Air Pollutant Deposition and Its Effects on

Natural Resources in New York State

Air Pollutant Deposition
and Its Effects on
Natural Resources in New York State

Timothy J. Sullivan

Comstock Publishing Associates
a division of
Cornell University Press
Ithaca and London

First published 2015 by Cornell University Press

First printing, Cornell Paperbacks, 2015
Printed in the United States of America

Library of Congress Cataloging-in-Publication Data

Sullivan, Timothy J., 1950 July 17– author.
Air pollutant deposition and its effects on natural resources in New York State / Timothy J. Sullivan.
    pages cm
  Includes bibliographical references.
  ISBN 978-1-5017-0060-6 (cloth : alk. paper)
  ISBN 978-0-8014-5687-9 (pbk. : alk. paper)

1. Air—Pollution—Environmental aspects—New York (State) 2. Atmospheric deposition—New York (State) 3. Natural resources—Environmental aspects—New York (State) 4. Environmental degradation—New York (State) I. Title.

TD883.5.N7S88 2015
551.57'7109747--dc23

2015014817

Cornell University Press strives to use environmentally responsible suppliers and materials to the fullest extent possible in the publishing of its books. Such materials include vegetable-based, low-VOC inks and acid-free papers that are recycled, totally chlorine-free, or partly composed of nonwood fibers. For further information, visit our website at www.cornellpress.cornell.edu.

Cloth printing   10 9 8 7 6 5 4 3 2 1
Paperback printing   10 9 8 7 6 5 4 3 2 1

This book is dedicated to all of the air pollution effects scientists I have worked with over the past three decades to better understand the environmental impacts of air pollution in New York. You are too numerous to list and I would undoubtedly omit some inadvertently. You know who you are. I have learned a great deal from our collaborations and interactions and it has been great fun.

Thank you!

Contents

Color plates follow page 120.

FIGURES

PLATES

1. Locations of Class I areas that receive maximum protection by the Clean Air Act

2. General vegetation types in the Adirondack and Catskill Parks

3. Locations of vegetation types in the Adirondack and Catskill Parks that are expected to contain red spruce or sugar maple trees

4. Locations of wetlands in the Adirondack Park

5. Total annual emissions per square mile of sulfur dioxide by county

6. Total wet plus dry sulfur deposition

7. Total annual emissions per square mile of nitrogen oxides by county

8. Wet plus dry deposition of total (oxidized plus reduced) nitrogen

9. Interpolated wet mercury deposition

10. Spatial distribution of lakes and streams in and around New York with measured surface-water ANC ≤ 100 µeq/L

11. Estimated target load of sulfur deposition to protect lake ANC to 50 µeq/L

12. Exceedance classes for Adirondack lakes

Tables

Preface

A power plant somewhere between Pittsburgh and Chicago releases gasses and particulates into the air; they contain mercury, oxides of sulfur, and oxides of nitrogen. A busy nearby highway is lined with cars on their morning commute; each vents nitrogen from its tailpipe. A livestock feeding operation smells faintly of ammonia. Tens, or maybe hundreds, of miles downwind a fish lays dying in a streambed after a large rainstorm. Sugar maple in a northern hardwood forest look unhealthy, show signs of canopy dieback, and are no longer producing seedlings. The forest is changing. A loon is suffering from neurotoxicity, the result of eating too many fish containing too much mercury. Atmospheric pollutant emissions and eventual biological damages are linked. We know this. The scientific community has been studying it for more than 30 years. But between the gas and particle emissions and the unhealthy trees and fish, stuff happens. It involves transportation and transformation, meteorology, physics, chemistry, biology, and ecology. Soils and rocks buffer atmospheric acidity, nutrient cycles are disrupted, inorganic mercury is converted to its highly toxic methylmercury form that is prone to bioaccumulation, and aluminum is mobilized from soil to drainage water, where it poisons tree roots and fish. This book is all about that stuff in the middle, between cause and effect.

Air pollution in New York and elsewhere in the eastern United States began to increase in a serious way more than a hundred years ago. The amounts of sulfur, nitrogen, and mercury that we put into the air increased rather steadily, reaching peak values during the last quarter of the twentieth century. As the scientific community and then the general public have come to understand the human health and ecological price we were paying for the right to pollute, emissions have declined, especially in response to federal legislation such as the Clean Air Act and its amendments. For the most part, these pollutant levels have continued to decline. Nevertheless, legacy effects remain. As chemical and in some cases biological conditions improve, the scientific community is trying to better understand how low pollution emissions need to go. How good is good enough? What are the ecological, economic, and societal trade-offs? How should we manage our resources? This book will help readers develop an understanding of the complexities and begin to answer such questions.

Air pollution in the form of sulfur, nitrogen, and mercury is emitted from a wide variety of sources. These include power plants, motor vehicles, agriculture, incinerators, and industrial facilities. These pollutants are carried with the prevailing winds and are eventually deposited to the earth's surface in the form of precipitation, air particles and gasses, mountain clouds, and fog. Once deposited to vegetation or soil surfaces, they move into the soil water, where they can become adsorbed and stored on soil, taken up by plants and soil microbes, or leached to surface waters. Effects in the soil, vegetation, and surface-water ecosystem compartments are varied. A multitude of chemical and biological transformations occur, altering the chemical characteristics of the deposited substances and their behavior in the environment. Key processes relating to elemental cycling, toxicity, and bioaccumulation in the food chain vary with contaminant type. Ecological impacts are diverse. Depending on the severity of impact, trees may die, plant species composition may change, fish may be eliminated from a particular body of water, estuaries may become overenriched with nutrients, or a potent neurotoxin may bioaccumulate in fish, which in turn might be consumed by and have adverse impacts on fish-eating wildlife and humans. The costs to individuals and society are diverse and often substantial.

Adverse effects on lakes and streams and their watersheds occur throughout sensitive regions of New York. Pollutants can impact water quality and harm many species of aquatic biota. They also alter the chemistry of watershed soils, a process that has potential impacts on plant roots and other terrestrial life forms. Scientists study the biogeochemistry of the entire landscape. The chemistry of drainage water integrates a host of terrestrial and aquatic processes that interact with water as it moves from the atmosphere as precipitation through the soil and into the groundwater and eventually to streams, lakes, rivers, and estuaries. Thus, study of the water can elucidate processes on the terrestrial side.

This book describes these interactions and others. Here I summarize the collective experience of researchers who have been studying the effects of air pollutants on soils, waters, and associated biota in New York and across the United States for the past three decades. This book is targeted to students and practitioners of environmental science, water and air pollution, soil science, biogeochemistry, water resources, and aquatic and terrestrial ecology and to water resource professionals and other scientists and natural resource managers.

A host of questions revolve around and depend upon the study of air pollution and its effects. What lakes are acid-sensitive or acid-impacted? Is a stream limited in its primary productivity by nitrogen or something else? Have mayflies, zooplankton, or fish been impacted by too much acidity? If so, which species? Are forest tree distributions and forest health changed by atmospheric nitrogen or sulfur input? Has mercury bioaccumulation affected piscivorous wildlife such as loons, eagles, and mink? What is the critical load of air pollution that sensitive downwind resources can tolerate without unacceptable damage? What level of damage is acceptable? What have been the effects of air pollution on aquatic and terrestrial food webs? Are water-quality, soil, and forest conditions getting better, staying the same, or getting worse over time? The questions are endless. If you have questions such as these or if you want answers that you can base policy or management on, this book can be of assistance.

The focus is on air pollution effects in New York. I highlight, in particular, research results generated in the Adirondack and Catskill Parks, Long Island Sound and associated coastal estuaries, and the Great Lakes region. However, I also draw extensively from research conducted elsewhere in the northeastern United States, throughout the country, and overseas. The principles described in this book are applicable to the study of air pollution and its effects globally.

The various examples that are provided here deal with studies of environmental effects from atmospheric deposition of acidifying, eutrophying, and neurotoxic air contaminants. However, the principles that are developed and illustrated here also apply to the study of other environmental issues besides atmospheric deposition of sulfur, nitrogen, and mercury. The cycles, processes, and transformations discussed in this book can also inform the study of agricultural, silvicultural, and urban pollutants; climate change; and other aspects of nonpoint- and point-source pollution. A reader who grasps the materials presented in this book will be well equipped to design, implement, and interpret many kinds of pollution effects studies.

I hope that the information presented here will help you design, conduct, and interpret environmental effects studies that will help all of us to better understand the impacts of human activities on ecosystem health. Armed with high-quality data and appropriate analyses, we can collectively move forward to reduce unacceptable human-caused impacts in an economically responsible fashion and protect and improve the quality of our natural resources for future generations.

Acknowledgments

Preparation of this book was supported by funding from the New York State Energy Research and Development Authority (NYSERDA) through a contract with E&S Environmental Chemistry, Inc. (E&S). Project management was provided by Gregory Lampman. Charles Driscoll and Gregory Lampman assisted in developing an outline for this effort. Todd McDonnell and Deian Moore contributed data analyses and prepared maps and figures. Jayne Charles and Deian Moore prepared the manuscript. Douglas Burns, Gregory Lampman, and two anonymous reviewers provided by the American Association for the Advancement of Science (AAAS) offered very helpful comments and suggestions based on review of earlier draft manuscripts.

This book was prepared by Timothy J. Sullivan in the course of performing work contracted for by E&S and sponsored by NYSERDA. The opinions expressed in this book do not necessarily reflect those of NYSERDA or the state of New York, and reference to any specific product, service, process, or method does not constitute an implied or expressed recommendation or endorsement of it. Further, NYSERDA, the state of New York, and E&S make no warranties or representations, expressed or implied, as to the fitness for particular purpose or merchantability of any product, apparatus, or service or the usefulness, completeness, or accuracy of any processes, methods, or other information contained, described, disclosed, or referred to in this book. NYSERDA, the state of New York, and E&S make no representation that the use of any product, apparatus, process, method, or other information will not infringe privately owned rights and will assume no liability for any loss, injury, or damage resulting from or occurring in connection with the use of information contained, described, disclosed, or referred to in this book.

Acronyms and Abbreviations

AcidBAP	acid biological assessment profile
ADRP	Acid Deposition Reduction Program
Al	aluminum
Al^{3+}	elemental aluminum ion
Al_i	inorganic aluminum
Al^{n+}	aluminum ions, expressed as the sum of all cationic aluminum ions
Al_o	organic aluminum
$Al(OH)_3$, $Al(OH)_2^+$, and $Al(OH)^{2+}$	aluminum hydroxides
ALS	Adirondack Lakes Survey
ALSC	Adirondack Lakes Survey Corporation
ALTM	Adirondack Long Term Monitoring Program
AMNet	Atmospheric Mercury Network
ANC	acid-neutralizing capacity
AQRV	air-quality related values
ASSETS	Assessment of Estuarine Trophic Status
BAF	bioaccumulation factor
BGC	biogeochemical
C	carbon
Ca	calcium
Ca^{2+}	calcium ion
CAA	Clean Air Act
CAAA	Clean Air Act Amendments
$CaCO_3$	calcite
CAIR	Clean Air Interstate Rule
CASTNet	Clean Air Status and Trends Network
Cl	chlorine
Cl^-	chloride
CMAQ	Community Multiscale Air Quality
CO_2	carbon dioxide
CWA	Clean Water Act
Δ, delta	difference; change
ELS	Eastern Lakes Survey
EMAP	Environmental Monitoring and Assessment Program
EPA	U.S. Environmental Protection Agency
ERP	Episodic Response Project
Fe	iron
F-factor	fraction of the change in mineral acid anions in solution that is neutralized by base cation release from the soil
H	hydrogen; hydrogen atom

H^+	proton, hydrogen ion
ha	hectare
HCO_3^-	bicarbonate anion
Hg	mercury
$HgCl_2$	mercury chloride
Hg(II)	divalent mercury
HNO_3	nitric acid
IA	integrated assessment
K	potassium
K^+	potassium ion
kg	kilogram
L	liter
LISS	Long Island Sound Study
LOAEL	lowest observed adverse effect level
LTM	Long Term Monitoring Project of the U.S. EPA
m	meter
MAGIC	Model of Acidification of Groundwater in Catchments
MDN	Mercury Deposition Network
MEA	Millennium Ecosystem Assessment
MeHg	methylmercury
meq	milliequivalent
μeq	microequivalent
μg	microgram
μmol	micromole
Mg	magnesium
Mg^{2+}	magnesium ion
N	nitrogen
n	number of observations
N_2	molecular nitrogen; nonreactive nitrogen
N_2O	nitrous oxide
NA	not available; insufficient data
Na	sodium
Na^+	sodium ion
NAAQS	National Ambient Air Quality Standards
NADP	National Atmospheric Deposition Program
NAPAP	National Acid Precipitation Assessment Program
NH_3	ammonia
NH_4^+	ammonium ion
NH_x	atmospheric reduced nitrogen; includes NH_3 and NH_4^+
NH_y	total reduced nitrogen
NO_2	nitrogen dioxide
NO_3^-	nitrate

NOAA	U.S. National Oceanic and Atmospheric Administration
NO_x	atmospheric oxidized nitrogen; sum of NO and NO_2
NPDES	National Pollutant Discharge Elimination System
N_r	reactive nitrogen
NRC	National Research Council
ns	nonsignificant
NSWS	National Surface Water Survey
NTN	National Trends Network
NYSBAP	New York State Biological Assessment Profile
NYSERDA	New York State Energy Research and Development Authority
O_2	molecular oxygen
O_3	ozone
OEC	overall eutrophic condition
OHI	overall human influence
P	phosphorus
pH	relative acidity
PHREEQC	model for soil and water geochemical equilibrium
PIRLA	Paleoecological Investigation of Recent Lake Acidification
pK_a	acid dissociation constant
PnET	Photosynthesis and EvapoTranspiration
PnET-BGC	Photosynthesis and EvapoTranspiration–Biogeochemical
PnET-CN	Photosynthesis and EvapoTranspiration–C, water, and N balances
PO_4^-, PO_4^{3-}	phosphate
ppm	parts per million
PSD	prevention of significant deterioration
$RCOO_s^-$	strongly acidic organic anions
RHR	Regional Haze Rule
S	sulfur
SAA	sum of mineral acid anion concentrations
SBC	sum of base cation concentrations
Si	silicon
SO_2	sulfur dioxide
SO_4^{2-}	sulfate ion
SO_x	sulfur oxides
SPARROW	SPAtially Referenced Regressions On Watershed Attributes
TIME	Temporally Integrated Monitoring of Ecosystems
WATERSN	Watershed Assessment Tool for Evaluating Reduction Strategies for Nitrogen
ww	wet weight
YOY	young of the year (fish)
yr	year

Air Pollutant Deposition and Its Effects on

Natural Resources in New York State

CHAPTER 1

Background and Purpose

1.1. ATMOSPHERIC DEPOSITION IN NEW YORK

New York is home to a wide range of plant and animal species that occupy a multitude of ecosystems, from the high peaks of the Adirondack Mountains to the southern Great Lakes and the estuaries and coastal waters of the Atlantic Ocean. New York's forests support wildlife and timber production and contribute clean water for human consumption and aquatic ecosystem health. Lakes and streams support fish and the life forms on which they feed and the predators that feed on them. Coastal waters and estuaries that include Long Island Sound, the Hudson River Estuary, and Raritan Bay provide diverse habitats for aquatic species and support fisheries economies. In addition to the ecologic and economic values and the ecosystem goods and services New York's ecosystems provide, they also attract millions of visitors to the mountainous, Great Lakes, and coastal regions each year.

The Adirondack and Catskill Mountains regions of New York contain many protected lakes and streams that are affected by air pollution. The Adirondack Park, in particular, has been the focus of extensive research and monitoring efforts for more than 30 years so that we can better understand the effects of air pollution. Surveys have been conducted of hundreds of streams and more than a thousand lakes. There have been investigations of short-term changes in water chemistry during periods of rain and snowmelt, studies of acidification processes, the development and application of mathematical models to predict the rate of future recovery as air pollution levels decline, studies to estimate the critical loads of atmospheric deposition required to protect sensitive resources, and periodic sampling of dozens of lakes for more than 30 years to document changes over time. Much is known about aquatic and terrestrial resource sensitivity and damage in these regions from atmospheric deposition of air pollutants. However, much remains to be learned.

Air pollution in the form of atmospheric deposition of sulfur (S), nitrogen (N), and mercury (Hg) has caused substantial damage to sensitive and valuable resources in New York. Additional damage might occur in the future if air pollution continues at relatively high levels. However, federal and state efforts over the past several decades to curb air pollution emissions from power plants, industry, and motor vehicles have resulted in a pronounced decrease in air pollution and atmospheric deposition of airborne contaminants in New York and elsewhere in the eastern United States. Hopes have risen among environmental scientists and policy makers that damaged resources will recover, and indeed some recovery has been documented.

Resources considered highly sensitive to the effects of atmospheric deposition of S, N, and Hg are not evenly distributed across New York. Some areas contain extensive sensitive resources; other areas contain few. Resource sensitivity also varies with respect to the types of effects. Resources sensitive to acidification are clustered largely in the Adirondack and Catskill Mountains. Acid sensitivity has also been documented in the Shawangunk region of the state. Sensitivity to nutrient enrichment from atmospheric N deposition occurs statewide, but concern is most heavily focused on coastal areas and the Great Lakes region. Sensitivity to Hg methylation occurs statewide but is most prevalent in areas containing abundant wetland vegetation. Figure 1.1 shows the locations in New York where some of these sensitive resources are located.

Resources in the Adirondack and Catskill Mountains have been damaged by acidic deposition caused by both S and N air pollutants. In fact, these regions are among the most sensitive and damaged in the United States. Sulfur and N have been contributed from the atmosphere to soil and drainage water, lowering the pH and causing chemical changes that affect the suitability of the soil and water for supporting sensitive species of algae, plants, and animals. Some of the species affected are especially important to the citizens of New York, including trout and other sport fish and sugar maple (*Acer saccharum*) and red spruce (*Picea rubens*) trees. Estuaries and marine coastal waters have been affected by overenrichment with nutrient N. Some of that N is deposited from the atmosphere, but other important sources include agricultural runoff and wastewater treatment facilities. Added N often acts to stimulate algal growth in estuarine and marine waters, leading to a cascade of deleterious impacts on coastal waters and the plants and animals that live there.

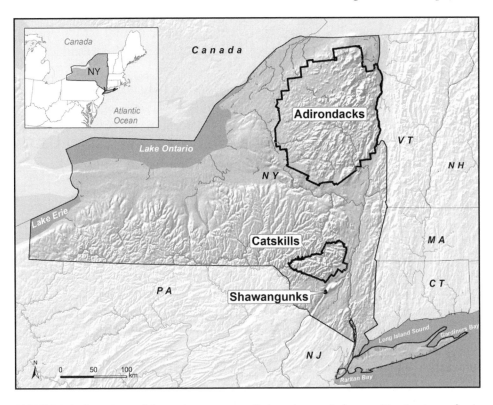

FIGURE 1.1. Locations of the major resources that are known to be sensitive to atmospheric inputs of sulfur, nitrogen, and mercury in New York: the Adirondack, Catskill, and Shawangunk regions; the Great Lakes; and coastal estuaries around Long Island.

Nitrogen deposition, especially to wetlands, meadows, and alpine environments, can alter competitive relationships among plant species and increase the establishment of nonnative species at the expense of some of the rare native species. Species shifts are thought to occur at N deposition levels as low as 5 to 10 kilograms per hectare per year (kg N/ha/yr) for raised and blanket bogs (Achermann and Bobbink 2003) and at even lower levels for alpine plant communities (Sverdrup et al. 2012). Some wetland plants are well adapted to low-N environments, including some species in the genera *Sphagnum* and *Isoetes* and some insectivorous plants, such as the green pitcher (*Sarracenia oreophila*) and the roundleaf sundew (*Drosera rotundifolia*). The pitcher plant (*Sarracenia purpurea*), a native of nutrient-poor peatlands in the eastern United States, has been proposed as an indicator of high atmospheric N supply (Ellison and Gotelli 2002).

Atmospheric deposition also contributes Hg to natural ecosystems in New York. When converted into a methylated chemical form, Hg bioaccumulates in food chains and is toxic to humans and wildlife predators. The latter include large fish, river otters (*Lontra canadensis*), mink (*Neovison vison*), loons (*Gavia* spp.), and bald eagles (*Haliaeetus leucocephalus*). There are also complex interactions between S and Hg deposition, because sulfate (SO_4^{2-})-reducing bacteria are believed to be the dominant source of methyl Hg (MeHg) in many natural ecosystems. By methylating Hg, these bacteria make it biologically available and facilitate biomagnification.

In response to atmospheric deposition of S, N, and Hg on New York ecosystems, fish, plants, and other life forms have suffered varying degrees of damage from acidification, nutrient enrichment, and toxicity. Citizens of New York have shown great concern about air pollution damage to brook trout (*Salvelinus fontinalis*) and other fish species. The general public may be less aware of effects on other life forms.

1.2. AIR QUALITY MANAGEMENT

1.2.1. Clean Air Act

The Clean Air Act (CAA) of 1970 was enacted to protect public health and welfare from the harmful effects of human-generated air pollution. Criteria pollutants are those for which the U.S. Environmental Protection Agency (EPA) has established National Ambient Air Quality Standards (NAAQS) as directed by the CAA. Standards were established for selected pollutants that are emitted to the atmosphere in significant quantities throughout the country and that may endanger public health and welfare. These include sulfur dioxide (SO_2) and nitrogen oxides (NO_x). Although reduced N (NH_x) also contributes to these effects, standards have not been set for it. The primary NAAQS are designed to protect human health, while the secondary NAAQS are designed to protect public welfare from the adverse effects of pollutant(s). The CAA defines public welfare effects to include, but not be limited to, "effects on soils, water, crops, vegetation, manmade materials, animals, wildlife, weather, visibility and climate, damage to and deterioration of property, and hazards to transportation, as well as effects on economic values and on personal comfort and well-being."

The CAA as amended in 1977 also established the Prevention of Significant Deterioration (PSD) program. The primary objective of the PSD provisions is to prevent substantial degradation of air quality in areas that comply with NAAQS and yet maintain a margin for industrial growth. A PSD permit from the appropriate air regulatory agency is required to construct a new pollution source or substantially modify an existing source (Bunyak 1993). A permit application must demonstrate that the proposed polluting facility will (1) not violate national or state ambient air quality standards; (2) use the best available control technology to limit emissions; (3) not violate PSD increments for SO_2, nitrogen dioxide (NO_2), or particulate matter (PM); and (4) not cause or contribute to adverse impacts to air quality related values (AQRVs) in any Class I area. Class I areas include certain national parks and wilderness areas that receive the highest level of federal protection from air pollution damage. The PSD increments are allowable pollutant concentrations that can be added by industrial development to baseline concentrations.

The values Congress chose as PSD increments were not selected on the basis of concentration limits causing impacts to specific resources. Therefore, it is possible that pollution increases that exceed the legal Class I increments may not cause damage to Class I areas. It is also possible that resources in a Class I area could be adversely affected by pollutant concentrations that do not exceed the increments.

The following questions may be addressed when reviewing PSD permit applications:

- What are the identified sensitive AQRVs in each Class I area that could be affected by the new source?
- What are the air pollutant levels that may affect the identified sensitive AQRVs?
- Will the proposed facility result in pollutant concentrations or atmospheric deposition that will cause the identified critical level to be exceeded or add to levels that already exceed the critical level?
- If the critical level of the sensitive indicator is exceeded, what amount of additional pollution is considered "insignificant"?

The first two questions are largely land management issues that should be answered on the basis of management goals and objectives for the protected area. The last two are technical and policy questions that must be answered on the basis of analyses of projected emissions from the proposed facility and predictions of environmental response to given pollutant concentrations (Peterson et al. 1992).

In Title IV of the 1990 Clean Air Act Amendments (CAAA), Congress called for decreases in annual emissions of SO_2 and NO_x from utilities that burn fossil fuels. The legislation specifically required utilities to reduce (from 1980 levels) annual emissions of SO_2 by 10 million tons and annual emissions of NO_x by 2 million tons by the year 2010. As a consequence of Title IV, emissions and deposition of NO_x, especially sulfur oxides (SO_x), have declined substantially since 1990. The Clean Air Interstate Rule (CAIR) of 2005 further reduced S and N emissions and deposition, but that rule has gone through legal challenges.

The CAIR emissions control rule focused primarily on emissions controls on coal-fired electricity generating plants for the purpose of attaining NAAQS for particulate matter and ozone (O_3). However, the CAIR was challenged in court. The Cross State Air Pollution Rule (CSAPR) was scheduled to replace CAIR in 2012, but it too was litigated and vacated. In April 2014, the U.S. Supreme Court revised the earlier District of Columbia circuit court opinion that had previously vacated CSAPR. At the time of this writing, CAIR remains in effect but will probably be replaced by something different. It is not clear at this time what exactly will replace it.

1.2.2.   Regional Haze Rule

More recent reductions in S and N emissions and deposition have been driven, in large part, by the need for states to comply with regulations aimed at improving visibility in Class I areas. The EPA promulgated regulations in 1980 to address visibility impairment that is "reasonably attributable" to one or a small group of sources. Congress subsequently added section 169B to the CAAA to focus attention on regional haze issues. On July 1, 1999, the EPA promulgated the Regional Haze Rule (RHR), which requires states (and tribes that choose to participate) to review how pollution emissions in the state affect visibility at Class I areas across a broad region (not just Class I areas in the state). These rules also require states to make "reasonable progress" in reducing any effect this pollution has on visibility conditions in Class I areas and to prevent future impairment of visibility. The rule requires states to analyze a pathway that takes the Class I areas from current conditions to "natural conditions" in 60 years. "Natural conditions," a term used in the CAA, means that no human-caused pollution can impair visibility. This program is aimed at Class I areas, which are not found in New York but occur in surrounding states (Plate 1). Thus, efforts to curtail emissions that impact Class I areas in nearby states also affect air pollutants that are emitted in or transported to New York. The RHR is improving regional visibility throughout the country and is noteworthy because the requirement to improve visibility will result in further decreases in S and N deposition. This is because ammonium sulfate is typically the principal contributor to haze at most locations in the eastern United States.

1.2.3.   Federal Water Pollution Control Act

The Federal Water Pollution Control Act, commonly known as the Clean Water Act (CWA), was promulgated in 1972, and significantly amended in 1977, 1987, and 1990. The primary purpose of the act is to protect and restore the physical, chemical, and biological quality of the nation's waters. The act established the goals of making all navigable waters fishable and swimmable and eliminating the discharge of pollutants into the nation's waterways. Like the CAA, the CWA provides an additional tool to help states meet pollution control mandates. The impaired streams and anti-degradation sections of this law are pertinent to air pollution effects assessment because streams acidified by S and/or N deposition may qualify to be listed as impaired streams on what is known as the 303(d) list.

States manage and protect water quality under the CWA through the development and enforcement of ambient water-quality standards. Water quality standards are composed of three interrelated parts: (1) designated beneficial uses of a water body, such as contact recreation or cold-water fishery; (2) numerical or narrative criteria that establish the limits of physical, chemical, and biological characteristics of water sufficient to protect beneficial uses; and (3) an anti-degradation provision to protect water quality that exceeds criteria and to protect and maintain water quality in in waters designated as Outstanding National Resource Waters, an EPA category. States comply with water-quality standards by controlling the type and quantity of point-source pollutants entering waters through the National Pollutant Discharge Elimination System (NPDES) and by implementing best management practices for nonpoint sources of pollution. Section 303(d) of the act requires states to also formally identify waters that do not currently meet water-quality standards and bring them into compliance through the development and implementation of total maximum daily loads, which establish the maximum loadings of pollutants that a water body can receive from point and nonpoint (including atmospheric deposition) sources of pollution without exceeding the standards.

1.2.4. Other Legislation

New York State has enacted statewide emissions regulations for coal-fired power plants. The original law, the State Acid Deposition Control Act, was passed in the 1980s. A second law, the Acid Deposition Reduction Program (ADRP), was passed in 2004 that will require fossil fuel–fired electric generators in New York State to reduce NO_x and SO_2 emissions. Affected sources must reduce SO_2 emissions to 50 percent below the levels allowed by Phase 2 of the federal acid rain program. Affected sources must reduce NO_x emissions during the non-O_3 season (October–April) to a level that corresponds with the NO_x reductions that were achieved starting on May 1, 2003, through the implementation of the CAIR NO_x Ozone Season Trading Program for the O_3 season (May–September). Although these regulations do not address emissions from upwind states, they do contribute to the overall pattern of regional emissions controls that influence atmospheric deposition of S, N, and Hg at the locations of sensitive ecosystem receptors in New York.

1.3. ECOSYSTEM FUNCTIONS AND SERVICES

Ecosystem services refer to the fundamental value of ecosystems to human welfare. Ecosystems provide many goods and services that are critical for the functioning of the biosphere and provide the basis for the delivery of tangible benefits to human society. These include food, materials, pharmaceuticals, ecosystem processes and cycles, recreation, relaxation, and spiritual enrichment. Terminology regarding ecosystem services, human value, ecosystem function, and social benefit can be confusing (Table 1.1). ICSU-UNESCO-UNU (2008), The Millennium Ecosystem Assessment (MEA; 2005), and the EPA (2008) have defined ecosystem services to include supporting, provisioning, regulating, and cultural services:

- Supporting services support the provision of other ecosystem services through such actions as production of biomass, production of atmospheric oxygen, soil formation and retention, nutrient cycling, water cycling, and provision of habitat.
- Provisioning services include products such as food, fiber, and medicinal and cosmetic products (Gitay et al. 2001).
- Regulating services include carbon sequestration; climate and water regulation; protection from natural hazards such as floods, water and air purification; and disease and pest regulation.
- Cultural services satisfy human spiritual, educational, and aesthetic needs and foster appreciation of ecosystems and their components.

TABLE 1.1. Key terms that are central to the ecosystem service concept based on the example of acidification effects on recreational fishing

Term	Description	Examples
Final ecosystem service	End product component of nature that yields human well-being	Sport fishery, surface water
Intermediate service	Intermediate product needed to support final ecosystem services	Water quality needed to support a sport fishery
Value	Importance to people, expressed in monetary or nonmonetary terms	Opportunity to fish in an aesthetically pleasing location that contains suitable sport fish
Function/process	Intermediate step that contributes to the service	Nutrient cycling, cleansing of drainage water as it flows through soil, microclimate regulation
Social benefit or source of well-being	Arises from the human use of an ES, often in combination with other conventional goods and services	Recreation, spiritual enrichment, relaxation, natural biodiversity maintained

In evaluating the effects of air pollutants on ecosystem services, each service should be geographically referenced and expressed where possible in generally comparable units of measure (Gimona and van der Horst 2007; Naidoo et al. 2008). This allows comparison of loss or gain in ecosystem services across space and across time. Services that can be measured and stacked to facilitate prioritization of key regions or watersheds are especially important to human well-being because key areas such as the Adirondack and Catskill Parks provide a multitude of services in proximity to centers of human population and provide services that are highly valued by the citizens of New York.

Regulating services include natural cycles and processes and the ways they benefit people. It is difficult to estimate their value. Cultural services include a variety of emotional, psychological, and spiritual benefits that humans derive from natural ecosystems; these are also difficult to quantify. Cultural services also include benefits related to outdoor recreation and ecotourism, which are important components of rural economies in New York. Such services include fishing, hunting, hiking, swimming, boating, and wildlife viewing. Sources of human well-being, ecosystem services, and the social benefits of ES are closely related (Figure 1.2).

However, many of the supporting and regulating services MEA (2005) and others have identified are inconsistent with a definition of ecosystem services that is measurable, mappable, and capable of being stacked and valued. There are also difficulties of double-counting of intermediate products or services when the value of one ecosystem service is embedded in the value of another ecosystem service. For example, supporting services, such as nutrient cycling, constitute intermediate services that should not be valued directly because they are already included in the process of assessing impacts on a final ecosystem service, such as provision of a trout fishery (Sullivan 2012).

Boyd and Banzhaf (2007) stress the importance of separating intermediate and final services for economic valuation. It is these final ecosystem services that satisfy the need to measure, stack, map, and assign value such services. Interim products, functions, processes, and cycles are intermediate to or contribute to final ecosystem services but are not themselves final services (Sullivan 2012). Given the importance of these considerations, and in keeping with the economic issues raised by Boyd and Banzhaf (2007), a final ecosystem service can be defined as follows:

> an endpoint component of nature that can be enjoyed, consumed, or used by people to generate human well-being and that can be measured, stacked, mapped, and valued using a common currency. (Sullivan 2012)

Resource management and public policy should focus in large part on these final ecosystem services, which are determined by the processes, cycles, and intermediate services that are the focus of ecological research.

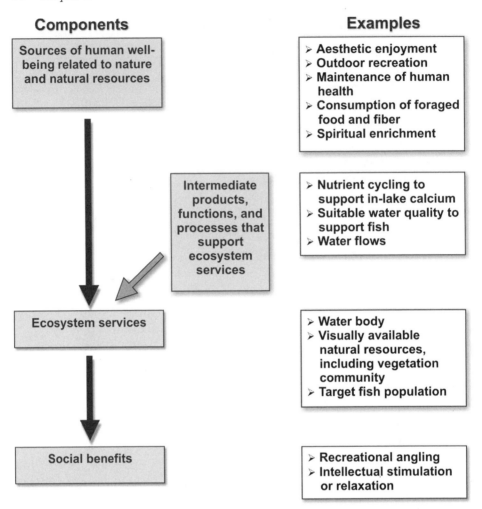

FIGURE 1.2. Relationships among sources of human well-being, ecosystem services, and social benefits that accrue from ecosystem services, using the example of recreational fishing. Adapted from Boyd and Banzhaf 2006; Sullivan and McDonnell 2012.

A decrease in ecosystem services is often reflected in economic loss. Determination of the net impact of changes in ecosystem services requires a complete accounting of costs and benefits using a common currency for valuation of ecosystem services. This is important because society and makers of environmental public policy will be less likely to ignore the consequences of air pollution and the benefits of mitigation if the economic costs and damages are clearly defined and are expressed in terms of monetary value. Restoration of ecosystem services that have been lost due to air pollution damage can affect a wide range of potential benefits to society (Table 1.2).

Calculation of a change in ecosystem services caused by an environmental stress such as acidification, Hg biomagnification, or nutrient enrichment must be based

TABLE 1.2. Anticipated ecosystem service benefits to be realized by moving from a state of critical load exceedance to non-exceedance in the Adirondack Mountains

Ecosystem service	Anticipated benefits	Notes
Provisioning services		
Production of maple syrup and related products	Continued or increased production of food products	Important regional benefit
Catch of brook trout and other game fish in sport fishery	Continued or increased catch of sport fish	Important regional benefit, especially in the Adirondack and Catskill Parks
Production of maple wood for furniture and other wood products industry	Continued or increased wood production	Limited benefit
Production of spruce wood for wood products	Continued or increased wood production	
Provision of wildlife habitats	Continued provision of habitat for species associated with sugar maple or red spruce trees or surface waters	Difficult benefit to quantify
Regulating services		
Climatic regulation	Decreased greenhouse gas production or increased carbon sequestration can reduce potential for climate warming impacts	Difficult benefit to quantify
Water regulation	Improved tree health in some habitat types can maintain or enhance water storage, reducing the impacts of flooding and providing increased stream flow during low flow periods	Difficult benefit to quantify
Erosion regulation	Decreased effect on vegetation cover can limit possible increases in erosion during heavy precipitation events	Difficult benefit to quantify

TABLE 1.2—continued

Ecosystem service	Anticipated benefits	Notes
Supporting services		
Primary production	Effects on primary production and biomass in N-limited ecosystems can be reduced	Difficult benefit to quantify
	Decreased evapotranspiration and increased groundwater recharge	Potentially significant based on changes in hydrology predicted in various climate change scenarios
Nutrient cycling	Decreased soil acidification can limit the depletion from the soil of Ca and other nutrient base cations, which are important for healthy sugar maple, brook trout, and other terrestrial and aquatic species	Important regional benefit
	Decreased mineralization, nitrification, and nitrate leaching can limit soil acidification, Al mobilization, and base cation depletion in base-poor terrestrial and aquatic ecosystems	Interactions are complex
Trophic interactions	Effects of acidification on aquatic invertebrates, such as mayflies and caddisflies, can be reduced, thereby increasing food sources for brook trout and other sport fish	Unclear whether alternate food sources will be used
Cultural services		
	Improved brook trout and other sport fisheries can lead to increased recreational activities and ecotourism	Important regional benefit

Ecosystem service	Anticipated benefits	Notes
TABLE 1.2—continued		
	Improved sugar maple health can enhance autumn foliage color in forests that contain sugar maple	Important regional benefit
	Aesthetic qualities of plant communities can be enhanced through improved health of red spruce, sugar maple, and other acid-sensitive species	Important, but subtle benefit
	Iconic species, such as brook trout and sugar maple, can increase in abundance	May be of substantial monetary value to the public
	Maple syrup production cottage industry can improve	Important regional benefit

on marginal (rather than total) change in value. The marginal value is the change that occurs in response to a modest change in the supply of services. For example, air pollution is more likely to cause incremental degradation of environmental conditions than it is to cause a complete destruction of the resource (Kareiva and Marvier 2011). Thus, it is important to estimate the value of services lost or gained as a consequence of differing levels of air pollution emissions controls. The total value of a given service is less relevant. The incremental change in that value that occurs in response to a change in atmospheric deposition needs to be considered when evaluating policies that regulate air pollution or its abatement.

1.4.   GOALS AND OBJECTIVES

The goal of this book is to summarize and synthesize recent scientific research findings on the topic of the effects of atmospheric deposition of S, N, and Hg on sensitive ecosystems in New York. It explores both chemical and biological effects. It mainly highlights research findings for the period 1990–2013, although it also discusses some earlier and some more recent studies.

The effects I am concerned with in this book are mainly acidification of soil and water (caused by S and N deposition with associated aluminum [Al] mobilization), nutrient enrichment (also called eutrophication) of aquatic and terrestrial ecosystems (caused by N deposition and interactions with non-atmospheric inputs of N and other key nutrients), and toxicity to biota (caused by Hg methylation and acidification attributable to S and N deposition). Each of these effects of concern can be caused by both natural processes and inputs of pollutants by humans.

Acidification effects in New York are confined mainly to the Adirondack and Catskill Mountains regions. Nutrient enrichment effects have been most thoroughly studied for estuaries such as Long Island Sound; potential nutrient effects on the Great Lakes and terrestrial ecosystems are also of concern. Some smaller lakes, especially those in remote locations, may also be sensitive to eutrophication from atmospheric N deposition, depending on the relative magnitudes of N versus phosphorus (P) inputs from all sources in the lake watersheds. Effects of Hg deposition and methylation in New York have been mostly studied in the Adirondack Mountains region, although such effects are of interest statewide.

This book is organized into 13 chapters. These include text and in some cases tables and figures. Color maps are presented as plates and are grouped together in one place. This first chapter gives general background material and explains the purpose of the book. The next chapter describes resource sensitivity to adverse effects of atmospheric deposition of S, N, and Hg. This is important because the level of air pollution and associated atmospheric deposition tell only part of the story. Some ecosystems and ecosystem elements are sensitive to a given level of pollution and others are not. To obtain a clear picture of impacts, you must consider the extent of inherent resource sensitivity. Chapter 3 describes the principal stressors addressed in the book. Although the major focus is on S, N, and Hg, these pollutants interact with other natural and human-caused ecosystem stressors and these are also described. They include watershed disturbances, especially those related to fire, land use, and the introduction of nonnative and invasive species. They also include the effects of biomagnification of Hg and the effects of climate change and climatic variability.

Chapter 4 provides an overview of the various chemical effects of the pollutants under consideration here. These are manifested as a consequence of a wide variety of chemical processes and interactions, especially those associated with S, N, base cations, Al, Hg, and C. The chapter also describes processes that include atmospheric deposition effects in terrestrial, wetland, and surface water environments.

In Chapter 5, the focus shifts to biological effects of S, N, and Hg deposition. These are described for terrestrial, freshwater aquatic, and near-coastal aquatic communities.

Effects of atmospheric deposition can perhaps best be understood in the context of past changes in resource condition. Chapter 6 addresses such changes as they have been assessed using paleoecological techniques focused on analysis of lake sediment cores and process model hindcasts of preindustrial chemical and biological conditions. The chapters examines observed recent changes in conditions; the major focus is on long-term monitoring data sets.

Chapter 7 addresses the extrapolation of various site-specific studies to the regional landscape. This is important for placing research findings in a regional context. Resource management decisions require information on effects across broad spatial scales. Adverse impacts to one site or a few sites are generally considered insufficient as the basis for decision making about pollution abatement.

In some cases, resource conditions have been projected into the future using dynamic models. Such projections typically consider multiple levels of future atmospheric emissions controls. Chapter 8 describes results where they exist for New York.

Chapter 9 discusses critical and target loads of S and N deposition. These specify the pollutant loading rates below which adverse impacts are not expected to occur. The critical load describes the sustained pollutant loading at an ecosystem's steady state, which may not occur for many years or decades. The target load represents what can be considered an acceptable pollutant loading rate for protecting sensitive ecological elements at a particular point in time, such as, for example, the year 2050 or 2100.

Chapter 10 explores climate linkages further. Changes in future temperature, water chemistry, and water quantity (hydrology) will modify how air pollutant deposition impacts sensitive ecosystem elements now and in the future.

Calculation of the critical load informs evaluation of pollutant effects on ecosystem services. Chapter 11 describes recent research that illustrates the improvements in ecosystem services that may occur as a consequence of reducing atmospheric deposition to levels that are below the critical load.

Existing and past air pollutant deposition has in some cases caused ecosystem damage that cannot be reversed simply by reducing or eliminating that deposition. In such cases, additional human intervention might be required to restore species composition, diversity, and functionality to affected ecosystems. Chapter 12 outlines options for active intervention.

Chapter 13 summarizes major findings and addresses some of the remaining gaps in data that would help us understand the effects of air pollutant deposition in New York. It also provides recommendations for further research.

CHAPTER 2

Resource Sensitivity to Atmospheric Deposition

Moderate- to high-elevation watersheds with steep topography, exposed bedrock, deep winter snowpack, and shallow base-poor soils tend to be most sensitive to acidification in response to acidic deposition (Sullivan 2000). However, the chemistry of a lake or stream is always changing. Water chemistry is variable on both intra-annual and inter-annual time scales in response to changes in rainfall, snow accumulation and melting, temperature, and other climatic variables. Several factors govern the sensitivity of terrestrial and aquatic ecosystems to acidification from S and N deposition. Some of the major factors that govern ecosystem sensitivity to acidification are outlined below.

2.1. GEOLOGY

One of the most important variables that governs acid sensitivity is geology. Geologic composition plays a dominant role in influencing the chemistry of drainage waters and thus their sensitivity to acidification. Bedrock geology is the basis of a national map of the acidification sensitivity of surface water compiled by Norton et al. (1982) and has subsequently been used in numerous regional and subregional acidification studies (e.g., Dise 1984; Bricker and Rice 1989; Stauffer 1990; Stauffer and Wittchen 1991; Vertucci and Eilers 1993; Sullivan et al. 2007b).

Most of the major concentrations of low acid-neutralizing capacity (ANC) surface waters, including those in the Adirondack and Catskill Mountains, are located in areas underlain by bedrock that is resistant to weathering, especially bedrock in which quartz is a dominant mineral.

2.2. SOILS

Soil acidification, which occurs in response to natural factors, timber harvesting, and human-caused air pollution, involves the loss of base cations such as calcium (Ca), magnesium (Mg), potassium (K), and sodium (Na) and the accumulation of acidic cations such as hydrogen (H) and inorganic Al (Al_i) in the soil. Soil acidification is partly a natural process that is caused by the production of carbonic and organic acids and by cation uptake by vegetation (Charles 1991; Turner et al. 1991). Mineral acids in soil water, such as nitric and sulfuric acids, result from air pollution containing NO_x and SO_x, respectively. Effects attributable to the addition of mineral acids to acid-sensitive soils have been pronounced in the northeastern United States and portions of the Appalachian Mountains in both hardwood and coniferous forests.

Uptake of nutrient cations by vegetation can also generate acidity in the soil. In addition, a considerable amount of natural organic acidity is produced in the upper O_a soil horizon through the partial decomposition of organic matter. This process can decrease the pH of soil water in the O_a horizon well below the lowest pH values commonly measured in acidic deposition (Krug et al. 1985; Lawrence et al. 1995). Natural acidification of soil is particularly pronounced in coniferous forests. The O_a-horizon soils under coniferous vegetation are strongly acidified by organic acids and are therefore unlikely to have experienced a substantial further lowering of pH as a result of acidic deposition (Johnson and Fernandez 1992; Lawrence et al. 1995). Soils that are influenced by the growth of hardwood tree species tend to have surface soil horizons that are naturally less acidic and are therefore more susceptible to decreased pH in the O_a horizon from acidic deposition. Several studies have documented declines in soil pH and base saturation in the O_a/A horizons and the upper B mineral soil horizon in sensitive regions of the United States over the past several decades (Johnson et al. 1994a; Johnson et al. 1994b; Drohan and Sharpe 1997; Bailey et al. 2005; Sullivan et al. 2006b; Warby et al. 2009). Thus, naturally occurring soil acidification can be exacerbated by acidic deposition. Despite recent decreases in acidic deposition and some improvement in surface water acid-base status, there are widespread observations of ongoing decreases in base cations that are exchangeable in soil (Bailey et al. 2005; Sullivan et al. 2006a; Sullivan et al. 2006b; Warby et al. 2009). Only the study of Lawrence et al. (2012) suggests the beginning of soil recovery from acidification in New York.

Several chemical indicators provide information that aids in interpretation of soil acid-base status, including cation exchange capacity, exchangeable Ca, exchangeable Mg, and base saturation. Base saturation, the quantity of exchangeable bases (Ca, Mg, K, sodium [Na^+]) as a percent of the total cation exchange capacity (which includes exchangeable H and Al), is often most useful. If base saturation is less than about 15% to 20%, soil-exchange chemistry is dominated by Al. Atmospheric inputs of sulfuric and nitric acid to soils having low base saturation are buffered mainly by release of Al_i from the soil to soil solution (Reuss 1983) and only partial neutralization of atmospherically deposited acidity is realized. Where soil base saturation is higher, acids contributed from the atmosphere are buffered mainly by release of base cations and such acid neutralization is generally complete, or nearly so.

Quantities of exchangeable bases are usually higher in the upper organic horizons than in the mineral B horizon of the soil, and the organic horizons tend to have higher base saturation. This occurs despite the lower pH of the organic horizons caused by the prevalence of naturally occurring organic acids. The B-horizon base saturation is particularly sensitive to acidification caused by base cation loss in response to leaching by SO_4^{2-} and nitrate (NO_3^-) contributed by acidic deposition. Little work has been done to relate changes in soil base saturation to stand or forest health, but Cronan and Grigal (1995) suggested that B-horizon base saturation values below about 15% could lead to Al_i stress to vegetation. Base saturation values less than 10% are common in the B horizon of soils in the areas where soil and surface water acidification from acidic deposition have been most pronounced, including the Adirondack Mountains (Sullivan et al. 2006a). In addition, the growth, regeneration, and canopy condition of sugar maple in the southwestern Adirondack Mountains is correlated with the presence of low soil base saturation and high acidic deposition (Sullivan et al. 2013). Effects on sugar maple are most pronounced at B-horizon soil base saturation ≤ 12%.

Release of base cations from soil into soil water through weathering, cation exchange, and mineralization contributes to the neutralization of acidity (van Breemen et al. 1983). If the acidity is associated with anions that are highly mobile in the soil environment, such as SO_4^{2-} (which is mobile in previously glaciated areas such as upstate New York) and NO_3^- (which is mobile to varying degrees in New York watersheds), cations can be leached to surface waters. The limited mobility of anions associated with naturally derived organic and carbonic acids control the rate of base cation leaching under conditions of low acidic deposition. In the absence of acidic deposition, anion concentrations derived from organic and carbonic acids in acid-sensitive watersheds are generally low except in watersheds where wetland area is substantial. Much of the organic acidity generated in soils is adsorbed deeper in the soil profile, further limiting anion mobility. Inputs of acidic deposition can supply the anions SO_4^{2-} and NO_3^- in relatively high concentrations and these anions tend to be mobile, especially in previously glaciated soils. These mineral acid anions can therefore accelerate base cation leaching (Cronan et al. 1978). Depletion of base cations, mainly Ca, from the soil can damage acid-sensitive plants and those that require substantial Ca.

There are four issues that are potentially important with respect to the terrestrial effects of atmospheric S and N deposition:

1. Toxicity of Al_i to plant roots and/or foliage,
2. Depletion of Ca and other nutrient base cations from soil,
3. N saturation (whereby N supply exceeds N demand), and
4. Nutrient enrichment effects, especially as they influence competitive interactions among plant species.

2.3.   FOREST VEGETATION

Vegetation in the Adirondack and Catskill Parks is mainly hardwood and mixed hardwood and conifer forest types (Plate 2). Relatively pure conifer stands are generally scattered and occur mainly at higher elevations. Agricultural lands are found mainly around the periphery of the parks.

Mineral acidity is deposited to the terrestrial ecosystem through wet, dry, and occult deposition processes. Only a minor component of the atmospherically deposited S is taken up from plant leaf surfaces through the stomata into aboveground plant tissue or from the soil into plant roots. In contrast, N can readily be transported from leaf surfaces to the interior of leaves through the leaf stomata and potentially washed with rainfall from plant surfaces into the soil. Once the deposited N moves into plant tissues or soil, it can cause several kinds of ecological effects, including nutrient enrichment and acidification effects.

There are many transformations of the deposited N that take place in forest soils, often facilitated by bacteria and fungi. The N form can change rapidly. Different plants and algae vary in their needs for N nutrition. Some prefer oxidized forms, some prefer reduced forms, and some can use small organic N molecules. The N that leaches from the soil into drainage water is mostly in the form of NO_3^-.

Two tree species (red spruce and sugar maple) are known to be highly susceptible to damage from acidic deposition. Red spruce occurs in New York throughout the Adirondack and Catskill Mountains (Plate 3). Sugar maple is more broadly distributed statewide. Although soil acidification effects in New York are expected to be especially pronounced in the plant communities that include these tree species, the same kinds of effects might occur in other vegetation types. For example, basswood (*Tilia americana*) accumulates base cations and contributes to relatively high Ca concentrations in the upper mineral soil (Fujinuma et al. 2005).

Resulting high soil Ca, in turn, can increase N mineralization, nitrification, and leaching (Page and Mitchell 2008). Effects on other tree species are not well documented in New York. However, research in Quebec by Duchesne and Ouimet (2009) suggested that base cation depletion caused partly by soil acidification is a major factor favoring regeneration of American beech (*Fagus grandifolia*) in sugar maple stands.

Red spruce trees in the eastern United States died at a rapid pace during the 1980s and 1990s. This mortality was linked to exposure of foliage to acidic cloud water and an increase in the amount of dissolved Al_i compared to the amount of dissolved Ca^{2+} in soil water. Some of the red spruce decline occurred at high-elevation sites that frequently experience cloud cover. Much of the total atmospheric S and N deposition at such locations probably comes in the form of cloud deposition, which is often more acidic than acid rain.

The leaching of atmospherically deposited SO_4^{2-} to surface waters controls soil acidification and Al_i toxicity to plants at most acid-impacted areas in New York. Nitrate mobility is also important in some watersheds, but the dominant mobile strong-acid anion under base-flow conditions is usually SO_4^{2-}. During snowmelt and to a lesser extent during rainstorms, NO_3^- is the dominant mobile anion in many Adirondack and Catskill surface waters.

Acidification effects have not been as thoroughly studied for sugar maple as they have for red spruce. Nevertheless, several studies, mainly in Pennsylvania (cf. Long et al. 1997; Herlihy et al. 2000; Juice et al. 2006; Horsley et al. 2008), have indicated that sugar maple decline is linked to the occurrence of relatively high levels of acidic deposition to base-poor soils and Ca depletion from soil exchange sites. A recent study of sugar maple in the Adirondacks (Sullivan et al. 2013) concurred with these previous results from Pennsylvania.

The health of sugar maple trees is strongly influenced by the availability of Ca in the soil. Other nutrient base cations (e.g., Mg^{2+}, K^+) might also be important at some locations. Trees growing on soils having low base cation supply are more susceptible to damage from defoliating insects, drought, and extreme weather. In response, mature trees can die and there can be poor regeneration of seedlings.

Depletion of soil base cations may contribute to sugar maple mortality on sites having marginal soils. Sugar maple dieback at 19 sites in northwestern and north-central Pennsylvania and southwestern New York was shown to be correlated with combined stress from defoliation and soil deficiencies of Mg and Ca (Horsley et al. 1999). Dieback occurred predominately on ridgetops and upper slopes, where base cation availability was much lower than on middle and lower slopes (Bailey et al. 1999).

The primary factors that determine the sensitivity of forest ecosystems to acidification-related effects of S and N deposition are (1) the distribution of plant species on the site; and (2) the size of the pool of exchangeable base cations in the soil that are available to neutralize acidic atmospheric inputs. Some plant species are more sensitive than others, although relative sensitivities are poorly known. The effects of acidification on plants are mostly governed by Al toxicity and deficiencies in the nutrient base cations. These two factors are closely related. Inorganic Al is toxic to tree roots. In response to high Al_i concentrations in soil solution, plants often exhibit reduced root growth, which restricts uptake of water and nutrients, especially Ca, from the soil (Parker et al. 1989). Calcium is an ameliorant for Al_i toxicity to roots in soil solution and so its loss from soil and soil water over the long term could potentially exacerbate Al_i toxicity. Magnesium, and to a lesser extent Na and K, may also mitigate Al_i toxicity.

The nutrient base cations (Ca^{2+}, Mg^{2+}, K^+) are taken up through plant roots in dissolved form. However, most base cation supplies in rocks and soils are bound in minerals and are unavailable to plants. The available base cations in the soil are adsorbed to negatively charged soil exchange sites. These exchangeable base cations can exchange for the acidic cations H^+ and Al^{3+} and enter the soil solution.

Weathering slowly breaks down rocks and minerals, releasing the base cations to the soil in dissolved form and making them available to plant roots. This process contributes to the pool of adsorbed exchangeable base cations on the soil. These base cation reserves are gradually leached from soil with drainage water, but they are constantly resupplied through weathering and to a lesser extent through atmospheric base cation deposition. Elevated leaching of base cations from the soil by acidic deposition can deplete the soil of exchangeable bases faster than they are resupplied through weathering and deposition input (Cowling and Dochinger 1980).

2.4.   HYDROLOGY AND HYDRODYNAMICS

Water is provided to the watershed in the form of rain, snow, and cloud-water inputs. This water is either lost back to the atmosphere via evapotranspiration or it flows downhill in response to gravity. Drainage water contacts various soil horizons and geologic materials en route to streams, rivers, and lakes. The pathway the drainage water follows and its duration in the soil largely determine the extent of acid neutralization provided by the soils and bedrock to the water that flows into streams and lakes.

Streams or lakes in the same setting can vary in their sensitivity to acidification depending on the relative contributions of near-surface drainage water and deeper groundwater (Eilers et al. 1983; Chen et al. 1984; Driscoll et al. 1991). Acidic deposition that falls as precipitation directly on the lake or stream surface does not interact with watershed soil, but it may be neutralized by in-lake reduction processes that are largely controlled by hydraulic residence time (Baker and Brezonik 1988).

Water routing determines the degree of contact between drainage water and acidifying or neutralizing materials in the watershed. Surface-water ANC varies depending in part on the proportion of the flow that has contact with deep versus shallow soil horizons and the chemical makeup of those horizons. In general, the more subsurface contact and greater exposure to the products of mineral weathering, the higher the drainage-water ANC (Turner et al. 1990). This pattern can be attributed to higher soil base saturation, lower organic acidity, and in some watersheds greater SO_4^{2-} adsorption in subsurface soils. The accumulation in upper soil horizons of acidic material derived from atmospheric deposition is also important (Lynch and Corbett 1989; Turner et al. 1990). The depth and composition of soils, talus, and colluvium in combination with the watershed slope influence the residence time of water in the watershed, the extent to which runoff interacts with soils and geologic materials that can neutralize deposited acids, S adsorption on soils, and the extent of NO_3^- uptake by lichens, microbes, plants, and other biota. Thus, hydrological conditions and water-flow paths are critical in regulating both mineral acid anion leaching and acid neutralization in the watershed. Hydrology is an important controlling factor for acidic deposition effects in all settings (Turner et al. 1990), but this is especially the case in mountainous ecosystems such as those found in the Adirondack and Catskill Mountains regions.

2.5. WETLANDS

Wetlands are common in many parts of New York that contain acid-sensitive surface waters. For example, wetlands constitute about 14% of the land surface in the large Oswegatchie/Black River watershed in the southwestern Adirondack Mountains (Ito et al. 2005), the portion of New York that has been most affected by surface-water acidification from acidic deposition. The locations of wetlands that have been mapped by the Adirondack Park Agency are shown in Plate 4.

Two-thirds of the wetlands in the Adirondack Park (10.1% of the surface area of the Adirondack Park) are classified as being palustrine (Cowardin et al. 1979). This class includes nontidal wetlands dominated by trees, shrubs, or emergent vegetation and wetlands that are small (< 8 hectares [ha]) and shallow (< 2 meters [m] at the deepest location) and that lack vegetation. The palustrine designation groups together vegetated wetland types often called marsh, swamp, bog, fen, or prairie and the small shallow water bodies called ponds. Other wetlands in the Adirondacks can be broadly classified as lacustrine or riverine. The former include both wetlands and deep-water habitats that are larger than 8 ha, are located in a topographic depression, and have vegetative coverage of less than 30%. The latter include wetlands and deep-water habitats contained in a channel, usually with flowing water. Riverine wetlands are subdivided into Lower Perennial (low gradient, slow water velocity, mainly sand and mud substrate, occasional O_2 deficit), Upper Perennial (high gradient, fast water velocity, mainly coarse substrate with occasional sand, dissolved oxygen concentration generally near saturation), and Intermittent (flowing water for only part of the year).

The topic of acidification effects on wetlands is not well represented in the literature. No studies have documented the extent or magnitude of acidification effects of S and N deposition on wetland ecosystems in New York. Research in Ontario, Canada, has shown, however, that SO_4^{2-} reduction in wetlands removes SO_4^{2-} from solution and thereby increases the ANC of drainage waters, at least temporarily. Because levels of natural organic acidity tend to be high in wetland soils and water, these ecosystems have likely not been acidified to a large extent by the levels of acidic deposition that have occurred to date. It is more likely that atmospheric deposition has affected wetlands in New York via nutrient N enrichment pathways and via stimulation of Hg methylation. These topics are discussed in subsequent chapters of this book. When wet conditions follow drought in wetland-influenced surface waters, concentrations of SO_4^{2-} and NO_3^- typically increase and ANC decreases with increasing flow levels (Tipping et al. 2003; Watmough et al. 2004; Schiff et al. 2005).

Wetlands modify the response of watersheds to atmospheric deposition of S, N, and Hg in multiple ways. Wetland soils influence element cycling so as to alter the supply of mobile anions (SO_4^{2-}, NO_3^-) to downslope surface waters. Naturally occurring organic acids acidify the surface waters that pass through wetlands and alter the speciation of Al, affecting its toxicity. Wetlands also provide an anoxic environment with high levels of labile organic matter that is conducive to methylation of inorganic Hg, increasing its bioavailability and concentration in fish and other aquatic biota and in some species of terrestrial wildlife, including especially those that prey on fish. The methylation of inorganic Hg is well known to be carried out most readily by sulfate-reducing bacteria, which are obligate anaerobes. Since their metabolism directly involves the chemical reduction of SO_4^{2-} to sulfide, the rate of Hg methylation is regulated partly by the SO_4^{2-} concentration in surface (including wetland) waters.

At relatively low loads of S deposition, the rate of S input versus atmospheric Hg deposition input may ultimately control Hg bioaccumulation. At higher ambient levels of S deposition, the rate of SO_4^{2-} reduction may be saturated and additional S enrichment would be expected to have only a small effect on Hg methylation.

Ombrotrophic bogs receive their nutrients primarily from precipitation and atmospheric input. Such wetlands form in locations where precipitation exceeds evapotranspiration and there is a barrier that impedes drainage of the surplus water input. Bogs are typically dominated by *Sphagnum* spp. mosses and may or may not be forested. Over time, the *Sphagnum* builds a thick layer of peat, which raises the bog above the surrounding landscape, preventing water from runoff or from groundwater flux from entering. Ombrotrophic bogs are especially sensitive to the effects of N deposition. However, such effects appear to mostly be due to nutrient enrichment rather than acidification processes (Bobbink et al. 2010).

Wetlands are closely tied to a number of important biogeochemical processes that regulate watershed response to acidic deposition and atmospheric inputs of Hg. Wetlands often contain acidic soils, primarily due to the presence of large amounts of naturally occurring organic matter. Fulvic and humic acids are formed during the natural breakdown of organic matter and contribute organic acidity to soil and surface waters in wetland environments. The soils of ombrotrophic bogs and nutrient-poor fens, in particular, tend to exhibit a paucity of exchangeable base cations; this would buffer both organic and mineral acidity.

2.6.   SURFACE WATER

2.6.1.   Streams and Lakes

2.6.1.1.   *Acid-Base Chemistry*

The sensitivity of surface waters to acidification and the degree of acidification or recovery that occurs over time are both commonly measured by the ANC, which reflects the ability of water to neutralize strong acids added in the form of SO_4^{2-} or NO_3^-. Each can be contributed to a watershed by air pollution in the form of S or N deposition. ANC values can be positive or negative. Waters that have ANC values below 0 µeq/L during the summer or fall index period when it has not been raining are defined as chronically acidic. Surface waters with ANC ≤ 50 to 100 µeq/L are generally considered to be potentially sensitive to acidification. Those with ANC values over 50 or 100 µeq/L are generally considered less sensitive or insensitive. When the ANC is low and especially when it is negative, stream-water pH is also low (less than about 5 to 6). Low pH indicates that acidity levels are high, which can have adverse effects on fish and other aquatic species. A decrease in ANC over time is called acidification. The capacity of a watershed to resist decreases in ANC is determined mainly by the amounts of base cations relative to the amounts of acidic anions in the water. The base cations include Ca^{2+}, Mg^{2+}, K^+, and Na^+; they are mostly derived from the soils and ultimately from the rocks that break down to form those soils. The acidic anions include mainly SO_4^{2-}, NO_3^-, and chloride (Cl^-) and in some cases also organic acid anions and are mostly derived from atmospheric deposition of S and N and wetland and organic soil influence. Exceptions include coastal areas having high Cl^- from ocean spray and areas impacted by geologic S or applications of road salt.

The most common indicators of surface-water acidification are ANC, pH, and the concentration of Al_i. Recently, the base-cation surplus index was developed to better account for the complicating effects of natural organic acidity (Lawrence et al. 2007). Acidification of surface waters is most often assessed on the basis of loss of ANC, which is often accompanied by a decrease in pH and base-cation surplus and an increase in Al_i.

Lakes and streams in the Adirondack and Catskill Mountains regions have heightened sensitivity to acidic deposition (Sullivan 2000; Lawrence et al. 2008a). Many Adirondack lakes and streams and many Catskill streams have been acidified by atmospheric deposition of both S and N (Driscoll et al. 1991; Sullivan et al. 2006a; Lawrence et al. 2008b). As a consequence, many of these surface waters are chronically low in ANC. Of the 1,489 lakes surveyed by the Adirondack Lakes Survey (ALS) in the mid-1980s, 24% had summer pH values below 5.0, 27% were chronically acidic (ANC ≤ 0), and an additional 21% were probably susceptible to episodic acidification (ANC between 0 and 50; Kretser et al. 1989; Driscoll et al. 2007a). However, surveyed lakes were not statistically selected and therefore cannot be used for making regional population estimates. In addition, some sampled lakes are smaller than 1 ha in area. Such small lakes may in many cases grade into wetlands with high natural organic acidity.

The Shawangunks region of New York has also been identified as an area characterized by acid-sensitive streams. Located just southeast of the Catskill Mountains, this region is underlain by the Shawangunk formation (orthoquartzite, conglomerate, and sandstone) and Martinsburg shale. There is also a shallow ≤ 60 cm layer of glacial till. The main stream in the area, Coxing Kill, shows a steep pH gradient and distinct changes in Al chemistry with elevation (Schultz et al. 1993).

As discussed in the previous sections, the sensitivity of waters to acidification from acidic deposition is determined mainly by the types of rocks found beneath the drainage waters, the characteristics of watershed soils, and the flow paths drainage water follows through the watershed. Effects can be complicated. As an example, in very general terms, the geology controls soil characteristics that interact with precipitation, topography, and air pollution to determine soil-water and surface-water chemistry, which affect trees and fish. If the underlying geology is poor in base cations and water drains through the soil very quickly following shallow flow paths, the soil and water in the watershed will tend to have poor ability to neutralize acids deposited from the atmosphere.

Watershed processes control the extent of ANC contribution from soils to drainage waters as acidified water moves through terrestrial systems. The concentration of acid anions in solution, including mainly SO_4^{2-}, NO_3^-, and organic acid anions, is an important driver of these processes. In upland systems, organic acid anions are largely produced in upper soil horizons, but they precipitate out of solution as drainage water percolates into the deeper mineral soil horizons and they become adsorbed to soil particles. This process limits the influence of organic acids on the ANC of drainage waters, except in wetlands, where dissolved organic carbon (DOC) concentrations can be quite high.

Soil acidification and neutralization processes reach a threshold at some depth in the mineral soil (Turner et al. 1990), and drainage waters below this depth generally have relatively high ANC and are not prone to acidification. Acidic deposition allows soil acidification and cation leaching to occur at greater depth and causes water that is rich in SO_4^{2-} or NO_3^- to flow into streams and lakes. If these acid anions are charge-balanced by H^+ or Al^{n+} cations, the water will have low pH and could be toxic to aquatic biota. If they are charge-balanced by base cations such as Ca^{2+}, the base cation reserves of the soil can become depleted but the surface water will not be acidified (Sullivan 2000).

Organic acids in surface waters originate from the degradation of (mainly plant) biomass in upland areas, wetlands, riparian zones, the water column, and stream and lake sediments (Hemond 1994). High concentrations of dissolved organic matter often derive from extensive wetlands and/or organic-rich riparian areas in the watershed (Hemond 1990; Sullivan 2000). Organic acids exert a strong influence on surface-water acid-base chemistry, particularly in dilute waters having moderate to high (greater than about 400 µmol/L) DOC concentrations. Organic acids typically include a mixture of functional groups having both strong and weak acid characteristics. Some streams and lakes are naturally acidic as a consequence of high concentrations of DOC. Organic acids also provide buffering to minimize pH change in response to inputs of SO_4^{2-} and NO_3^- from acidic deposition. Dissolved organic carbon forms complexes with Al_i, reducing or eliminating its toxicity. Many streams and lakes in the Adirondack Mountains are chronically acidic or low in ANC due to the presence of organic acids.

The National Acid Precipitation Assessment Program (NAPAP; 1991) concluded that about one-fourth of all acidic lakes and streams surveyed in the National Surface Water Survey (NSWS; Linthurst et al. 1986; Kaufmann et al. 1988) conducted in the 1980s were acidic largely as a consequence of organic acids. A survey of about 1,400 lakes in the Adirondack Mountains by the Adirondack Lakes Survey Corporation (ALSC; Kretser et al. 1989) included many small lakes and ponds (1 to 4 ha in size) with high DOC. More than one-third of those had pH < 5 that was judged to be mainly due to the presence of organic acids (Baker et al. 1990b).

Specification of the acid-base character of dissolved organic acids is uncertain, and acid-base behavior cannot be described using a single H^+ dissociation constant (pK_a). Organic acids in surface waters comprise a complex mixture of acidic functional groups, some of which are quite strong. Some ionization occurs at pH values well below 4.0 (Driscoll et al. 1994; Hemond 1994). Various modeling approaches have been used to estimate the acidity of organic acids in surface waters, often as simple organic acid analogs that have different pK_a values (Oliver et al. 1983; Perdue et al. 1984; Driscoll et al. 1994).

Leaching of base cations by acid anions in acidic deposition depletes the soil of exchangeable bases. The importance of this response is indicated by the observation that most streams and lakes in the northeastern United States are not exhibiting increases in ANC and pH that are equivalent to recent decreases in S deposition. This limited recovery of the acid-base chemistry of water can be at least partially attributed to base cation depletion from soil and the consequent decreased base cation fluxes from soil to surface water.

The scientific research community's understanding of the importance of the base cation response developed slowly. During the 1980s, the generally accepted paradigm of watershed response to acidic deposition was analogous to a large-scale titration of ANC (Henriksen 1980). Atmospheric inputs of acidic anions were believed to result in movement of those anions through soils into drainage waters with a near-proportional loss of surface-water ANC. This view was modified when Henriksen (1984) suggested that up to about 40% of the SO_4^{2-} contributed by acidic deposition could be balanced by increased movement of base cations from soils into surface water and the remaining 60% to 100% of the added SO_4^{2-} resulted in loss of surface-water ANC. During the late 1980s, it became increasingly clear that more than 40% of the added SO_4^{2-} was typically neutralized by base cation release and that historical ANC and pH of surface waters were not as high as believed earlier, meaning that the magnitude of acidification was not great. This understanding developed largely from paleoecological studies (e.g., Davis et al. 1988; Charles et al. 1990; Sullivan et al. 1990) that suggested that past changes in lake-water pH and ANC had been small relative to estimated increases in lake-water SO_4^{2-} concentrations since preindustrial times (Sullivan 2000). The expectation that changes in acidic deposition were accompanied mainly by changes in ANC and pH was replaced by the realization that past changes in SO_4^{2-} and NO_3^- concentration were also accompanied by substantial changes in base cation concentration. Lakes and streams have not been acidified by historical deposition to the extent that was earlier believed. As a consequence, surface-water ANC and pH should not be expected to show a large increase in response to reduced emissions and deposition of S and N. The magnitude of the base cation response has clearly limited the extent of surface-water acidification caused by acidic deposition. However, this same response has contributed to base cation deficiencies in some soils, with associated adverse terrestrial effects and delayed aquatic recovery from acidification (Sullivan 2000).

The Great Lakes and their watersheds are so large that they are not sensitive to acidification from atmospheric deposition. Rather, impacts on surface waters in the Great Lakes region occur on smaller inland lakes and soils in acid-sensitive portions of the region.

2.6.1.2. *Nutrients*

Nutrient enrichment is a suite of environmental changes that can occur when the availability of a key nutrient is increased. Nutrient enrichment caused by atmospheric N deposition and/or other point or nonpoint sources of N pollution can cause excessive growth of algae in N-limited surface waters. Ecological responses to atmospheric N input depend to a large extent on the nutrient limitation status of the water and the dynamics of the invertebrate grazer populations. Comparisons of nutrient and grazer regulation of phytoplankton have focused heavily on differences in nutrient loading and the structure and dynamics of the food web (Carpenter et al. 1991; Elser and Goldman 1991; Saunders et al. 2000). For example, Saunders et al. (2000) conducted in situ enclosure experiments on 18 Adirondack lakes to investigate responses to macrozooplankton grazing and N and P nutrient limitation. Results depended on differential hydrologic flow paths. Phytoplankton of clear-water drainage lakes were P-limited. Phytoplankton in seepage lakes responded to additions of either N or P or N + P with increased chlorophyll *a* (chl *a*). For many drainage-lake phytoplankton assemblages, crustacean grazing was as important as nutrient limitation in driving primary production (Saunders et al. 2000).

The extent to which atmospheric N deposition alters the trophic state of fresh waters in New York is poorly known. Sensitivity to such effects is largely determined by the extent to which algal and plant growth is limited by the availability of N. Bergström and Jansson (2006) suggested that the majority of lakes in the northern hemisphere may have originally been N-limited and that atmospheric N deposition has contributed enough N to change the stoichiometric balance of N and P in lakes. As a consequence, P limitation is more commonly observed today. Nevertheless, a meta-analysis of 990 freshwater field experiments to determine the patterns of autotrophic nutrient limitation found that N limitation of stream benthos, lake benthos, and phytoplankton was as common as P limitation from sites in all biomes worldwide (Elser et al. 2007). Thus, addition of N to aquatic systems that are N-limited or co-limited by both N and P (Axler et al. 1994) is expected to stimulate primary production and modify the freshwater biological community.

2.6.2. Estuaries and Near-Coastal Marine Waters

Estuaries and near-coastal marine waters are highly sensitive to N input, which contributes to eutrophication. The largest sources of N for New York's estuaries and other coastal waters are probably human activities related to food production and consumption, followed by atmospheric N deposition. Production and consumption of food result in leaching of N and other nutrients from agricultural lands to surface waters and human waste contribution to streams and rivers, commonly processed through centralized wastewater treatment systems. Fertilizer runoff and runoff from livestock operations are the most important agricultural sources of N to drainage water. Wastewater treatment plants contribute nutrient (including both N and P) pollution to coastal waters through permitted and accidental wastewater discharge. Septic systems can also be significant nutrient (especially N) sources in rural areas.

The response of estuaries to nutrient inputs is regulated to some extent by climatic factors, including precipitation to the watershed and temperature, both of which influence thermal stratification. Climatic influence on eutrophication has been pronounced in portions of Long Island Sound. Unlike western Long Island Sound, the Hudson River–Raritan Bay has not shown a significant trend in water temperature or temperature stratification during the latter several decades of the twentieth century (O'Shea and Brosnan 2000). Changes in wind frequency and duration may also alter the extent of turbulent mixing and hydrodynamics in large lakes and estuaries.

CHAPTER 3

Principal Stressors

The effects of atmospheric deposition of S, N, and Hg air pollutants are governed by three different kinds of processes. The first concerns emissions of air pollutants into the atmosphere, largely from electricity-generating facilities, agriculture, and transportation. The second is transport within and deposition from the atmosphere. Emitted pollutants are transported with the prevailing winds, modified by physical and chemical reactions, and eventually deposited to the earth's surface through atmospheric deposition. The third kind of process involves the transformations that occur as these constituents move through the soil and water in watersheds and the resulting effects on ecosystem structure and function.

3.1. SULFUR, NITROGEN, AND MERCURY EMISSIONS AND DEPOSITION

A variety of air pollutants, including S, N, and Hg, are emitted into the atmosphere by energy-generating facilities, industry, motor vehicles, agriculture, and other sources. These substances can be chemically modified in the atmosphere and transported short or long distances, where they can be deposited and adversely affect humans and sensitive resources.

Airborne NO_x, NH_x, and SO_x particles, gasses, precursors, and transformation products are removed from the atmosphere by wet, dry, and occult deposition. This transfer is commonly called "acid rain", although rain accounts for only part of the transfer. Pollutants move from the air to the ground surface in rain, snow, and clouds and as dry particles and gases. The overall transfer process is called atmospheric deposition. This deposition can fall on plant foliage, bare rock, soil, water, or snow. Locations of wet- and dry-deposition monitoring stations in New York that are operated by the National Acid Deposition Program (NADP) and the Clean Air Status and Trends Network (CASTNet) are shown in Figure 3.1. Precipitation and snowmelt combine with moisture that fog or clouds transfer to the terrestrial ecosystem to transport atmospherically deposited substances to soils. This transfer process lowers the long-term buildup of these pollutants in the atmosphere and thus moderates the potential for direct human health effects caused by their inhalation. However, deposition also transfers atmospheric pollutants to other environmental compartments where they can alter the structure, function, diversity, and sustainability of complex terrestrial, wetland, and aquatic ecosystems (U.S. EPA 2008).

Acidic deposition is commonly measured in units of mass of the deposited substance per unit of ground area over the course of one year. Common units of deposition measurement include kilograms per hectare per year (kg/ha/yr) or milliequivalents per square meter per year (meq/m²/yr). Several networks operate monitoring sites throughout the United States to measure the atmospheric concen-

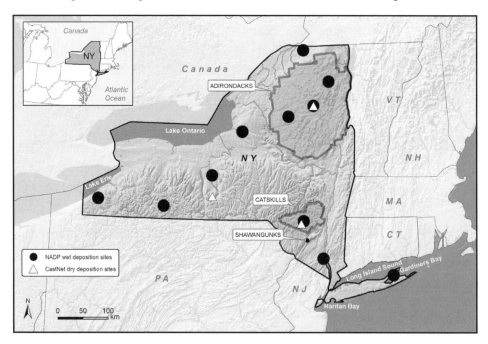

FIGURE 3.1. Locations of NADP/NTN wet and CASTNET dry deposition monitoring sites in New York. Data source: http://nadp.sws.uiuc.edu/.

trations and wet or dry deposition of pollutants. Wet-deposition monitoring data from the National Acid Deposition Program/National Trends Network (NADP/ NTN) for a recent three-year period are summarized in Figure 3.2. Monitoring sites are present at sufficient density to allow spatial interpolation of wet-deposition estimates across New York and the entire eastern United States.

Dry and occult deposition are both more difficult to measure than wet deposition. Dry deposition can be estimated from measurements or model projections of pollutant concentrations in the air, assuming a rate of transfer from the air to the ground. CASTNet operates a relatively sparse network of dry deposition monitors. Some of these monitoring stations are located in New York, including in the Adirondack and Catskill Mountains (Figure 3.1).

Occult deposition has been measured at only a few locations. It tends to be high at coastal (fog-influenced) and high-elevation (cloud-influenced) locations. In general, cloud deposition in the eastern United States is assumed to occur primarily at elevations above about 1,000 m and to be quantitatively important above about 1,500 m. Cloud deposition can constitute as much as half of the total deposition in high mountain areas, although few watersheds in New York have appreciable landscape at such a high elevation.

The estimates for cloud ion chemistry under clean air conditions at Whiteface Mountain in the Adirondacks suggest a pH somewhere in the range of about 5.0

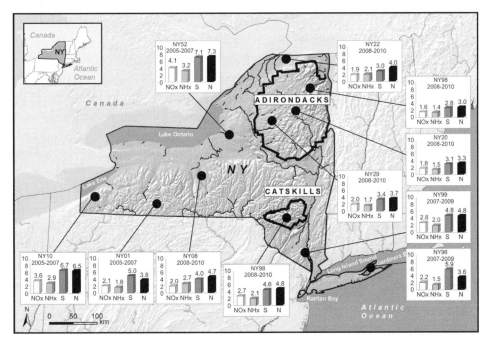

FIGURE 3.2. Measured values of wet atmospheric deposition of oxidized nitrogen (NO$_x$), reduced nitrogen (NH$_x$), total inorganic nitrogen (N), and total sulfur (S), expressed as a three-year average annual measurement at each of the NADP/NTN deposition monitoring sites in New York. Units of deposition are kg/ha/yr of S or N. Data source: http://nadp.sws.uiuc.edu/.

to 5.25 under conditions that are relatively unimpacted by air pollution. This is based on measurements for the cleanest clouds that are commonly intercepted by the mountain at the monitoring location. The pH, in turn, is linearly related to the cloud-water concentrations of SO_4^{2-} and NO_3^- (Dukett et al. 2011). Using these relationships, Dukett et al. estimated that cloud concentrations of SO_4^{2-} and NO_3^- in 1994 were about 27 and 13 times higher, respectively, than concentrations in clear air. By 2009, these multipliers had decreased to about 4 and 3 times the clean air values. Thus, there has been a substantial decrease since 1994 in the cloud concentrations of SO_4^{2-} and NO_3^- at Whiteface Mountain.

Coniferous trees have greater leaf surface area than do deciduous trees, shrubs, grasses, or forbs. In addition, needles remain on coniferous trees and intercept aerosol pollution year round. As a consequence, dry deposition to conifer needles is generally higher than dry deposition to other vegetative surfaces. Dry deposition to hardwood forest vegetation is generally higher than dry deposition to nonforested areas.

Some emissions are deposited to the ground in proximity to the source. When the emitted S or N is transported vertically upward by convection to the middle and upper troposphere, however, it can be transported long distances from the emission source areas before depositing to the earth's surface. Atmospheric transport varies with chemical and meteorological conditions along the path from the emissions source to the location where pollutants are deposited from the air to the ground.

3.1.1. Sulfur Emissions and Deposition
3.1.1.1. *Sulfur Emissions into the Atmosphere*

Sulfur is emitted into the atmosphere mainly as SO_2, which is released when coal or other S-containing fuel is burned. This SO_2 is readily oxidized to SO_4^{2-}. Amounts of S emitted into the atmosphere vary across the airshed that contributes pollutants to New York (Plate 5). The highest emissions occur from coal-fired power plants and from human activities in and around major population centers, especially from centers of energy, agricultural, and industrial development.

Atmospheric deposition of S in New York originates from emissions that occur largely outside New York's boundaries. Emissions of S into the atmosphere are generally higher in neighboring upwind states, especially in the Ohio River Valley, than they are in New York (Figure 3.3). New York residents consume some of the electricity generated in the Ohio Valley, however. The problem of S air pollution is regional to national in scope and involves landscapes well beyond New York's borders. Emissions of SO_2 have decreased substantially since 2005, largely in response to a move within the energy production sector of the United States toward less reliance on coal.

3.1.1.2. *Sulfur Deposition*

Both total S and total N deposition in New York likely increased nearly tenfold during the twentieth century due to energy, agricultural, industrial, and transportation development. Patterns of deposition have been complex and variable due to the influence of meteorology, atmospheric transport, atmospheric chemistry, landscape topography, precipitation, and vegetation type. In complex mountainous terrain in New York, the total amount of deposition per unit time can vary severalfold over relatively short distances.

Wet deposition of S measured by the NADP/NTN was mapped for the eastern United States for this book, using Grimm and Lynch's (1997) approach, which uses a statistical interpolation procedure that corrects for the effects of changing elevation on precipitation amount (data provided by J. Grimm). Estimates of dry deposition for 2006

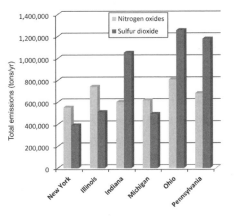

FIGURE 3.3. Estimated total annual emissions of sulfur dioxide and nitrogen oxides in New York and the five principal upwind states. These estimates were generated by the U.S. EPA for the year 2005. For each state, the nitrogen oxide bar is on the left and the sulfur dioxide bar is on the right. Source: EPA National Emissions Inventory.

were added to interpolated wet deposition for the analysis presented here, using output from the Community Multiscale Air Quality (CMAQ) atmospheric transport model (data provided by R. Dennis, U.S. EPA). The model provides estimates of total wet plus dry S deposition (Plate 6), although uncertainty is especially high for estimating dry deposition. These estimates of wet plus dry deposition are relatively coarse (12-km CMAQ grid cells) and do not fully capture spatial variation, especially in the mountainous areas. They do, however, provide a reasonable approximation of patterns in total S deposition that potentially impacts sensitive ecosystems. There may be additional cloud deposition at the highest elevation locations in some areas.

Wet deposition of major ions, including SO_4^{2-}, is relatively well characterized throughout the northeastern United States, largely because of the extensive networks of monitoring stations, especially in the NADP/NTN network. Highest values of S wet deposition in the eastern United States are generally found in a band from the southern Appalachian Mountains to New York and New England. Measured values of wet S deposition have decreased markedly in New York since attainment of peak values in the 1970s and 1980s (Figure 3.4). Wet N deposition, in contrast, remained high and relatively constant during the last two decades of the twentieth century but subsequently declined (Figure 3.4).

Dry deposition inputs of S and N are poorly known and cannot be directly measured. In general, dry deposition is highest in close proximity to pollution emission sources and varies significantly with differences in vegetative canopy and meteorological conditions. Networks for monitoring atmospheric chemistry,

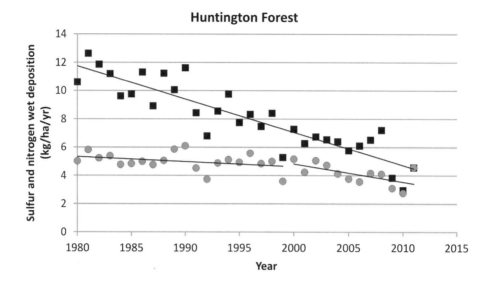

FIGURE 3.4. Total wet sulfur (squares) and nitrogen (circles) deposition at the Huntington Forest NADP/NTN monitoring site in the Adirondack Mountains over the period of record, 1980–2011. Data Source: http://nadp.sws.uiuc.edu/.

which provide the basis for estimating dry deposition, are not as well developed as the networks for monitoring wet deposition.

The environmental impacts of atmospheric S on soil, plants, lichens, or aquatic organisms are dependent on deposition from the atmosphere to the ground, water, or vegetation surface. Inputs of all forms of deposition first interact with the vegetative canopy unless they fall directly on the water or other unvegetated surfaces. This interaction can occur from a few centimeters above the ground (in grasslands) to tens of meters above the ground (in forests). In the canopy, deposited pollutants can be taken up by the foliage or by organisms that live in the canopy or on the leaf surface. Because N uptake occurs in the canopy, only a portion of the dry-deposited N is transported from the canopy to the soil in throughfall and stemflow. In contrast, most of the deposited S moves through the canopy to the soil, where it can be adsorbed on soil surfaces or transported with drainage water into surface water. Total S deposition varies substantially with elevation. For example, Lovett et al. (1999) observed a pronounced elevational gradient in the Catskill Mountains, due in part to differences in the tree canopy.

3.1.2. Nitrogen Oxide and Ammonia Emissions and Deposition

Urban, industrial, and agricultural development have significantly changed the amount of N released by human activities into the environment (Galloway et al. 1994; Vitousek et al. 1997). Emissions of N are high (> 20 tons/mi^2/yr) in counties in southern New York in the vicinity of New York City and Long Island.

County-level N emissions are also relatively high (5 to 20 tons/mi^2/yr) in the Great Lakes portion of the state. Emissions are generally lower in and around the Adirondack and Catskill Parks (Plate 7). Increased human population and development in the coastal zone, in particular, have contributed to the movement of point and nonpoint sources of N (and P and other pollutants) into estuaries and near-shore coastal environments.

3.1.2.1. *Nitrogen Emissions into the Atmosphere*

Global emissions of reactive N (N_r) to the atmosphere have increased significantly over the past century. This has largely been due to three causes: (1) increased cultivation of legumes, rice, and other crops that promote conversion of molecular N (N_2) to organic N through biological N fixation; (2) combustion of fossil fuels, which converts both atmospheric N_2 and fossil N to reactive NO_x; and (3) synthetic N fertilizer production via the Haber-Bosch process, which converts nonreactive N_2 to N_r to sustain food production and some industrial activities (Galloway and Cowling 2002; Galloway et al. 2003).

The oxidized forms of N (primarily NO_2) are emitted mainly by motor vehicles, power plants, and industrial facilities. The reduced forms (primarily ammonia, NH_3) are emitted mainly by agriculture, through volatilization of N contained in animal manures and fertilizers.

Quantities of NO_x and NH_x emitted into the atmosphere vary across the landscape; the highest emissions are in and near major human population centers and centers of energy, agricultural, and industrial development. Total NO_x emissions sources in and near New York are mapped by county in Plate 7, based on U.S. EPA emissions estimates for the United States in 2008 (Western Regional Air Partnership 2013). Emissions of NO_x have been especially high to the west and southwest of the location of the major sensitive resource receptors in New York. Highest upwind emissions are from Ohio, Indiana, and Pennsylvania (Figure 3.3). These N emissions are transported with the prevailing winds.

3.1.2.2. *Nitrogen Deposition and Other Watershed N Sources*

Atmospheric N deposition in New York is higher than in many other areas of the United States. Golden and Boyer (2009) estimated spatial patterns across New York in NO_3^- and NH_4^+ wet deposition using data from 31 stations that monitor the volume and chemistry of precipitation. These data had been collected by NADP/NTN, the Atmospheric Integrated Research Monitoring Network, New York State Acid Deposition Monitoring Network, and sites operated by the Institute for Ecosystem Studies. Based on measurements for 2002 to 2004, total inorganic N wet deposition in New York ranged from about 4.7 to 10.5 kg N/ha/yr. Hot spots of relatively high wet N deposition were identified according to location and elevation. In general, oxidized inorganic N deposition was higher than reduced inorganic N deposition, although NH_x deposition may have been underestimated by up to about 15% because of loss of NH_4^+ from the deposition-monitoring collectors during sample storage (Meyers et al. 2001; Grimm and Lynch 2005).

Relatively high levels of total inorganic N deposition in New York were estimated to occur in the plateau regions east and southeast of the Great Lakes, the western Adirondack Mountains, and the western Catskill Mountains. Deposition generally decreases regionally in these areas from southwest to northeast (Ollinger et al. 1993; Ito et al. 2002; Golden and Boyer 2009). The estimates of total inorganic wet N deposition Golden and Boyer (2009) reported for the period 2002–2004 were about 20% higher than estimates based only on data from the nine NADP/NTN monitoring sites in New York.

Temporal patterns in wet N deposition at one monitoring site (Huntington Forest) in the Adirondack Mountains indicated that N deposition was high for several decades but has decreased in more recent years (Figure 3.4). Model estimates from CMAQ of dry N deposition are available at the time of this writing for 2006. These simulations of dry N deposition were combined with interpolated wet deposition for 2006 to yield a continuous coverage of total deposition estimates for 2006 (Plate 8). Atmospheric N deposition in New York is highest (> 12 kg N/ha/yr) in the southern part of the state, in and around New York City and much of Long Island, and in the Great Lakes region (Plate 8). These high levels of N deposition correspond with the locations of many nutrient-sensitive estuarine, marine, and freshwater resources. Acid-sensitive resources are more common in the Adirondack and Catskill Parks, where N deposition is somewhat lower (< 12 kg N/ha/yr).

In upland locations, deposited N that is not taken up in the tree canopy falls to the ground as throughfall and/or stemflow, where plants, bacteria, and fungi compete for it. This competition for deposited N plays an important role in determining the extent to which N deposition will stimulate plant growth or cause acidification and the degree to which added N is retained in the ecosystem (U.S. EPA 1993). Landscape features such as elevation, aspect, and forest edge can play important roles in creating high variability in total deposition rates in complex terrain (Weathers et al. 2000).

Humans increased the average flux of N to coastal waters of the United States over the last century by fourfold to fivefold, and increases were even larger in some areas (Howarth et al. 2000a; National Research Council [NRC] 2000). These N fluxes are attributed to both atmospheric and non-atmospheric sources and may increase further in the future in response to human population growth, land development, and intensified agriculture (Scavia and Bricker 2006), adding further to the N flux received by coastal waters from atmospheric and non-atmospheric sources. The effects of N deposition to near-coastal areas in New York are of particular concern because near-coastal surface waters are often N-limited, because N loading from large river watersheds funnel into them, and because many species of estuarine biota are especially sensitive to effects of eutrophication.

Alexander et al. (2001) modeled N loading in 40 coastal water bodies throughout the conterminous United States using the Spatially Referenced Regression on Watershed Attributes (SPARROW) model. This mechanistic model establishes correlations between the stream N load and the spatial distributions of N sources, landscape characteristics, and stream properties. Model results suggested a wide range of atmospheric contributions to the total N load delivered to the study estuaries, from 4 to 326 kg N/km²/yr, with a median of 20% contributed from the atmosphere (range 4 to 35%). The largest modeled N source was agriculture, which accounted for more than 33% of the total N load in most watersheds. Municipal and industrial point sources represented the largest proportion of the total input (35 to 88%) in half of the North Atlantic study watersheds, including Hudson River–Raritan Bay (40%). The largest proportion of N input was estimated to be derived from agricultural fertilizer in Gardiners Bay (38%) and from atmospheric deposition in Long Island Sound (35%).

Source contributions to the total N export from watersheds to major estuaries suggested that the atmospheric input accounted for a higher percentage of the total N input in Long Island Sound than any other major estuary in the conterminous United States. Hudson River–Raritan Bay was also estimated to have a relatively high proportional contribution of atmospheric sources of N (26%). The estimate for Gardiners Bay was substantially lower, 11% of the total N loading.

Castro et al. (2003) estimated N inputs to 34 estuaries along the Atlantic and Gulf Coasts of the United States, including Long Island Sound and Hudson River–Raritan Bay. The major sources of inorganic N contributed to the watersheds of these estuaries by human activities were assumed to include fertilizer application to crops and lawns, N fixation by crops and pastures (mainly legumes), import of food into the watershed for human and livestock consumption, and atmospheric deposition of N. Riverine and watershed N sinks were estimated. For 11 of the 34 estuaries selected for study, urban sources of N were estimated to comprise the majority of the total N load contributed to the estuary. This was the case for both of the study estuaries that were located in New York: Long Island Sound and Hudson River–Raritan Bay. Human sewage was estimated to contribute 56.7% and 65.5% of the N exported to these two estuaries, respectively. For the 11 apparently urban-dominated watersheds, the estimated total human-caused N inputs were highly correlated ($r^2 = 0.88$) with human population density. On average, each human in these watersheds contributed 5.6 kg N/ha/yr to the urban-dominated estuaries. This estimate is consistent with earlier estimates of per capita human N excretion (4.4 to 5.2 kg/N/yr; Howarth et al. 1996). The atmospheric contributions of N to Long Island Sound and Hudson River–Raritan Bay were estimated at 17% and 16.4% of the totals, respectively. Atmospheric N deposition to the watershed was estimated to be dominant in only one northeastern study watershed, Barnegat Bay, New Jersey, where atmospheric N deposition accounted for an estimated 51% of the total estuarine N loading. This high estimate for the atmospheric N contribution was largely because sewage effluent generated in the Barnegat Bay watershed was discharged to the ocean offshore, bypassing the estuary (Castro et al. 2003).

Dry deposition amount varies with the form of N (e.g., oxidized, reduced, organic) and the morphology of the receiving surfaces. It depends on the concentrations of N species in the atmosphere and the affinity of the N-containing gas or aerosol for deposition to exposed environmental surfaces (reflected in the deposition velocity). The deposition velocity to open freshwater surfaces is substantially (perhaps 2 to 5 times) lower than to terrestrial vegetation; over open saltwater, dry deposition of NO_3^- particles can be much higher, of the same order of magnitude as wet NO_3^- deposition (Paerl et al. 2002a). Thus, dry deposition rates are different for oxidized versus reduced N and for freshwater versus saltwater surfaces. Dry deposition of oxidized N to terrestrial surfaces may approximate wet deposition, whereas dry deposition of reduced N is more commonly less than wet deposition (Clarke et al. 1997). Model analyses suggested that the gradient of dry deposition of oxidized N across estuaries with a salinity gradient will roughly track salinity (Paerl et al. 2002a). The delivery of atmospheric oxidized N to the upper, freshwater portions of an estuary is expected to be dominated by wet deposition. Dry deposition is proportionately more important in delivering N to the lower estuary. The major source of reduced N to an estuary is often deposition of reduced N to the lower portions of the estuary.

3.1.3. Mercury Emissions and Deposition

3.1.3.1. *Mercury Emissions into the Atmosphere*

Mercury is emitted to the atmosphere primarily as elemental Hg or divalent Hg (Hg[II]) from a variety of sources, both natural and human-caused. Some Hg degasses naturally from the earth's crust; natural emissions vary from location to location. The most important anthropogenic sources have included coal-fired power plants, mining, incineration of hazardous wastes, cement manufacturing, production of chlorine, and the breakage and spillage of products that contain Hg. Globally, mining is the greatest source of Hg emissions (U.S. EPA 2015). Coal combustion is the second largest source globally but the largest source in the United States.

In 2012, the U.S. EPA established national Hg emissions standards. This led to new emissions control strategies at power plants. The EPA was litigated on the Hg rule and the outcome of that litigation is not clear at the time of this writing (Haeuber 2013). In January 2013, the international community agreed on the previsions of a global treaty (the Minamata Convention) to regulate use and release of Hg.

3.1.3.2. *Mercury Deposition*

Atmospheric Hg deposition in many areas has probably increased with industrialization by one order of magnitude or more (Steinnes et al. 2005). Cores collected from peat deposits and lake sediments provide evidence that industrial emissions and deposition of Hg over the twentieth century contributed to Hg accumulation in sensitive watersheds (Rood et al. 1995; Engstrom and Swain 1997). However, the complexities of atmospheric Hg chemistry and generally low atmospheric concentrations make quantification of Hg deposition difficult (Lynam and Keeler 2005).

Mercury deposition is very much a global issue because elemental Hg has a long atmospheric residence time of many months. As a consequence, a sizeable component of the Hg deposited in New York may have originated in Asia or other locations far from New York's borders. Atmospheric deposition of Hg is bidirectional. Once deposited, Hg(II) can re-volatilize from the earth's surface back into the atmosphere. Conversely, it can be methylated in anaerobic soils and microaerophilic zones or transported by runoff and leached to a water body. Methylation also occurs in the water body in sediments and in anoxic bottom waters. The spatial extent and duration of anoxia (absence of dissolved oxygen) in water zones is highly variable depending on seasonal weather patterns, short-term wind mixing and runoff events associated with storms, lake productivity (i.e., trophic state), and the morphometry of the lake basin (embayments, shallow littoral areas, sediment characteristics, etc.). Whether Hg is methylated in these anoxic zones may also depend on the concentration of SO_4^{2-} in soil and in sediment interstitial water (pore water). Inputs of SO_4^{2-} may stimulate Hg methylation in low SO_4^{2-} zones where SO_4^{2-} availability limits the growth of sulfate-reducing bacteria. Either Hg(II) or MeHg can be volatilized from the water back into the atmosphere. Mercury incorporated into the organic horizon of the soil can be retained until it is re-emitted to the atmosphere with burning or volatilization (Dicosty et al. 2006).

The spatial patterns of the sensitivity of the ecosystem to potential damage from Hg exposure across the United States can be evaluated in part using maps of interpolated wet Hg deposition. Wet deposition data are available from the NADP's Mercury Deposition Network (MDN) at monitoring sites across the United States. In general, wet deposition of Hg is highest in Florida, the Gulf states, and north into the heartland. It is lower in the West, Upper Midwest, Northeast, and mid-Atlantic regions (Plate 9). Dry Hg deposition data are less available regionally and are more uncertain. The NADP has assembled a monitoring network, the Atmospheric Mercury Network (AMNet; NADP 2014), of gaseous and particulate Hg measurements at multiple sites across the United States. These data may be helpful in the future for estimating dry Hg deposition fluxes. However, at the present time, there is considerable uncertainty in the methodology used in modeling the deposition velocity needed to make dry deposition estimates.

The current scientific understanding is that elemental Hg moves through plant stomata into leaf tissue and accumulates in the leaf throughout the growing season. Long-term measurements of the Hg content of litterfall and throughfall in forests provide an indication of changes in dry and total Hg deposition. This is because dry Hg deposition is transferred to the soil with litterfall. The amount of litterfall is generally considered to approximate the amount of dry deposition (Risch et al. 2012). After Hg is transferred to the forest floor with litterfall, some of the deposited Hg is re-emitted; the remainder is mostly incorporated into the soil.

Monitoring of wet Hg deposition occurs at four sites in New York within the MDN: Huntington Forest (Adirondacks), Biscuit Brook (Catskills), Rochester, and the Bronx. An additional site was in operation at West Point until 2010. The Huntington Forest site has been in operation since 1999 and the Biscuit Brook site since 2004. Sites in Rochester and the Bronx were added in 2008. The annual average concentration of Hg in precipitation showed statistically significant downward seasonal Kendall trends over the period of record at Huntington Forest and Biscuit Brook (Levine and Yanai 2012). Trends were also downward but not significant at the other sites that have been in operation for a more limited time period. In contrast, estimates of wet deposition of Hg (which is equal to the concentration in precipitation times the amount of precipitation) did not show significant trends at any of the sites because of annual variability in precipitation amount.

3.2. WATERSHED DISTURBANCE

Effects of S, N, or Hg deposition on sensitive ecosystems do not occur in isolation. They interact with a host of other natural and human-caused perturbations. Changes in land use or other watershed disturbance and associated changes in vegetative structure influence how an ecosystem responds to external stressors such as acidic deposition, nutrient enrichment, forest harvest, land disturbance, and climatic conditions. Some of these effects are described below.

3.2.1. Timber Harvest and Fire

Watershed export of N in drainage water typically increases after logging or other major ground disturbance, often reaching peak NO_3^- concentrations in stream water within a few years after the disturbance, then returning to background concentrations after about 5 to 10 years (Likens et al. 1978; Bormann and Likens 1979; Eshleman et al. 2000). Short-term increased NO_3^- leaching has been well documented after logging (Martin et al. 1984; Dahlgren and Driscoll 1994; Yeakley et al. 2003) and forest insect infestation (Eshleman et al. 1998, 2004).

Burns and Murdoch (2005) studied the effects of clear-cut logging of a northern hardwood forest in the Catskill Mountains on net mineralization and nitrification rates over a six-year period. Although NO_3^- leaching caused an increase in stream NO_3^- concentrations of 1,400 μeq/L within five months of harvesting, the in situ rates of N mineralization and nitrification remained essentially unchanged. Instead, the increased NO_3^- leaching largely reflected reduced N uptake by plant roots.

McHale et al. (2007) investigated the interactions among acidity and the concentrations of NO_3^-, Al, and Ca in stream water, soil water, and groundwater subsequent to clear-cutting. Their focus was on effects on the speciation, solubility, and concentrations of Al_i post-harvest. Export of Al_i increased fourfold during the first year after harvest. Concentrations of Al_i and NO_3^- in B-horizon soil water were highly correlated ($r^2 = 0.96$). Groundwater inflow into the stream contained high concentrations of base cations and relatively low Al_i; this spring water mixed with acidic, high-Al_i upper-mineral-soil water, yielding intermediate concentrations of Al_i in the stream after harvest. Within five years of harvest, soil water NO_3^- concentrations decreased below pre-harvest levels due to N demand for plant regrowth. However, groundwater NO_3^- concentrations remained higher than pre-harvest levels because groundwater has a long residence time. McHale et al. (2007) speculated that NO_3^- and Al_i increased in this study more than in previous studies of forest harvesting and that the recovery was also slower because the watershed had previously been acidified by acidic deposition. Pulses of high NO_3^- and Al_i occurred during high-flow conditions after the first growing season. There was 100% mortality of caged brook trout during snowmelt episodes the first spring season after harvesting; mortality decreased to 85% the second spring season and subsequently returned to pre-harvest levels (Baldigo et al. 2005).

Timber harvest removes nutrients from the watershed with the harvested trees and increases leaching losses of NO_3^- and nutrient base cations (Bormann et al. 1968; Mann et al. 1988). Logging contributes, in particular, to loss of N and Ca^{2+} from the soil (Tritton et al. 1987; Latty et al. 2004). The extent of nutrient loss depends in large part on the intensity of the logging and whether or not it is accompanied by fire (Latty et al. 2004). Fire is sometimes followed by establishment of N-fixing plant species that provide substantial new supplies of N_r to the soil (Johnson 1995; Johnson et al. 2004). The effects of logging and fire on nutrient cycling are influenced by subsequent shifts that occur in species composition and the degree to which C and N pools are altered in the mineral soil and the forest floor. Different tree species vary in their N cycling properties, especially in their influence on litter mass and quality (Finzi et al. 1998; Ferrari 1999; Ollinger et al. 2002).

Watershed disturbance by logging and fire also disrupts the normal flow of water through a watershed and can cause an increased flux of N acidity or increased contact between runoff water and soil surfaces. Forest harvesting affects post-harvest forest N demand because young forests grow rapidly.

Harvesting also reduces the likelihood of future N saturation, even if atmospheric N inputs are relatively high. Forest management affects soil erosion, nutrient cycling, and the buildup of organic material in soils. These, in turn, can influence the availability of base cations for acid neutralization.

Drainage water quality is affected by harvesting in multiple ways. Atmospheric dry deposition of S and N decreases because the canopy has been reduced or removed. Leaching of NO_3^- increases and in some cases causes a pulse of surface-water acidification. Base cations tied up in wood are transported off site with the harvested timber. Further loss of base cations accompanies the erosion that often follows logging, especially on steep slopes. Erosion depletes topsoil, including its base cation content. Regrowth of the forest may further affect drainage water quality through increased vegetative uptake of N and nutrient base cations by rapidly growing young trees, which accumulate base cations to a greater extent than anions. In order to balance the charge discrepancy, roots release an equivalent amount of protons, which acidifies the soil.

In areas such as the Adirondack and Catskill Mountains that experience relatively high levels of acidic deposition and consequent base cation leaching, there may be concern about sustainable timber production. Increased water flux through the soil after tree removal further increases the leaching loss of base cations from the soil (U.S. EPA 2008). Harvest-induced leaching losses have been estimated to range from 28 to 48 kg/ha/yr of Ca^{2+} and 7 to 16 kg/ha/yr of Mg^{2+} (Federer et al. 1989). C. Johnson et al. (1991) measured effects of logging on the acid-base chemistry of soil over a three-year period at Hubbard Brook Experimental Forest in New Hampshire, a forested ecosystem that shares many similarities with acid-impacted forests in the Adirondack and Catskill Mountains. Base saturation of the mineral soil B_h horizon decreased from 14% to 11% and pH decreased by 0.24 pH units.

Using the Photosynthesis and EvapoTranspiration-Biogeochemical (PnET-BGC) model, Chen and Driscoll (2004) simulated the response of five forested watersheds in the Adirondack and Catskill Mountains to changes in atmospheric N deposition and land disturbance. Simulation results suggested that forest harvesting caused increased leaching of base cations and NO_3^- from the watersheds. These changes also affected model projections of future pH and ANC of surface water. Both were lower in response to forest cutting compared to undisturbed conditions.

Because fire can cause a short-term increase in the concentrations of NO_3^- and SO_4^{2-} in drainage water (cf. Chorover et al. 1994; Riggan et al. 1994), fire suppression might contribute to the development of N saturation by allowing N to accumulate in soil and in the forest floor and by maintaining mature stands with low N demand (Fenn and Poth 1998). Nitrogen leaching can be many times higher in burned watersheds than in unburned watersheds, and the amount and concentration of N release may be related to fire intensity (Riggan et al. 1994). Nitrate concentrations have been shown to increase and remain elevated in stream water for several years after burning (Chorover et al. 1994).

Extensive past logging appears to have considerable and long-lasting effects on nutrient cycling (Goodale and Aber 2001; Fisk et al. 2002). Latty et al. (2004) compared soil nutrient pools and N cycling among three types of forest stands in the Adirondack Mountains: old growth, selectively logged, and selectively logged and then burned. Results suggested that even the combination of relatively light logging and burning may influence the extent of subsequent N limitation over periods of decades to centuries (Latty et al. 2004).

Watershed disturbance might also impact Hg cycling and its relationship to S deposition. In a study of 20 watersheds in Quebec, Canada, Garcia and Carignan (2000) found that the average Hg concentration in 560 mm northern pike (*Esox lucius*) was significantly higher in lakes whose watersheds had recently been logged (3.4 µg/g) than in reference lake watersheds (1.9 µg/g), which had remained undisturbed for at least 40 years. Since both inorganic and MeHg tend to adsorb to particulate material (both inorganic and organic sediment and soil), increased erosion associated with disturbance from urbanization, road building, or forest practices can lead to increased Hg runoff from watersheds (Ruzycki et al. 2011). Fire can contribute to volatilization loss of Hg from the watershed (Wiedinmyer and Friedli 2007).

3.2.2. Land Use Change

Land use and vegetation patterns have been changing in New York for more than a century. Some land-use activities contribute to acidification of soil and surface waters; other activities decrease acidity (Sullivan et al. 1996b). In addition to loss of base cations through erosion and timber harvesting, land use can also influence watershed sensitivity to acidification through land disturbance and consequent exposure of S-bearing minerals to oxidation and through change in N status of the forest in response to timber management. These land-use activities can influence the relative availability of mobile mineral acid anions (SO_4^{2-} and NO_3^-) in soil solution and base cations on the soil cation exchange sites. The balance between mineral acid anions and base cations, in turn, affects the ANC and pH of soil water and surface water. Reforestation subsequent to extensive timber harvesting in New York in the late nineteenth and early twentieth centuries undoubtedly contributed to depletion of Ca in the soil and exacerbated soil acidification caused by acidic deposition.

Regional acidification of surface waters in New York and other portions of North America and Europe has been attributed mainly to acidic deposition.

This regional acidification cannot be attributed to changes in land-use practices or landscape characteristics. Nevertheless, it is clear that changes in land use and disturbances in the drainage basins of lakes and streams modify acid-base water chemistry (Sullivan 2000). Land-use history has an especially important influence on N leaching from forested watersheds (Pardo et al. 1995; Aber and Driscoll 1997; Goodale et al. 2000; Lovett et al. 2000). Watershed disturbance often causes an increase in the ANC and pH of drainage water and contributes to problems with water quality other than acidification. These include excessive sedimentation causing turbidity and suspended sediment effects on aquatic macroinvertebrates and fish and increased fertility from nutrient stimulation of algal and higher plant growth. Large changes in water chemistry due to land use may make it difficult to quantify the effects of acidic deposition at some locations.

Urban land use also influences the acid-base chemistry of drainage water. Runoff is transmitted more efficiently to surface waters in the urban environment than in a more natural vegetation community, altering the magnitude of discharge and the flux of contaminants from the urban landscape to surface water (Arnold and Gibbons 1996; Montgomery and Buffington 1998). Increased pavement, rooftops, and other impervious surfaces increase surface runoff and reduce the infiltration rate of precipitation into soil. Roadside ditches route precipitation directly from roads into streams. Urbanization also results in alteration of the land surface in ways that affect other nonpoint-source pollution processes. Urbanization is generally associated with substantial clearing of vegetation and compaction of soil, thereby increasing storm-water runoff and erosion (Poff et al. 1997; Burges et al. 1998; Jones et al. 2000; Trombulak and Frissell 2000; Alberti et al. 2007). Degraded riparian and wetland conditions and functionality diminish the ability of the urban watershed to remove contaminants from runoff, including atmospherically deposited N (Peterjohn and Correll 1984). Many studies have documented correlations between water quality or ecological conditions and various measures of the extent of urbanization, such as human population density or the percent of the watershed covered by impervious surfaces (Hachmoller et al. 1991; Charbonneau and Kondolf 1993; Johnson et al. 1997; Thorne et al. 2000; Alberti et al. 2007).

3.2.3.   Invasive Species

Invasive plant species are opportunistic and may be better able than some of the rare native plant species to take advantage of added N. This may especially be the case in ecosystems that have developed under relatively low N supply. Thus, there may be a link between atmospheric N deposition and the prevalence of invasive nonnative plant species in some plant communities.

Nonnative plant and animal species introductions have substantially altered communities of phytoplankton, zooplankton, benthos, and fish in the Great

Lakes. The largest changes are attributable to introductions of quagga (*Dreissena rostriformis bugensis*) and zebra (*D. polymorpha*) mussels (collectively dreissenid mussels; Vanderploeg et al. 2002). These species introductions have altered the response of the aquatic ecosystem to atmospheric N input. Mida et al. (2010) compared changes in concentrations of lake nutrients between the periods 1983 to 1999 (pre-quagga influence) and 2000 to 2008 (post-quagga influence) in southern Lake Michigan. They observed significant changes in lake productivity, which they attributed largely to dreissenid mussel filtering. Dreissenids have only recently (about 2005) colonized offshore portions of the lake as the populations of offshore quagga mussels have increased (Nalepa et al. 2009). Consequently, dreissenid impacts have been apparent only in the most recent years of monitoring (Mida et al. 2010). In response to large-scale removal of algae from the water column by mussel filtration, summer algal uptake of N declined significantly in recent years, causing an increase in NO_3^- concentration in lake water. The offshore pelagic zone of the southern portion of Lake Michigan has changed from its former mesotrophic condition to more closely resemble the oligotrophic Lake Superior. During the post-quagga period, total P concentration decreased to only 2.9 µg/L and chlorophyll *a* concentrations decreased to only 0.9 µg/L, both reflective of oligotrophic conditions. Midea et al. (2010) suggested that these changes in the trophic state of southern Lake Michigan will likely have important implications for the food web that supports Lake Michigan fisheries. However, the large quagga populations may not be sustainable and mussel densities may decline in the future (Strayer and Malcom 2006; Nalepa et al. 2009). Additional monitoring will be needed to ascertain the long-term impact of this nonnative species introduction on the water quality and biological communities of Lake Michigan and their responses to N inputs. Dilute Adirondack lakes are likely less sensitive to dreissenid invasion because these mussels tend to be limited by low concentrations of Ca^{2+}, which are common in the Adirondacks.

3.2.4. Other Disturbances

Other watershed disturbances such as road building, agriculture, insect infestation, plant disease, mining, blowdown, and road salt application can also alter the biogeochemistry of an ecosystem. Such disturbances influence the hydrologic budget, base cation mobilization from soil to drainage water, routing of drainage water through the watershed, atmospheric nutrient input to surface waters, and S, N, and Hg cycling. Changes in these processes can affect the acid-base chemistry and nutrient dynamics of soils and drainage waters (Sullivan et al. 1996b). The effects of such disturbances can greatly modify the biogeochemical responses of a given watershed to atmospheric inputs of S, N, and Hg.

Erosion can reduce the size of the soil base cation pool in the watershed and limit the capacity of soils to neutralize atmospheric acidity. Tree defoliation caused by insects can have large effects on the N cycle of forest ecosystems. Nitrogen contained in the foliage consumed by insects is deposited on the forest floor in insect feces, greenfall, and insect biomass. Some of this deposited N is subsequently taken up by tree roots and soil microbes. However, a sizable component of this N can also be leached in drainage water, which has substantial nutritional consequences for the site and effects drainage water chemistry (Lovett et al. 2002a). Low N supply can limit the population growth of defoliating insects (Mason et al. 1992) and increase the trees' chemical defenses (Hunter and Schultz 1995). The concentration of NO_3^- in drainage water may be partly related to the extent of defoliation and tree mortality that occurs; it may also be influenced by the patterns and amount of precipitation that occur immediately after the defoliation (Lovett et al. 2002a).

Road salt applications during winter months can acidify drainage water via the neutral salt effect. This mechanism entails ion exchange on the soil surface whereby a base cation present in the added salt (typically Na^+ or Mg^{2+}) exchanges for H^+ on the soil, releasing H^+ acidity to the drainage water. Acidification of surface water after salt application is primarily an episodic, rather than a chronic, influence on the chemistry of drainage water. Salt application can also influence microbial N transformation processes, mainly by increasing soil pH (Green et al. 2008).

3.2.5.   Multiple Stress Response

As discussed in the preceding sections, natural ecosystems are typically subjected to multiple stressors, including nutrient enrichment from atmospheric deposition of N and acidification in response to atmospheric deposition of S and N. Additional stressors that can also be important include exposure of plants to atmospheric O_3, climatic variation, natural and human disturbance, the occurrence of invasive nonnative plants or animals, native and nonnative insect pests, and disease. Atmospheric deposition interacts with these other stressors to affect ecosystem patterns and processes in ways that scientists are only beginning to understand (U.S. EPA 2008).

3.3.   MERCURY BIOACCUMULATION AND BIOMAGNIFICATION

Mercury bioaccumulation is the process whereby the concentration of Hg increases over time in a given organism. This occurs because Hg concentrates in fatty tissue of animals. The progressive accumulation of a chemical with an increase in trophic level is called biomagnification (LeBlanc 1995). It results in chemical accumulation from lower to higher levels of the food web. This is a particular concern with Hg because it not only biomagnifies but also is a potent neurotoxin.

Pollutants that biomagnify accumulate in body fat, which allows them to become more concentrated at each level of tropic transfer. Mercury in its methyl form biomagnifies to increasingly higher concentrations in aquatic and terrestrial food webs. This is because organisms can efficiently assimilate MeHg and it is slowly eliminated (Reinfelder et al. 1998; Croteau et al. 2005).

The northern Great Lakes region and eastward into northern New York is an area that is especially sensitive to Hg biomagnification. This increased susceptibility is due in part to moderately high Hg deposition and in particular to watershed and lake characteristics that exacerbate Hg transport, methylation, and bioaccumulation (Evers et al. 2011a).

Mercury methylation and biomagnification are important concerns in northern New York in large part because of the common occurrence of wetlands. The DOC contributed by wetlands to drainage water is an important parameter affecting Hg bioavailability and transport through watersheds in part because Hg binds readily to organic matter (Grigal 2002). Methylation is a critical step in the Hg cycle with respect to effects of Hg on aquatic biota. Increased S deposition has been shown to increase rates of Hg methylation in freshwater wetlands (Galloway and Branfireun 2004; ICF International 2006; Jeremiason et al. 2006). This is because it appears that most methylating bacteria require SO_4^{2-} to carry on their metabolic activities. Other bacteria (iron [Fe] reducers) have been shown to methylate Hg as well, so SO_4^{2-} reducers are not necessarily required, but they are believed to be the dominant methylators in most environmental settings. Methylation is also inversely correlated with surface water ANC and pH (Wiener et al. 2006; Driscoll et al. 2007b) such that water acidification stimulates Hg biomagnification.

3.4. CLIMATE CHANGE

The Earth has warmed by nearly 1 °C over the past century and is predicted to continue to warm by another 1.4 to 5.8 °C during the twenty-first century (IPCC 2007). Along with climate warming, a large number of other climate-related variables are also projected to change, and virtually all of these will interact in one way or another with the effects of atmospheric S, N, and Hg deposition (Burns et al. 2011). Potential effects of anticipated climate change on natural resources in New York were summarized in an extensive review and analysis reported by Rosenzweig et al. (2011).

Climate influences watershed biogeochemistry, largely through effects on hydrology and nutrient cycling. Rain and snowmelt events alter acidification and neutralization processes, as does extended drought (Webster et al. 1990). Watersheds that are most sensitive to acidification from S and N deposition tend to be those that receive substantial precipitation input and those that build up a substantial winter snowpack. These characteristics may change in the future under climate warming. Atmospheric deposition of nutrient N interacts with climate change, affecting the diversity of plant species that occur at a site, especially in wetland environments (Bobbink et al. 2010) and at high elevations (Sverdrup et al. 2012).

The effects of future changes in N availability will exacerbate contemporaneous shifts in species distributions that occur in response to changes in temperature, moisture availability, and invasive species occurrences. There may be pronounced shifts away from cold-adapted to warm-adapted aquatic and terrestrial species (Durance and Ormerod 2007; McKenney et al. 2007; Lenoir et al. 2008). Thus, climate change will (and in some cases already does) constitute an additional ecosystem stress that will interact with the acidification and nutrient enrichment stressors associated with S and N deposition.

Both intra- and inter-annual variation in climatic conditions and long-term changes in climate can have substantial effects on the acid-base and nutrient dynamics of soils and surface waters. Soil and air temperature, precipitation amounts and patterns, and snowpack development and melting all influence key biogeochemical processes and cycles. Sensitive terrestrial and aquatic receptors respond to climatic factors, and these responses can complicate interpretation of biological impacts or recovery in response to changing levels of atmospheric deposition. For example, fish and other aquatic biota respond to ambient water levels, water temperature (and associated concentrations of dissolved oxygen), and the occurrence of hydrologic events such as snowmelt and storm flow. Plants respond to changes in air temperature and soil temperature as well as water availability. The amount of snowpack influences soil temperature during winter, winter dieback of fine roots, and the availability of water throughout the early portion of the growing season.

Climate warming and associated changes in moisture availability affect a host of biogeochemical processes and cycles. These include weathering, nitrification, and mineralization (Dalias et al. 2002; Campbell et al. 2009; Wu and Driscoll 2010). There are so many interactions that predicting the direction and magnitude of likely impacts is highly uncertain (Spranger et al. 2008). It will be important to continue ongoing monitoring efforts to track changes in discharge, soil condition, surface-water chemistry, and other ecosystem elements as these climate-related impacts play out in New York and elsewhere.

Winter temperatures have a large influence on snowpack development, snowmelt hydrology, and freeze-thaw cycles of soil. Burns et al. (2007) documented earlier snowmelt in the northeastern United States in recent years. There has also apparently been an increase in winter rain throughout the eastern United States (Hodgkins et al. 2003) and an increase in large rainstorms (Murdoch et al. 2000). Climate model predictions suggest the likelihood of future reductions in snowfall, resulting in smaller snowpacks and a lesser role for snowmelt in the hydrologic cycle of north temperate regions such as are found in New York (IPCC 2007).

Diminished snowpack development and earlier snowmelt can cause lower stream flows and warmer water during summer, with consequent impacts on cold-water fish (Mohseni et al. 2003). These changes will also impact soil freezing (Burns et al. 2011). Freeze-thaw behavior of soils in turn contributes to fine root damage and consequent changes in the amount of NO_3^- leaching (Fitzhugh et al. 2003).

Conditions are constantly changing in response to seasonal and inter-annual cycles and processes. Climatic variations influence the amount and timing of precipitation, snowmelt, plant growth, depth to groundwater, and water loss via transpiration. Such variables influence the chemistry of drainage water and interactions between S, N, and Hg deposition and the sensitive aquatic and terrestrial receptors present on the site.

Climate change can alter ecological resources in estuaries. Primary production has increased significantly in Hudson River–Raritan Bay during summer periods of low water discharge from the watershed. Low discharge increases water residence time and thermal stratification in the estuary and deepens the photic zone, each of which can alone or collectively contribute to increased primary production (Howarth et al. 2000b). Climate change models predict lower freshwater input to estuaries during summer in the northeastern United States in the future. Thus, New York estuaries may become more susceptible to eutrophication in the future, even under stable levels of nutrient loading.

3.4.1. Influence of Soil Freezing on N Cycling

Soil freezing affects the winter mortality of plant roots and can alter N and other nutrient cycling. Increased soil freezing caused by unusually cold temperature and/or reduced winter snowpack can contribute to substantial mobilization of N from soil and plant roots to surface waters. The results can include increased concentrations of NO_3^- in drainage water and associated effects on base cation leaching and Al mobilization.

3.4.2. Extreme Events

The projected effects of ongoing and future climate change include increased magnitudes and frequencies of events related to extreme weather, including drought, flood, and fire. These increases in extreme events will likely have more substantial effects on the future biogeochemistry of ecosystems than changes in average conditions (Dale et al. 2001; Jentsch et al. 2007). Acid-sensitive ecosystems in New York may experience more pronounced episodes of acidification driven by large rain events, especially those preceded by drought.

Extreme climatic events can affect nutrient-phytoplankton interactions and the effects of nutrient inputs on the productivity of lakes, rivers, and estuaries.

For example, nutrient enrichment and estuarine eutrophication are influenced by drought and the frequency and intensity of large storms. Fast-growing diatoms have been shown to be favored during periods of high discharge and short residence time in estuary water in Chesapeake Bay (Paerl et al. 2006). High rates of discharge from the watershed, such as are associated with storm activity, can reduce water salinity, increase the sediment load, and contribute nutrients from the watershed to the estuary (U.S. EPA 2008). All of these changes in the aquatic environment can influence primary productivity and eutrophication processes.

Extreme events affect the biogeochemistry of freshwater and terrestrial ecosystems (Mitchell et al. 1996b). For example, root mortality increases N mineralization and subsequent N leaching, potentially contributing to acidification or eutrophication of drainage water. Extreme drought followed by storm flow mobilizes previously stored S and related acidity from wetlands, contributing to episodic acidification of down-gradient receiving waters (Kerr et al. 2011).

Climate models generally predict increased severity and duration of drought in the northeastern United States (IPCC 2007). Drying and re-wetting of soils can have a large influence on the storage of S and N and subsequent release from soils to drainage waters in response to increased precipitation following drought. Post-drought episodic acidification of streams is an important concern (Tipping et al. 2003; Eimers et al. 2007; Burns et al. 2011).

CHAPTER 4

Chemical Effects
of Atmospheric Deposition

Studies that assess relationships between atmospheric N and S deposition loading and the estimated or expected extent, magnitude, and timing of ecosystem effects often employ a weight-of-evidence evaluation (see U.S. EPA 1995; van Sickle and Church 1995; NAPAP 1998). This approach was followed by NAPAP in its Integrated Assessment Report to Congress (NAPAP 1991). NAPAP used several lines of evidence to assess the extent and magnitude of acidification in sensitive regions of the United States, including the Adirondack and Catskill Mountains in New York. These included the following:

1. watershed simulation models
2. empirical biological dose/response models
3. observed relationships between surface-water chemistry and ambient atmospheric deposition
4. process studies
5. trend analyses
6. paleolimnological reconstructions of past water chemistry
7. whole-watershed or whole-lake acidification or deacidification field experiments

Each type of study contributed to overall scientific understanding of acidification and acid neutralization processes and their effects on the chemistry and biology of lakes and streams.

4.1. SULFUR

4.1.1.   Upland Sulfur Cycling Processes

In some ecosystems (mainly in some of the unglaciated portions of the Appalachian Mountains), much of the incoming S deposition is adsorbed to the more highly weathered soils that are found in these areas and therefore cannot immediately contribute to acidification of soils or drainage water. In previously glaciated ecosystems, such as those found in the Adirondack and Catskill Mountains, however, SO_4^{2-} acts as a mobile anion and readily moves through soil and into surface water. Because of its mobility, SO_4^{2-} typically contributes proportionately more to the acidification of soil, soil water, and surface water than does NO_3^-.

The cumulative depositional inputs of S from the atmosphere to the forested terrestrial environment are largely transferred from vegetation to the soil. Much of this transfer is accomplished by movement of water through the canopy as throughfall. In the soil, biogeochemical processes control the extent to which atmospheric S deposition affects ecosystem structure and function.

Sulfur budgets compiled for glaciated watersheds in the northeastern United States, including the Adirondack and Catskill Mountains, showed that S deposition inputs approximately equaled outputs in drainage water on an annual basis (Rochelle et al. 1987). Such results suggested that relatively little S retention on soil occurs at glaciated sites in mountainous portions of New York, compared to nonglaciated sites in the southern Appalachian Mountains, for example. This finding was due to relatively low SO_4^{2-} adsorption on the younger soils typical of glaciated sites. The observed balance between S inputs and outputs implied that decreases in atmospheric deposition of S would lead directly to decreases in SO_4^{2-} leaching. The strong correlation between decreases in atmospheric S deposition and decreases in SO_4^{2-} concentrations in surface waters throughout the Northeast during the last three decades is widely recognized as reflective of this linkage (Stoddard et al. 2003). Nevertheless, there is evidence to indicate that S inputs in glaciated ecosystems are not completely conservative. Deposited S is cycled through microbial and plant biomass (David et al. 1987; Alewell and Gehre 1999; Likens et al. 2002). In addition, substantial S is stored as organic complexes in the soil. Mitchell et al. (2011) showed that stream S export now generally exceeds inputs at many sites in the Northeast, which suggests that excess S stored in soil organic matter is bleeding out of soils as they re-equilibrate with new lower levels of S deposition. Thus, while northeastern soils are responsive to changes in S input, there are still lags in the system due to soil storage. Mitchell et al. constructed watershed S budgets for 15 sites in the northeastern United States and southeastern Canada, including two sites in New York (Arbutus Lake in the Adirondacks and Biscuit Brook in the Catskills). It appeared that mobilization of stored S from the soil contributed about 1 to 6 kg S/ha/yr to stream fluxes at the study sites.

Surface-water acidification is reflected in a decrease in ANC, usually a decrease in pH, and often an increase in the concentration of Al_i in surface waters. ANC is the most widely used water chemistry indicator for acidic deposition sensitivity and effects. It can be measured in the laboratory by Gran titration or defined as the difference between the measured base cation and mineral acid anion concentrations in water:

$$ANC = (Ca^{2+} + Mg^{2+} + K^+ + Na^+ + NH_4^+) - (SO_4^{2-} + NO_3^- + Cl^-) \quad (1)$$

Surface-water ANC reflects the end result of all of the chemical, physical, and biological interactions that occur as atmospheric deposition and precipitation move from the atmosphere into the soil and drainage water. ANC reflects the relative balance between base cations and strong acid anions in solution. If the sum of the base cation concentrations (SBC; expressed in equivalence units) exceeds the sum of the strong acid anions, the water will have positive ANC. Higher ANC is generally associated with higher pH and Ca^{2+} concentrations; lower ANC is generally associated with higher H^+ and Al_i concentrations and a greater likelihood of toxicity of the water for aquatic biota.

Surface-water ANC concentrations can be grouped into five major classes: acute concern (less than 0 µeq/L), severe concern (0 to 20 µeq/L), elevated concern (20 to 50 µeq/L), moderate concern (50 to 100 µeq/L), and low concern (greater than 100 µeq/L). Each range represents a probability of ecological damage (Cosby et al. 2006). Aquatic biota are generally not harmed when ANC values are above 100 µeq/L (U.S. EPA 2009). However, some surface waters have or had ANC well below 100 µeq/L in the absence of acidic deposition.

The pool of S stored in the soil is typically large, and S is mainly stored as organic complexes. David et al. (1987) estimated that total annual S deposition at a site in the central Adirondacks was only about 1% of the soil organic S pool. Similarly, Houle et al. (2001) estimated that annual S deposition at 11 sites in North America ranged from about 1% to 13% of the soil organic S pool. Organic S stored in soil is primarily unreactive, but it can be mineralized to SO_4^{2-} under the oxic conditions that are commonly found in well-drained soils (Johnson and Mitchell 1998).

Available data provide evidence of net loss of S from soils at a number of sites in the northeastern United States in recent years (Mitchell et al. 2011). Gradual removal from the soil of previously accumulated S apparently promotes some continued SO_4^{2-} leaching and soil acidification despite lower levels of S deposition input. Estimates of ecosystem fluxes such as weathering and dry deposition are uncertain, and there are complications in separating effects of S desorption from mineralization. For these reasons, it is difficult to predict when S outputs will no longer exceed inputs (Martinson et al. 2005; Mörth et al. 2005). In the future, increasing S mineralization because of a warming climate might further affect S retention and release (Knights et al. 2000; Driscoll et al. 2001a).

Leaching of atmospherically deposited S from soils to soil waters and eventually to surface waters is a dominant mechanism controlling soil acidification in New York. This process also largely controls Al mineralization and toxicity to plants and aquatic organisms, base cation depletion from soils, and surface-water acidification. Most aspects of ecosystem acidification in New York are controlled by SO_4^{2-} mobility. Nitrate mobility is also important at some locations and some times, especially under high-flow conditions (Wigington et al. 1996a). Nevertheless, the dominant mobile strong acid anion is SO_4^{2-} in most areas of New York where long-term chronic acidification of soil and water is an important concern.

4.1.2.   Wetland Sulfur Cycling Processes

Although S is generally mobile in upland soils in the northeastern United States, wetlands provide increased capacity for storage and release of S, especially during changing moisture cycles. Wetlands can serve as both sources and sinks of atmospherically deposited S, depending on water levels and water flux. Wetlands retain and release S in response to variations in hydrology, and S cycling reflects the balance between oxidation and reduction processes in wetland soils. This balance also affects Hg methylation in wetlands.

Ito et al. (2005) examined the influence of land-cover types on SO_4^{2-} fluxes in Adirondack lake watersheds. The concentration of SO_4^{2-} in drainage water decreased in association with increased wetland area in the watershed (adjusted $r^2 =$ 0.58, $p \leq 0.001$). Ito et al. attributed this finding to dissimilatory SO_4^{2-} reduction in anaerobic wetland soils. This is the same process that contributes to Hg methylation in watersheds.

Sulfur storage in saturated wetland soils prevents or delays acidification of downstream surface waters when there is a substantial flux of mineral acidity through the watershed from acidic deposition. However, the water table in wetland areas usually drops during drought conditions, allowing development of aerobic conditions in the upper layers of wetland soils. Under aerobic conditions, stored reduced S is reoxidized to SO_4^{2-}, which is then available to move to downstream surface waters under subsequent high-flow storm conditions. This process can contribute to episodic pulses of SO_4^{2-} and acidity in surface waters that receive drainage water from wetlands. Wetlands buffer downstream surface waters against chronic contributions of mineral acidity during low-flow conditions, but they can also be significant sources of periodic episodes of short-term high levels of SO_4^{2-} and acidity (Kerr et al. 2011).

Elevated concentrations of SO_4^{2-} have been observed in surface waters throughout the northeastern United States and southeastern Canada following periods of substantial drying during summer months. Kerr et al. (2011) compared watershed responses to seasonal drying in 20 watersheds across the region. They calculated a seasonal dryness statistic as the number of days for each month of the dry season when discharge was below the 25th percentile of long-term flow conditions. A SO_4^{2-} response score was calculated for each of the study's watersheds based on linear regressions of the seasonal dryness statistic compared to either the annual SO_4^{2-} concentration or the residual of annual SO_4^{2-} as a function of time. The response score provided an estimate of the proportion of variation in SO_4^{2-} that could be explained by seasonal drying. Possible values ranged from 0 to 1. More than half of the study watersheds showed some degree of SO_4^{2-} response to seasonal drying (response score = 0.04 to 0.72). In addition, the response score was positively related to percent wetland area and percent saturated area. This suggested that wetland/saturated area is an important driver of variation in the SO_4^{2-} response to seasonal drying. Any future shift toward dryer summer conditions might impact SO_4^{2-} dynamics in these watersheds (Kerr et al. 2011).

4.1.3. Surface Water Sulfur Cycling Processes

Measurements of SO_4^{2-} concentration in surface water are useful for understanding acid-base chemistry. They provide information about the extent of cation leaching and S retention on soils. Effects of S deposition are primarily indirect, and are partly attributable to soil and water acidification. Assessments of acidic deposition effects conducted from the 1980s to the present have shown SO_4^{2-} to be the primary anion in most acid-sensitive waters throughout the northeastern United States (Driscoll and Newton 1985; Driscoll et al. 1988, 2001a). For example, in an analysis representative of over 10,000 acid-sensitive lakes in the Northeast, inorganic anions constituted the majority of negative charge in 83% of the lakes. In this group of lakes, 82% of the total negative charge was attributed to SO_4^{2-} (Driscoll et al. 2001a). In contrast, organic anions predominated in only 17% of the lakes (Driscoll et al. 2001a). Naturally derived organic anions represented an average of 71% of total negative charge in this latter group of lakes.

In regions of New York affected by acidic deposition, the concentration of SO_4^{2-} and (to a lesser extent) NO_3^- in surface waters has changed from preindustrial conditions. In response, concentrations of other ions in surface water must also have changed to maintain electroneutrality. Leaching of SO_4^{2-} to soil water and surface water does not directly cause adverse environmental impacts. It is the changes in concentrations of these other ions (including mainly Al_i, H^+, bicarbonate [HCO_3^-], Ca^{2+}, and Mg^{2+}) that cause ecological effects from water acidification. As SO_4^{2-} and (to a lesser extent) NO_3^- concentrations increased, other anions (mainly HCO_3^- and organic acid anions) must have decreased and/or cations (e.g., base cations, H^+, or Al^{n+}) increased to maintain the charge balance in solution.

In-lake processes that influence the cycling of S can have a large influence on S retention in lakes with long water residence times. This in-lake S reduction increases the ANC of such lakes, reducing the influence of acidic deposition. Many of the smaller lakes in the Adirondacks have relatively short residence times (days to weeks). In-lake S reduction is not very important for producing ANC in such lakes. In some of the larger lakes, however, retention times are longer and in-lake processes are more important to ANC regulation.

4.2. NITROGEN

Nitrogen is one of the most important nutrients in virtually all ecosystems on earth (Vitousek and Howarth 1991). Because of the potential importance of N as a growth-limiting nutrient, N deposition from air pollution has the potential to cause wide-ranging ecological impacts, increasing the growth of some species and causing others to be outcompeted.

Although N makes up 80% of the total mass of the earth's atmosphere as N_2, most atmospheric N is not biologically available. Nitrogen fixation converts atmospheric N_2 to biologically active forms. The term reactive N (N_r) refers to all of the biologically active N compounds (Galloway et al. 2003). These include inorganic reduced forms such as NH_3 and NH_4^+, inorganic oxidized forms (NO_x, HNO_3, N_2O, and NO_3^-), and organic N compounds such as urea, amines, proteins, and nucleic acids (Galloway et al. 2003). Forms of N that are contained in atmospheric deposition and that influence nutrient dynamics and ecosystem health include mainly the oxidized and reduced forms of inorganic N but also some organic N.

In response to atmospheric emissions of N associated with motor vehicles, power plants, agriculture, and other human activities, N_r has accumulated in the atmosphere, soil, and water on a global scale (Galloway 1998; Galloway and Cowling 2002; Galloway et al. 2003). This has had a variety of effects on natural ecosystems (Galloway 1998; Rabalais 2002; van Egmond et al. 2002; Townsend et al. 2003). The sequence of biochemical transformations and environmental effects is referred to as the N cascade (Galloway and Cowling 2002; Galloway et al. 2003).

The effects of N addition are varied. They can include eutrophication, changes in the composition of plant and algal communities, disruptions in nutrient cycling, increased emissions into the atmosphere of the greenhouse gas nitrous oxide (N_2O), accumulation of N compounds in the soil, increased availability of N_r to primary producers, soil acidification, and increased susceptibility of plants to stress factors (Aber et al. 1989; Aber et al. 1998; Bobbink et al. 1998; Fenn et al. 1998; Driscoll et al. 2003b; U.S. EPA 2008). The nutrient enrichment effects of N deposition are controlled by retention and release of N_r in terrestrial, wetland, and aquatic ecosystems and by interactions between N and other nutrients. How much N an ecosystem retains or loses is regulated mainly by biological processes. In contrast, how much S an ecosystem retains or releases is mainly governed by chemical processes (S reduction in anoxic sediments is one notable exception). This is partly why it has been more challenging to understand and model the effects of N than to do so for S deposition (U.S. EPA 2008).

4.2.1. Upland Nitrogen Cycling Processes

Both NO_3^- and SO_4^{2-} have the potential to acidify drainage waters and leach potentially toxic Al_i from watershed soil to drainage water. However, N is often limiting for plant growth, and thus in most upland watersheds in New York much of the N inputs are quickly incorporated into biomass as organic N with little leaching of NO_3^- into surface waters.

One principal indicator of N enrichment in forested watersheds is the leaching loss of NO_3^- in soil drainage water and/or surface water. High leaching loss during the growing season is indicative of N saturation (Stoddard 1994). The concentration of NO_3^- in stream water provides a useful index of the extent to which N is leaching from the soil compartment to water. Most of the N export typically occurs as NO_3^-. There is also some leaching loss of dissolved organic N. However, export of dissolved organic N is typically less than 2 kg N/ha/yr from most northeastern forested watersheds (Campbell et al. 2000; Goodale et al. 2000; Lovett et al. 2000; Aber et al. 2003).

Atmospherically deposited N is largely retained in the soil of most upland forested areas in the northeastern United States (Nadelhoffer et al. 1999). However, damage to terrestrial and aquatic life can occur in response to NO_3^- leaching (Driscoll et al. 2003b). Although an estimated 70% to 88% of atmospheric N deposition was retained in Catskill Mountain watersheds during the 1990s, fish populations in some of the more acid-sensitive streams could not be sustained because high episodic NO_3^- concentrations in stream water during high-flow events caused the concentrations of Al_i in the stream to increase to levels above toxicity thresholds (Lawrence et al. 1999).

Aber et al. (2003) analyzed data collected during the mid- to late 1990s from lakes and streams throughout the northeastern United States and concluded that nearly all N deposition is retained or denitrified in northeastern watersheds that receive less than about 8 to 10 kg N/ha/yr of atmospheric N deposition. The concentration of NO_3^- in surface water typically exceeded 1 μmol/L only in watersheds that receive more than about 9 to 13 kg N/ha/yr of atmospheric N deposition. Leaching rates were variable among the watersheds that received N deposition above that level.

Nitrate leaching from soil to drainage water is influenced by complex ecosystem processes and the N cycle is controlled by many factors (Aber et al. 1991, 1998); addition of N from atmospheric deposition is only one. Mineralization and nitrification processes in the terrestrial compartment play especially important roles in regulating leaching losses from the soil rooting zone (Reuss and Johnson

1986; Joslin et al. 1987; D. Johnson et al. 1991a, 1991b). The leaching of NO_3^- is mainly controlled by biological processes, and this causes large seasonal variability (van Miegroet et al. 1993).

The N cycle involves multiple steps, including fixation, assimilation, mineralization, nitrification, and denitrification. Each of these steps involves biochemical transformations by plants and microbes. In general, with the addition of N from outside the ecosystem, competition among plants and microbes for available N decreases, net nitrification increases, and if the N supply is sufficient, NO_3^- leaching from the ecosystem in drainage water increases (Aber et al. 1989, 2003).

Availability of N to plants is controlled mostly by the rate of N mineralization, the microbial conversion of organic N to simple amino acids and eventually to inorganic forms such as NH_4^+ (Schimel and Bennett 2004). The two-step aerobic process of nitrification is mediated by microbes and converts the NH_4^+ produced by mineralization to NO_3^-. Nitrification acidifies the soil, releasing 2 moles of H^+ per mole of NH_4^+ converted to NO_3^- (Reuss and Johnson 1986).

Canham et al. (2012) developed and parameterized a spatially explicit mass-balance watershed model to predict total N concentrations in Adirondack lakes, following the general approach of Canham et al. (2004). The approach was designed to predict average midsummer N concentrations in 252 lakes by balancing inputs to the lake, primarily from the contributing watershed, and net losses, mostly from in-lake degradation processes and output to discharge. Results of the modeling suggested that forests are on average the largest source of N loading to Adirondack lakes. On average, forests retained about 87% of deposited N. Direct N deposition to the lake surface accounted for some of the N loading to lakes. Overall, however, lake N concentrations mainly reflected N loading from forests and loss rates determined by water residence time and in-lake processes (Canham et al. 2012).

Nitrogen storage in forest ecosystems occurs primarily in the soil. The soil N pool in the forest often constitutes 85% or more of the total terrestrial ecosystem N pool (Bormann et al. 1977; Cole and Rapp 1981). Most of this soil N is bound in organic matter on the forest floor associated with humic material or other organic complexes that are resistant to microbial degradation. Soil organic N is not readily available for direct biological uptake by most plants and microbes and is also not very susceptible to leaching loss into ground water or surface water. Only the labile pool of N_r is biologically active (Aber et al. 1989). It often limits photosynthesis and controls net primary productivity (cf. Field and Mooney 1986). Plant roots and soil microbes compete for available soil N_r. Plants obtain N to sustain primary productivity by absorbing NH_4^+, NO_3^-, or simple organic N compounds through their roots. Additional N is taken up by symbiotic fungi, bacteria, and cyanobacteria that occur in and in association with plant roots (cf. Lilleskov et al. 2001; Schimel and Bennett 2004).

The tree species present on a site has a profound influence on N cycling. Lovett et al. (2004) measured aspects of the N cycle in small single-species plots of five common trees in the Catskill Mountains: sugar maple, American beech, yellow birch (*Betula alleghaniensis*), eastern hemlock (*Tsuga canadensis*), and red oak (*Quercus rubra*). Hemlock plots showed characteristics of slow N cycling, including low foliar N, low litter N, high soil C-to-N ratio, low soil extractable N pools, and low rates of potential net N mineralization and nitrification. At the opposite extreme was sugar maple, which had low C-to-N, high soil extractable NO_3^-, and high nitrification. Because of such differences among species, N cycling in forests can be patchy, depending on distributions of tree species. Thus, changes in the composition of forest species resulting from forest management, disease, or climate change might substantially change the N cycling on the site.

Mature forests in the Catskill Mountains can retain large amounts of deposited N, even in the absence of any appreciable biomass growth increment. This is apparently because large amounts of N are incorporated into soil organic matter (Magill et al. 2000; Lovett et al. 2004), although the mechanism(s) of incorporation are complex and not well understood (D. Johnson et al. 2000). The N retention capacity varies among forest stands (McNulty et al. 1996; Magill et al. 2000). One important control seems to be reflected in the chemical quality of the litter produced by the trees. High lignin-to-N ratio is associated with slow litter decomposition (Melillo et al. 1982) and low N mineralization (Lovett et al. 2004).

Leaching of NO_3^- from forest soils to stream water can deplete soils of nutrient base cations, especially Ca^{2+} and Mg^{2+} (Likens et al. 1998). Such base cation depletion acidifies the soil and increases mobilization of Al_i. Considerable available evidence links N deposition to acidification of soils. Much of this evidence comes from the northeastern United States, where increased accumulation of N in soil is suggested by a strong positive correlation between atmospheric deposition levels and total N concentration in the O_a soil horizon at sites in New York, Vermont, New Hampshire, and Maine (Driscoll et al. 2001a). Further evidence that atmospheric deposition has increased availability of N in soil is shown by a strong negative correlation between atmospheric deposition and the C-to-N ratio of the O_a horizon in the Northeast (Lovett et al. 2002a; Aber et al. 2003; Ross et al. 2004). Atmospheric N deposition has increased N_r availability in soils in mountainous areas of New York, which has caused increased nitrification and acidification of soil and soil water. Land-use history further contributes to varying relationships between N deposition and ecosystem N status.

Lovett and Goodale (2011) developed a new conceptual model of N saturation, based on recent studies and results of experimental fertilization of a mixed oak forest (*Quercus rubra* and *Q. prinus*) with several species of hickory (*Carya* spp.) in New York. The conceptual model Lovett and Goodale proposed was based on the mass balance of N. Added N was modeled as flowing simultaneously to all sinks; the fate of the added N depended on the relative strengths of the sinks. That movement of N in the ecosystem determined how N saturation was manifested. The observed pattern of forest response differed substantially from that hypothesized in the widely quoted N saturation model of Aber et al. (1989, 1998). Foliar N concentrations in the N-amended stands increased to about 20% above levels in the control stands and stayed constant. Nitrate leaching increased almost immediately in response to fertilization.

Lovett et al. (2004) found that the factors controlling N mineralization and nitrification in Catskill Mountain forest stands are different. For example, both N mineralization and nitrification are low in eastern hemlock stands. Sugar maple plots showed moderate N mineralization but very high nitrification. Thus, different aspects of the N cycle may vary independently, and it may not be appropriate to generalize a particular species as having a fast or slow N cycle.

4.2.2.   Wetland Nitrogen Cycling Processes

Bogs, peatlands, and coastal marshes are types of wetland ecosystems that appear to be highly sensitive to the effects of N addition (U.S. EPA 2008). Bogs commonly receive the majority of their nutrient loads from the atmosphere. If atmospheric N loading to a bog increases, plant productivity will likely also increase and may be accompanied by changes in species composition to favor species that are adapted to higher nutrient levels. The U.S. EPA (1993) reviewed field experiments involving N fertilization of wetlands. Results suggested that wetland vegetation is commonly limited by N availability. In a modeling study, Canham et al. (2012) estimated that essentially all atmospheric N deposition to wetland surfaces in the Adirondack Mountains was retained in the wetland and not discharged to down-gradient lakes.

Nitrogen cycling in wetlands differs from N cycling in terrestrial ecosystems. The periodic anaerobic conditions in saturated wetland soils change the relative importance of the various microbial N transformations. Under anaerobic conditions, the rate of decomposition of organic matter is reduced, which can contribute to peat formation. Denitrification is increased; NO_3^- is converted to N_2O or N_2 and released to the atmosphere. Ammonium can be transported from anoxic sediments to oxidized surface sediment or to the water column, where nitrification can occur. The NO_3^- formed by nitrification can then be transported to downstream locations or denitrified and released back into the atmosphere as N_2O. Both NH_4^+ and NO_3^- may also be readily assimilated by aquatic plants and by periphyton (attached algae). There is also some evidence that the relative amounts of NO_3^- and NH_4^+ may affect aquatic plant diversity in wetlands and shallow lakes (Camargo and Alonso 2006; Jampeetong and Brix 2009a, b, c).

Nitrogen deposition to wetlands can cause nutrient imbalances and increases in the sensitivity of shrubs to drought, frost, and insects (Heil and Diemont 1983). These stresses can result in gaps in the shrub canopy, which then can be invaded by grasses that gain a competitive advantage because they are more efficient in using the additional N than shrubs and forbs (Krupa 2003).

Deposited N that is denitrified leaves the ecosystem without contributing to acidification or nutrient enrichment effects on terrestrial and aquatic receptors. Unfortunately, denitrification is difficult to measure directly in a field setting. This is due in part to the difficulty of measuring small changes in the very large atmospheric N_2 pool and to the substantial spatial and temporal heterogeneity in denitrification that is commonly observed across the landscape (Davidson and Seitzinger 2006). However, a variety of approaches can be used to estimate the rate of or potential for denitrification. These include laboratory sediment incubations, in situ NO_3^- injection and recovery studies, and stable isotope studies.

Denitrification requires anaerobic conditions and sufficient NO_3^- and organic C to support denitrifying bacteria. Most denitrification in terrestrial and transitional environments occurs at spatially limited locations and times that exhibit favorable conditions. These are called hot spots and hot moments. They are typically intermittently wet places or times in anaerobic soil microsites (McClain et al. 2003; Seitzinger et al. 2006). Organic matter in the soil provides the needed C. Factors that typically limit rates of denitrification include an inadequate supply of NO_3^- and the occurrence of aerobic conditions (U.S. EPA 2008).

Wetlands can remove significant quantities of NO_3^- from drainage water that passes through wetland soil. Wetlands also provide supplies of C to support denitrification. When NO_3^- in drainage water from O_2-rich uplands passes through wet, C-rich, anoxic wetland soils, loss of N to the atmosphere via denitrification can be substantial. Ashby et al. (1998) used the acetylene block technique to estimate denitrification rates of Catskill Mountain soils. Higher denitrification per unit area was found in soils with higher organic matter content and water-filled pore spaces. Denitrification was stimulated by experimental amendments with glucose or glucose plus NO_3^-, suggesting potential limitation of the denitrification process by the availability of both labile C and NO_3^-.

Denitrification is often an important N cycling process in riparian zones and wetland ecosystems (Lowrance 1992; Pinay et al. 1993; Pinay et al. 2000; Watts and Seitzinger 2001; Hefting et al. 2003). This is at least in part because the denitrification rate is strongly affected by C availability and the presence of anoxia. Riparian and wetland soils that are both rich in organic matter and are at least intermittently anaerobic generally have high denitrification potential. In contrast, nitrification, rather than denitrification, can be the dominant process that produces N_2O at locations where soils are aerobic (Stevens et al. 1997).

The main pathway of N loss in aquatic systems has long been assumed to be heterotrophic denitrification, the conversion of organic N to N_2 gas by heterotrophic bacteria. More recently, autotrophic bacteria have been shown to oxidize NH_4^+ under aerobic conditions, yielding N_2. This process is called anammox (Arrigo 2005; Zhu et al. 2013). It appears to be responsible for the majority of N loss in marine ecosystems, but its importance in freshwater ecosystems is uncertain (Zhu et al. 2010; Zhu et al. 2013).

Research on the processes that remove NO_3^- has mainly focused on plant and microbial assimilation and denitrification by bacteria. The latter produces gaseous N_2 as a by-product of the oxidation of organic matter. In recent years, however, application of tracer techniques has indicated that there are a number of other microbially mediated NO_3^- transformation processes, including not only anammox but also dissimilatory reduction of NO_3^- to NH_4^+. Estimated denitrification rates vary substantially among environments and measurement techniques. Nitrate loss in soils and sediments is typically assumed to be mainly attributable to denitrification. However, measurements based on direct assays, such as the acetylene block method, commonly account for less than half of the total NO_3^- loss (Seitzinger 1988). One possible explanation for this discrepancy is that much of the NO_3^- removal can be attributed to other processes. Burgin and Hamilton (2013) reviewed evidence suggesting the importance of these alternative NO_3^- removal pathways in aquatic and wetland ecosystems. They argued that the possible prevalence of these alternative pathways has important implications for managing excess N contributions to natural ecosystems.

Removal of NO_3^- by some of these other processes can cause NO_3^- transformation to compounds other than N_2. Sulfur-oxidizing bacteria may take up some of the NO_3^-. Thus, SO_4^{2-} loading might increase NO_3^- removal (Burgin and Hamilton 2013). Current understanding of the importance of these processes in freshwater ecosystems is limited.

4.2.3.   Fresh Surface Water Nitrogen Cycling Processes

Nutrient enrichment of freshwater aquatic ecosystems can result from N input as a consequence of atmospheric N deposition. Many lakes and streams are P-limited, however, and would not be expected to increase primary productivity in response to increased inputs of N. Many fresh waters are also N-limited or N and P co-limited (cf. Elser et al. 1990; Fenn et al. 2003a; Tank and Dodds 2003; Baron 2006; Bergström and Jansson 2006). Atmospheric inputs of N would be expected to increase productivity and alter the phytoplankton and perhaps biota at other trophic levels if the fresh waters are N-limited.

Down-gradient transport of N provides an important source of N to downstream ecosystems, including larger rivers, lakes, and coastal waters. The transport of N is determined by the net balance of N inputs by atmospheric deposition, land-based sources, and leaching from uplands minus the biological uptake, sediment storage, and gaseous loss of N during downstream transport.

Moving in the downstream direction, atmospheric N typically combines with fertilizer N in agricultural areas and with N from wastewater treatment facilities and runoff from impervious surfaces in urban areas. The role of atmospheric deposition in contributing to the river N load in residential and urban ecosystems has rarely been directly addressed in a rigorous fashion. The extent to which atmospheric N deposition in areas of mixed land use actually contributes to the N load of large rivers and to estuaries remains uncertain (U.S. EPA 2008). Smaller upland streams appear to have greater relative in-stream losses of NO_3^- due to greater contact with stream sediment. This would tend to decrease the role of atmospheric N at larger river scales.

Concentrations of N in streams of upland forested watersheds of the northeastern United States tend to be lower than in streams that drain watersheds characterized by other, more intensive land uses. In a comparison of small watersheds in eastern New York, for example, concentrations of N were highest and most variable in a stream draining a watershed where the predominant land use was row-crop production. Total dissolved N concentrations in streams in sewered suburban and urban watersheds were somewhat lower and less variable than in streams draining the agricultural watershed. Streams in urban and suburban watersheds may also experience high episodic N loading caused by combined sewer overflows (Driscoll et al. 2003c).

The export of N from terrestrial and wetland ecosystems to aquatic ecosystems is affected by seasonal fluctuations in temperature and biological uptake and by episodic fluctuations in water movement associated with rainstorms and snowmelt. Uptake by plants, microbial transformation reactions, and cycling among ecosystem compartments all vary with season and climatic factors.

Lovett et al. (2000) measured the chemistry of 39 first- and second-order streams in the Catskill Mountains to examine patterns in N loss from forested watersheds. Variability was pronounced: some streams had chronically low NO_3^- concentrations, some had strong seasonal cycles, and some had relatively high NO_3^- concentration year round. Retention of atmospherically deposited N varied greatly, from about 49% to 90%.

Throughout most of the year, the main cause of acidification of most acid-sensitive Adirondack and Catskill Mountain streams and lakes is S. At the peak of snowmelt, however, the influence of N deposition becomes proportionately more important and in some waters is just as important as S acidity (Sullivan et al. 1997; Figure 4.1). The seasonal shift in the relative importance of S and N as drivers of surface-water acidity is related to the seasonal dynamics of plant and microbial growth and hydrological cycles associated with snowpack accumulation and melting.

In the absence of major sources of domestic or industrial wastewater discharge or inputs of agricultural runoff laden with livestock manure or fertilizer, N is generally tightly cycled in stream watersheds, especially small streams with large relative areas of water contact with benthic surfaces and hyporheic zones (Burns 1998). Hyporheic losses of NO_3^- may be controlled in large part by supplies of labile C. For example, Bernhardt and Likens (2002) found that adding 6 mg/L

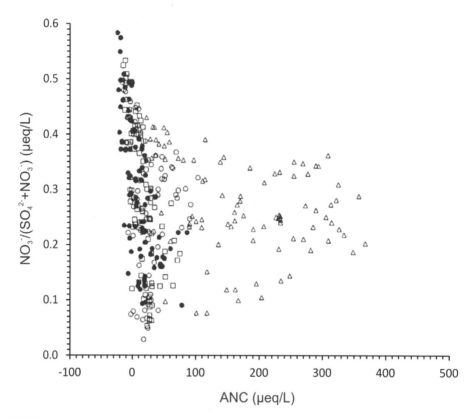

FIGURE 4.1. The ratio of nitrate:(sulfate plus nitrate) concentration to ANC in stream-water samples collected during rainfall and snowmelt hydrological episodes for the four streams included in the Adirondack region of the U.S. Environmental Protection Agency's Episodic Response Project. Site identification: ● = Buck Creek; ○ = Bald Mountain Brook; □ = Seventh Lake Inlet; Δ = Fly Pond Outlet. With kind permission from Springer Science+Business Media: Water, Air, and Soil Pollution, Increasing role of nitrogen in the acidifications of surface waters in the Adirondack Mountains, New York, 95(1997):330, T. J. Sullivan et al., Figure 8. Copyright 1997 Kluwer Academic Publishers.1997 Kluwer Academic Publishers.

of dissolved organic carbon as acetate to a small stream at Hubbard Brook, NH, reduced stream NO_3^- concentrations from ~5 to less than 1 mmol/L. In experimental mesocosms designed to mimic hyporheic flowpaths of a small river in the Catskill Mountains, Sobczak et al. (2003) found that adding 0.5 to 1.0 mg/L of dissolved organic carbon resulted in the net consumption of nearly all of the 40 mmol/L of NO_3^- in solution. Acetylene-block measurements of N_2O production by denitrification suggested that the majority of the NO_3^- loss was due to microbial assimilation rather than denitrification. This interpretation is consistent with results of isotopic tracer studies by Mulholland et al. (2004).

The C-to-N ratio is correlated with N transformation rates, especially nitrification. For example, nitrification typically increases markedly at a C-to-N ratio below about 22 to 24 (McNulty et al. 1991; Goodale and Aber 2001; Lovett et al.

2004). Lovett et al. (2002b) showed that more than half of the variation in stream NO_3^- concentration could be explained by the C-to-N ratio of watershed soils.

Water residence time is frequently identified as an important control on N loss from lakes (Howarth et al. 1996) and large river basins (Seitzinger et al. 2002). For example, Seitzinger et al. (2006) found that water residence time explained 56% of the variance in rates of N loss across lakes, rivers, estuaries, and continental shelves. Fast-flowing river reaches (residence time of hours) showed N losses ranging from about 0% to 15%. Large lakes that have long water residence (~ 100 years) eventually experience 80% to 100% N loss due to denitrification and sedimentation with permanent burial.

Driscoll et al. (1989) evaluated the relationship between measured wet N deposition and the concentration of NO_3^- in stream water for sites in North America (mostly eastern areas); this study was augmented by Stoddard (1994). The resulting data showed a pattern of N leaching at wet inputs greater than about 5.6 kg N/ha/yr. Stoddard (1994) presented a geographical analysis of patterns of watershed loss of N throughout the northeastern United States. He identified approximately 100 surface-water sites in the region with sufficiently intensive data to determine their N status. Sites were coded according to their presumed stage of N saturation. Sites ranged from Stage 0 (background condition) through Stage 2 (chronic N saturation effects). The geographic pattern in watershed N retention Stoddard (1994) depicted followed the geographic pattern of N deposition. At sites in the Adirondack and Catskill Mountains, N deposition was about 11 to 13 kg N/ha/yr and study sites were mostly characterized as Stage 1 (episodic effects) or Stage 2 (chronic effects). Sites in Maine, where N deposition was about half as high, were nearly all Stage 0. Sites in New Hampshire and Vermont, which received intermediate levels of N deposition, were identified as primarily Stage 0, with some Stage 1 sites. Based on this analysis, a reasonable threshold of N deposition for transforming a northeastern site from the "natural" Stage 0 condition to Stage 1 would correspond to the wet deposition levels found throughout New Hampshire and Vermont at the time of the study, approximately 6 to 8 kg N/ha/yr. This would likely correspond to total N inputs above 10 kg N/ha/yr, in general agreement with Driscoll et al.'s (1989) interpretation. This is the approximate level at which episodic aquatic effects of N deposition would likely become apparent in many watersheds of the Adirondack and Catskill Mountains.

Similarly, Aber et al. (2003) estimated the level of atmospheric deposition that causes release of NO_3^- to surface waters in the northeastern United States to be about 7 kg N/ha/yr. In watersheds that receive N deposition above this level, Aber et al. found that concentrations of NO_3^- in surface waters were positively correlated with atmospheric N deposition. Most watersheds that received N deposition less than about 7 kg N/ha/yr had little or no NO_3^- in their surface waters. The threshold value of 7 kg N/ha/yr was based on atmospheric deposition levels estimated at the base of forested watersheds. When scaled to include the higher deposition levels expected in the upper reaches of those watersheds, this value would probably be closer to about 10 kg N/ha/yr, similar to the European estimate of Dise et al. (1998).

Aber et al. (2003) reviewed studies throughout the northeastern United States and concluded that atmospheric total N deposition above about 9 to 13 kg N/ha/yr is generally required to cause substantial NO_3^- leaching for forest ecosystems in the Northeast. They found leaching to be generally negligible below those levels.

Eutrophication in response to nutrient addition is one of the most common water-quality problems in the United States affecting rivers, streams, wetlands, estuaries, and marine waters (Carpenter et al. 1998; Dodds et al. 2009). The most common and visible effects of increased nutrient (N, P) loading to aquatic ecosystems are increases in the growth of aquatic plants and algae. Nevertheless, environmental consequences of high nutrient loading are complex and far-reaching (Smith et al. 1999) and include loss of species and ecosystem services and accompanying economic effects (Dodds et al. 2009). The excessive accumulation of algal biomass to nuisance levels is readily observable and is often strongly related to nutrient loading (Camargo and Alonso 2006; Dodds 2006). Also of great consequence is the frequently observed shift toward increasing dominance of cyanobacteria in the phytoplankton community, some of which produce toxins (Dodds 2003; Howarth and Marino 2006).

Eutrophication of the Great Lakes, largely in response to P inputs (secondarily to N inputs) has been an important concern for many years, with the major focus on the lower lakes (Erie and Ontario) and the southern part of Lake Michigan. The Great Lakes Water Quality Agreement addressed loads, mainly point-source loads, of P. In response to management actions, summer total P concentration in Lake Erie declined between the 1970s and 1995 (Charlton et al. 1999). Nevertheless, the western basin of the lake has not met open-water nutrient guidelines intended to maintain algal growth below nuisance levels. Evaluation of the relative importance of atmospheric N deposition to the trophic state of Great Lakes waters is complicated by the strong influence of agricultural nutrient loading and the effects of other stressors such as the introduction of nonnative species and changing climatic conditions.

Microcystin toxins occur in the lower Great Lakes (Murphy et al. 2003; Boyer 2006; Makarewicz et al. 2006). These toxins, which are produced by cyanobacteria associated with eutrophic conditions, are neurotoxins and hepatotoxins linked to animal mortality, gastrointestinal disorders, and liver cancers (Ueno et al. 1996; Falconer 2005; Lehman 2007).

Eutrophication of the Great Lakes became an important concern in the 1950s and 1960s, when impaired water quality became apparent in Lake Erie (Schelske et al. 2006). Before that time, the Great Lakes were considered too large to be affected by anthropogenic nutrient addition (Hasler 1969) or affected only in harbors and near-shore areas (Beeton and Edmondson 1972).

Nutrient-driven changes (mainly P) in diatom production in the Great Lakes caused depletion of silica from the water column (Schelske et al. 1983; Schelske et al. 1986). The increased growth of diatoms, stimulated by nutrient additions to the lakes and their watersheds, contributed to silica sedimentation and eventually to silica-limited diatom growth.

Nitrate concentrations increased substantially in Lake Superior over the previous century, affecting the stoichiometric balance in the lake (Sterner et al. 2007). Earlier, it was believed that atmospheric N deposition could account for the observed increase in lake water NO_3^- concentration (Bennett 1986). However, more recent studies (Finlay et al. 2007; Sterner et al. 2007; McDonald et al. 2010) have suggested that the atmospheric N load has been insufficient to account for the full magnitude of the increase in NO_3^- concentration in lake water. Annual biological uptake of N appears to exceed the annual supply (Urban et al. 2005), as do measurements of seasonal NO_3^- decreases in the water column (Urban et al. 2005; McDonald et al. 2010). Mass balance model calculations by McDonald et al. (2010) suggested that the combination of atmospheric N deposition, tributary N loading, and changes in N burial in the sediment plus denitrification would reproduce the observed N accumulation in the lake. The model simulations further suggested that this long hydraulic retention time lake is approaching peak N concentration and is unlikely to show continued increases in lake NO_3^- concentrations in the future in the absence of major additional perturbations. The potential impacts of this accumulated NO_3^- in Lake Superior food webs remains unknown.

Lake Erie experienced the largest harmful cyanobacterial bloom (*Microcystis* sp. and *Anabena* sp.) of its history in 2011. It is likely that land use, agricultural activities, and meteorology all contributed to stimulating and exacerbating the cyanobacterial bloom (Michalak et al. 2013). By early September, the bloom was more than double the size of the previous largest bloom, which had been recorded in 2008. Michalak et al. (2013) hypothesized that severe precipitation events during the spring mobilized nutrients derived from agricultural practices and contributed them to the western basin of Lake Erie. Incubation and growth conditions were apparently optimized for bloom development by warm and calm conditions in late spring and summer. The researchers concluded that these conditions that are believed to have contributed to the severity of the cyanobacterial bloom are consistent with predicted future conditions under climate change (Michalak et al. 2013). More scientific information regarding harmful algal blooms in Lake Erie is available from the Experimental Lake Erie Harmful Algal Bloom Bulletin (see Great Lakes Environmental Research Laboratory 2014).

4.2.4.   Coastal Nitrogen Cycling Processes

The transport of N from soil to fresh surface waters has important implications beyond the potential for impacts on upland lakes and streams. Nitrogen export can contribute to down-gradient receiving waters and ultimately to the eutrophication of coastal ecosystems (Driscoll et al. 2003b; Camargo and Alonso 2006;

Howarth and Marino 2006). The impact of excess nutrient N loading is among the most significant coastal water pollution problems in the United States (Howarth et al. 2000a; NRC 2000; Scavia and Bricker 2006).

Because much of the atmospherically deposited N is retained in upland soils and vegetation or denitrified during export, only a small fraction of the N deposited on an estuary watershed typically reaches the estuary (Castro et al. 2000; Alexander et al. 2002; Seitzinger et al. 2002; van Breemen et al. 2002). However, inputs of large amounts of N to coastal waters from multiple sources, in addition to and including atmospheric sources, contribute to the development of hypoxia (reduced dissolved oxygen) and related water-quality problems in many East Coast estuaries, including those in New York. Atmospheric deposition contributes a portion of the N load that leads to coastal water hypoxia in many cases, but that portion is generally minor relative to the cumulative contribution from other sources (Paerl et al. 2001; Howarth and Marino 2006).

Nutrient loadings to estuaries in the northeastern United States have increased over the past century in response to human activities and are currently many times higher than natural background levels. This nutrient addition has been mainly due to agriculture, wastewater discharge, and atmospheric deposition. Eutrophication of coastal waters has become a major environmental problem worldwide (Hodgkin and Hamilton 1993; Joint et al. 1997; Ferreira et al. 2007). Significant environmental consequences of coastal water eutrophication include

- fish kills due to hypoxia or anoxia (Glasgow and Burkholder 2000)
- loss of submerged aquatic vegetation habitat (Twilley et al. 1985; Burkholder et al. 1992)
- loss of benthic fauna
- reduced abundance, diversity, and harvest of fish (Breitburg 2002)
- harmful algal blooms, of which toxins from cyanobacteria and their effects on humans and wildlife are but one element
- economic losses associated with tourism and seafood production (Anderson et al. 2000)

These environmental effects can potentially cause substantial impacts on the availability, edibility, and marketability of fisheries resources (Burkholder and Glasgow 1997; Smith et al. 1999; Camargo and Alonso 2006; Howarth and Marino 2006).

Atmospheric N deposition to the surfaces of estuaries and coastal marine waters provides direct N loading to these ecosystems. In addition, atmospheric deposition to the terrestrial watersheds can provide an indirect source of N to estuaries and near-coastal marine water to the extent that the deposited N leaches to drainage water and is then transported to the coastal zone with stream and river discharge. In the estuaries and coastal ecosystems that experience N overenrichment, the role of atmospheric deposition as a cause of eutrophication is determined by the relative contribution of atmospheric versus non-atmospheric sources of N that actually move into the coastal waters.

The relative importance of N sources varies from estuary to estuary. Atmospheric sources are proportionately more important to estuaries that have large surface area relative to watershed drainage area and estuaries that drain watersheds that are dominated by forested ecosystems rather than agricultural or urban lands (Boyer et al. 2002).

Even though only a relatively small percentage of the N that is atmospherically deposited on an estuary watershed would be expected to reach the estuary or coastal marine waters, a quantitatively important portion of the riverine N load can in some cases be derived from atmospheric N deposition (U.S. EPA 2008). It is difficult to estimate the percentage of atmospherically deposited N that leaches to estuaries and marine environments, especially under mixed land-use conditions. Nevertheless, the fraction that is exported from upland watersheds to estuaries in the eastern United States appears to be ecologically significant.

The relative contribution of atmospheric deposition to the total N loading to estuaries varies with local inputs of atmospheric N deposition, land use, watershed and estuary area, and hydrological and morphological characteristics that influence water retention time in the estuary (Paerl et al. 2001). Valigura et al. (2000) estimated that direct atmospheric deposition to the estuary surface generally constitutes at least 20% of the total N load for estuaries in the eastern states that occupy more than 20% of their watershed. Driscoll et al. (2003b) estimated annual net anthropogenic N inputs to eight large watersheds in the Northeast for the year 1997. Total estimated atmospheric plus non-atmospheric anthropogenic N inputs ranged from 14 kg N/ha/yr in the watershed of Casco Bay in Maine to 68 kg N/ha/yr in the watershed of Massachusetts Bay. In all eight study watersheds, the net import of N in food for humans was the largest anthropogenic input and atmospheric deposition was estimated to be the second largest anthropogenic N input. The range for atmospheric input was from 5 to 10 kg N/ha/yr, or 11% to 36% of the total N input.

The estimates of Driscoll et al. (2003c) are broadly consistent with those of Boyer et al. (2002), who used a similar N budgeting approach for 16 large river basins in the northeastern United States and reported that N deposition contributed, on average, about 31% of the total N load to these large river basins. Because Boyer et al. (2002) considered only the portions of each basin above U.S. Geological Survey gauging stations, which often were located upstream from large population centers, the Boyer et al. (2002) budgets included landscapes that have less human food consumption than the landscapes Driscoll et al. (2003c) considered. Boyer et al. (2002) estimated that across all basins, riverine export of N amounted to about 25% of total watershed N inputs; the range was from 11% to 40% among the estuaries that were studied. This result is also consistent with a similar analysis by Howarth et al. (1996), who estimated that basins draining to the north Atlantic Ocean exported approximately 25% of anthropogenic N inputs on average. The U.S. EPA (1999b) estimated that between 10% and 40% of the total N input to estuaries in the United States is typically derived from atmospheric deposition. The NRC (2000) judged that the EPA study underestimated the importance of atmospheric deposition as a contributor to the total N load. Across the various studies that have been conducted, atmospheric N deposition has been estimated to account from only a few percent to more than 40% of total external N loading to estuaries in North America and Europe (U.S. EPA 2008). These estimates of the relative importance of atmospheric deposition (compared to non-atmospheric sources of N) are highly uncertain. One source of substantial uncertainty in the development of detailed N budgets for coastal ecosystems concerns how much of the N deposited on the watershed is transferred through soils to drainage water and eventually to the estuary (U.S. EPA 2008). Such estimates for a given estuary commonly differ by more than a factor of two and sometimes differ by much more than that (see Table 4.1).

TABLE 4.1. Estimated percent of total nitrogen load to Delaware Bay and Hudson River/Raritan Bay contributed by atmospheric deposition

Study	Delaware Bay	Hudson River/Raritan Bay
Paerl 1985	44	--
Hinga et al. 1991	--	33
Scudlark and Church 1993	15	--
Paerl 1995	--	68
Jaworski et al. 1997	44	68
Alexander et al. 2001	22	26
Castro et al. 2000	--	10

Stacey et al. 2001
  Land-based
  SPARROW model
	
	16
	25
		
		10
		27

Castro and Driscoll 2002	20	17
Castro et al. 2003	23	18

There is also uncertainty and some controversy related to how much N is lost through in-stream processes prior to being transported by streams and rivers to estuaries.

It is difficult to quantify the magnitude of the various N sources that contribute to the load of N that enters an estuary. This difficulty stems in large part from (1) multiple agricultural and fuel combustion sources; (2) difficulty in quantifying dry atmospheric deposition to the estuary surface and to the watershed; (3) a lack of quantitative data on gaseous losses of NH_3 and NO_x compounds to the atmosphere; and (4) the complexity of N transport pathways through the watershed (NRC 2000; U.S. EPA 2008). Despite these sources of considerable uncertainty, atmospheric N is widely believed to be an important source of N to estuaries in the northeastern United States (U.S. EPA 2008).

Ultimately, the N loading to an estuary is a function of the human population in the estuary watershed. Turner et al. (2001) found a strong correlation ($r^2 = 0.78$) between human population density and the total N loading from watershed to estuary for coastal watersheds throughout the United States. This finding can partly be attributed to N emissions and deposition from automobiles in heavily populated areas, plus various non-atmospheric sources of N from human activities, especially from wastewater treatment plant effluent and accidental sewage releases.

Direct atmospheric deposition to the estuary surface becomes increasingly more important as a contributor to the total N loading to an estuary as the open-water surface area increases relative to the watershed area. However, estimation of direct atmospheric deposition to estuary surfaces has been hampered by uncertainties in the amount of dry deposited N. Many published studies have assumed that dry N deposition is equal to measured wet N deposition throughout the watershed (e.g., Fisher and Oppenheimer 1991; Hinga et al. 1991; Scudlark and Church 1993). Other studies have assumed lower dry N deposition rates for estuarine and near-coastal areas, equal to 40% (Jaworski et al. 1997) or 67% of wet deposition (Meyers et al. 2001). Dry deposition to open-water surfaces is probably much lower on a per-unit-area basis than dry deposition to vegetated terrestrial surfaces in the watershed. Paerl et al. (2001) estimated that dry deposition to open estuarine surfaces is three to five times lower. This difference has often not been considered in N-budgeting studies, and that oversight might have had a substantial impact on estimates of direct atmospheric loading to estuary surfaces, especially for estuaries having low ratios of watershed area to estuary surface area (U.S. EPA 2008).

Castro and Driscoll (2002) reported results of model simulations suggesting that large reductions (more than 25%) in atmospheric N deposition would be needed to substantially reduce the contribution made by atmospheric N deposition to the total N loads to estuaries in the northeastern United States. A simulated reduction in atmospheric deposition of 25% of ambient deposition reduced the contribution made by atmospheric deposition to the total estuarine N loads by only 1% to 6% (Castro and Driscoll 2002). Driscoll et al. (2003c) estimated that the implementation of aggressive emissions controls on both mobile N emissions sources and electric utilities would produce a reduction in estuarine N loading in Casco Bay, Maine, of only about 13%.

In the upper reaches of an estuary, the water is primarily fresh; discharge from the watershed via the river system(s) dominates the estuary hydrology, and atmospheric N deposition to the estuary surface is relatively unimportant as a contributor to the total N load compared to deposition to the larger watershed. Salinity levels are higher further downstream in the mesohaline portions of the estuary. In the upper estuary region, N derived from point and nonpoint sources in the watershed is largely assimilated by phytoplankton and plants. In addition, denitrification releases some N back to the atmosphere (Nixon 1986; Boynton et al. 1995; Paerl et al. 2002b). Atmospherically deposited N becomes increasingly important in the lower estuary (Paerl and Whitall 1999), downstream of the mesohaline zone of substantial nutrient uptake and loss to denitrification.

The response of an estuary to N input is highly variable, depending on the characteristics of the N sources and the morphology, hydrology, and biogeochemistry of the estuary. Water residence time in the estuary is an important controlling factor for eutrophication and is influenced by estuarine depth and surface area, water volume, and hydrologic flushing.

Estuaries vary substantially in their inherent sensitivity to eutrophication in response to nutrient input, especially N. The Hudson River–Raritan Bay estuary system is a drowned river valley of the Hudson River, with associated beaches, bays (including New York Harbor), and tidal straits (O'Shea and Brosnan 2000). Most of the freshwater input to this estuary is from the Hudson River. Raritan Bay is located in the lower harbor. It is an 81-km^2 open bay that receives freshwater inflow from the Hudson River at its northeastern end, from the Raritan River at the western end, and from the Hackensack and Passaic Rivers through the Arthur Kill tidal strait (National Oceanic and Atmospheric Administration [NOAA] 1988).

The various contributions of N to an estuary are commonly assessed by considering sources that contribute N to both the watershed and the airshed. The watershed that contributes drainage water to an estuary has boundaries that can be clearly defined based on topography, but the atmosphere lacks such clearly defined boundaries. Therefore, the contributing airshed is defined based on modeled climatology, the distribution and size of emissions sources, and the assumed deposition efficiency (Dennis 1997). The range of influence of a given emissions source is mapped as the distance at which the accumulated deposition from that source encompasses the majority of the total (often designated as more than 65%) (Paerl et al. 2002b). Contributing airsheds are typically much larger than contributing watersheds.

Results of a modeling study by Whitall et al. (2007) of 11 estuaries in the northeastern and mid-Atlantic coastal United States, using the Watershed Assessment Tool for Evaluating Reduction Strategies for Nitrogen (WATERSN) model, suggested that total N loading to the Long Island Sound watershed was about 10 kg N/ha/yr. Human sewage was estimated to be the largest source of N loading to all nine estuaries studied in the Northeast (contributing 36% to 81% of total N loading). Estimated runoff from atmospheric deposition of N represented an additional 14% to 35% of the estimated total N loading; for Long Island Sound and Raritan Bay, atmospheric N loading accounted for about 20% of the total (Whitall et al. 2007).

Not all N contributed to an estuary remains in the estuary. Denitrification is an important process that controls N loss from estuaries. The major environmental controls on the rate of estuarine denitrification likely include temperature and the availability of NO_3^-, O_2, and C (Seitzinger 1988; Rysgaard et al. 1994).

In response to growing concerns expressed in the 1990s about nutrient degradation of some U.S. estuaries, including extended hypoxia in Long Island Sound (Welsh 1991), NOAA conducted a nationwide assessment of estuary condition. The aim was to determine the magnitude, severity, and location of eutrophic estuaries in order to ascertain the spatial distribution and causes of eutrophication and to improve resource management. The National Estuarine Eutrophication Assessment (NEEA) included questionnaires about 16 nutrient-related water-quality variables (NOAA 1996, 1998), an assessment of overall trophic condition based on six key variables (Bricker et al. 1999), and an evaluation of probable causes of eutrophication and projections of future conditions (Scavia and Bricker 2006). Results varied from estuary to estuary.

Gardiners Bay, located on the northeastern-most extension of Long Island, is a moderate-sized (512 km²) estuary with largely (34%) urban land use and a moderate flushing time (~ one month). It was rated by Bricker et al. (2007) as having low overall eutrophic condition.

Great South Bay, which stretches along the southeastern shoreline of Long Island, is a coastal lagoon estuary. Estuaries of this type are enclosed by barriers that limit water exchange with the ocean. The estuary surface area is 383 km² and the estuary has a short water residence time of about two days (Bricker et al. 2007). Bricker et al. rated the overall eutrophic condition as moderate to high. This estuary is characterized by high chlorophyll *a*, high macroalgal abundance, and some nuisance/toxic algal blooms.

Hudson River/Raritan Bay has a surface area of 799 km² and a short water residence time of nine days. The watershed is urbanized (19% urban land use) and contains more than 12 million people. The Hudson River Estuary is considered to be only moderately sensitive to eutrophication (Malone 1977; Bricker et al. 1999), largely due to its deep water, light limitation, and short water residence time (Howarth et al. 2000b). However, the Hudson River estuary receives very high nutrient loading from municipal, agricultural, and atmospheric nutrient sources (Johnson and Hetling 1995; Jaworski et al. 1997), and it has shown a substantial response to drought (Howarth et al. 2000b). Because the watershed is large compared with the estuary surface, N inputs to the watershed per unit of estuary area are higher than most other estuaries in the United States (Bricker et al. 1999). Bricker et al. (2007) gave it an overall eutrophic condition rating of moderate. This estuary has a low ability for flushing and dilution of nutrients.

Long Island Sound is a large (~3,000 km²) estuary shared by New York and Connecticut; it also receives freshwater discharge from Massachusetts, New Hampshire, and Vermont. Most of its fresh water enters through the Connecticut River, near the eastern end of the estuary. It is moderately flushed, with a water residence time of about two months. The watershed is heavily developed and includes about 5 million people and 16% urban land use. Important sources of N to Long Island Sound include wastewater treatment plant discharge, non-point-source runoff from the watershed, and atmospheric deposition (Bricker et al. 2007). Bricker et al. ranked Long Island Sound as having moderate to high N input and moderate to high susceptibility to eutrophication. The presumed sensitivity of this estuary to eutrophication was due largely to low capacity to dilute nutrients with freshwater inflow and limited flushing of nutrients from the estuary to the ocean.

The causes and effects of hypoxia in Long Island Sound were intensively studied in the Long Island Sound Study (LISS), which began in 1988 as part of the National Estuary Program. It identified hypoxia as the most important water-quality problem affecting Long Island Sound (U.S. EPA 1994). The LISS documented bottom-water hypoxia in Long Island Sound affecting about 400 km² and persisting for about 2 to 11 weeks each year (Welsh et al. 1994; U.S. EPA 1998). Hypoxia has been a common occurrence in Long Island Sound since the 1980s or earlier (Figure 4.2) and has been most pronounced in the western portion of the sound where dissolved oxygen levels below 2 mg/L have been relatively common and where anoxia (dissolved oxygen < 1 mg/L) occurs during some years. Algal biomass is high in western Long Island Sound, with chlorophyll *a*

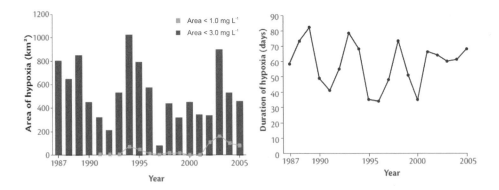

FIGURE 4.2. Areal extent and duration of Long Island Sound hypoxia, 1987–2005. The extent of hypoxia is expressed as the estuary area having dissolved oxygen less than 1 mg L⁻¹ and less than 3 mg L⁻¹. Source: Bricker et al. 2007.

concentration frequently higher than 30 μg/L (Cosper and Cerami 1996; O'Shea and Brosnan 2000). Nitrogen loads to the sound are trending slightly downward under a total maximum daily load agreement by New York and Connecticut (Bricker et al. 2007).

The Assessment of Estuarine Trophic Status (ASSETS) model was applied to Long Island Sound using water-quality data from more than 111 monthly samples collected at 7 stations in 1991 and 387 monthly samples collected at 17 stations in 2002. Between these sampling periods, about one-fourth of the wastewater treatment plants in Connecticut and New York that discharge into the Long Island Sound watershed were upgraded to include biological N removal, contributing to a 30% decrease in N loading from wastewater treatment plants (LISS 2003). The 90th percentile chlorophyll *a* concentration decreased from 18 to 9 μg/L between 1991 and 2002 for all Long Island Sound monitoring stations. In addition, the 10th percentile dissolved oxygen concentration increased from 3.9 to 6.4 mg/L during that time interval and the hypoxic area decreased from nearly 800 km^2 in 1987 to about 330 km^2 in 2002 (LISS 2003).

Submerged aquatic vegetation in Long Island Sound had been lost during the 1970s and 1980s in response to light shading by high algal concentrations in the water column (LISS 2003). The amount of submerged aquatic vegetation has increased slightly in more recent years (Ferreira et al. 2007).

In addition to the commonly recognized effects of eutrophication on water quality and aquatic biota, Cai et al. (2011) suggested that coastal eutrophication can also increase the acidification of coastal ocean water. The mechanism is described as follows: Eutrophication increases algal growth. Microbial consumption of excess algae lowers dissolved oxygen and increases carbon dioxide (CO_2), which in turn increases water acidity associated with carbonic acid. Thus, it is possible that coastal eutrophication increases the susceptibility of coastal waters to ocean acidification. This may become a more important research focus in the future.

4.2.5. Nitrogen Saturation

An undisturbed forest typically uses and stores almost all N received from atmospheric deposition in soil and vegetation. However, forests and other plant communities have a limited capacity to store N received from outside the watershed. This capacity is influenced by the plant species present and the history of disturbances such as logging, agriculture, and residential development that previously removed from the watershed some of the N that had been stored in the soil and trees. Nitrogen saturation refers to a condition whereby N inputs exceed the N storage capacity of a site. In this condition, more of the incoming N leaches as NO_3^- to soil water and eventually to surface waters. Increased leaching of NO_3^- can contribute to soil acidification, which has harmful consequences for plants. In the early stages of N saturation, trees may grow faster because they are being fertilized by N, which is typically the most important growth-limiting nutrient in temperate zone forests. During the latter stages of N saturation, tree health deteriorates and the forest may release large quantities of N to drainage water. Under conditions of advanced N saturation, tree growth declines and sensitive tree species can die in response to acidification, Al toxicity, and base cation depletion (U.S. EPA 2008).

Because forest growth in New York is generally limited by the availability of N (Aber et al. 1989), most forests take up the N provided by atmospheric deposition with few or no signs of N saturation. At some locations, however, decades of atmospheric N deposition has increased N supply in the soil to levels that promote increased plant growth. Excess N input at such sites increases net nitrification (formation of NO_3^- from NH_4^+ and organic N) and, consequently, NO_3^- leaching (Aber et al. 2003). Some terrestrial ecosystems in New York have become at least partly N-saturated, and high levels of N deposition have contributed to increased NO_3^- leaching losses in drainage water (Aber et al. 1989; Murdoch and Stoddard 1992; Murdoch and Stoddard 1993). Nitrogen saturation is a condition whereby the input of N to the ecosystem exceeds the nutritional requirements of terrestrial biota. A substantial fraction of the incoming N then leaches out of the ecosystem as NO_3^- in groundwater and surface water. Thus, atmospheric deposition of N has increased N availability in soils at some locations, which has led to increased nitrification and associated acidification of soil and soil water. The N retention capacity of soils is strongly dependent on land use history, however, so the relationships between N deposition and ecosystem N status are highly variable.

High concentrations of NO_3^- in soil solution may be largely responsible for peaks in toxic Al_i concentration in soil solution. Sulfate may also play a role by serving to elevate chronic Al concentrations (Eagar et al. 1996; U.S. EPA 2008). Studies in the southern Appalachian Mountains (cf. Joslin et al. 1992; Van Miegroet et al. 1992a; 1992b; Joslin and Wolfe 1994; Nodvin et al. 1995) found high concentrations of NO_3^- in soil water and stream water at high-elevation locations of spruce and fir forest. This NO_3^- leaching was believed to have been caused by high N deposition, low N uptake by forest vegetation, and inherently high N release from soils.

Nitrogen saturation occurs when N_r inputs become greater than the capacity of plants and soil microorganisms to use and retain N (Aber et al. 1989; U.S. EPA 1993; Aber et al. 1998). Under such high N loading, some resource other than N becomes growth limiting. This can be water, light, or another nutrient.

Nitrogen saturation develops gradually. As N availability increases beyond the capacity for N retention in the terrestrial system, there is increased loss of N gases to the atmosphere and NO_3^- to drainage water (Aber et al. 1989, 1998). A good indicator of ecosystem N saturation is the presence of high NO_3^- concentration in streams, lakes, and soil drainage water. Some forested ecosystems in the Adirondack and Catskill Mountains have exhibited minor to moderate symptoms of N saturation (Murdoch and Stoddard 1992; Murdoch and Stoddard 1993; Mitchell et al. 1996a). Other ecosystems in these forested regions may not reach saturation at current rates of deposition for many decades or centuries.

Under conditions of N saturation, soil microbial communities might change from being dominated by fungi to being dominated by bacteria. A meta-analysis of the effect of nutrient addition on mycorrhizae indicated a 15% decrease in mycorrhizal abundance in response to N fertilization across 16 studies at 31 sites, covering a range of grassland, shrubland, and temperate and boreal forest ecosystems (Treseder 2004). Loss of mycorrhizal function has been hypothesized as a key process contributing to reduced N uptake by vegetation and increased NO_3^- mobility from soil into drainage water as N saturation develops (U.S. EPA 2004).

A new conceptual model of N-saturation has recently been developed based on a mass balance of N instead of the temporal dynamics of N cycling indicators (Lovett and Goodale 2011). Under this paradigm, added N can flow simultaneously to the various N sinks and loss mechanisms. The movement of N to the N sinks and losses governs how N saturation is manifested.

4.3. DISSOLVED ORGANIC CARBON

4.3.1. Upland Processes

The N and C cycles are tightly coupled. Effects of nitrification, soil acidification, and cation leaching in response to atmospheric N deposition have been found primarily in soils having relatively low C and high N content and a C-to-N ratio below about 22 to 25 (Aber et al. 2003; Ross et al. 2004). The C-to-N ratio can be helpful in evaluating N-mediated soil acidification, largely because N mineralization and nitrification rates in the soil cannot be measured directly under natural conditions. All measurement approaches disturb the soil and break fine roots, contributing to artificially high estimated rates. Thus, such measurements provide only a relative index rather than a realistic rate (Ross et al. 2004). In addition, there is often high temporal and spatial variability in the measurements of N mineralization and nitrification. Measurements of total C and N are less variable, and the C-to-N ratio is a relatively straightforward measure for identifying forest ecosystems that are experiencing soil acidification and base cation leaching associated with high rates of nitrification in soils (U.S. EPA 2008).

In general, if the C-to-N ratio falls below about 22 to 25, nitrification is stimulated, resulting in increased N transport out of the soil as NO_3^- in drainage waters (Aber et al. 2003). In Europe, a C-to-N ratio of 24 was identified as the critical level below which increased nitrification occurred (Emmett et al. 1998). Although the C-to-N ratio of the forest floor can be altered by N deposition, it can be difficult to detect a modest change over time against substantial background spatial heterogeneity. Soil C-to-N ratio and nitrification data from 250 plots located across the Northeast showed statistically significant but weak correlations between the soil C-to-N ratio and annual N deposition (Aber et al. 2003).

4.3.2.   Wetland Processes

The presence of dissolved organic carbon in lakes and streams can often be attributed to the influence of wetlands on water flow and surface-water chemistry. Forested areas also contribute substantial dissolved organic carbon to drainage waters in the Adirondacks (Canham et al. 2004). High concentrations (higher than about 500 μmol/L) of dissolved organic carbon can substantially reduce the pH and ANC of surface waters, buffer surface waters against pH changes in response to added mineral acidity, and form stable complexes with dissolved Al. The latter process reduces Al toxicity to aquatic life. Thus, responses of surface waters to acidic deposition can be strongly influenced by upstream and shoreline wetland development and other sources of dissolved organic carbon to drainage water.

Wetlands serve as temporary (and sometimes also as long-term) sinks for atmospheric S and N (Baeseman et al. 2006; Sanchez-Andrea et al. 2012; Pound et al. 2013). Chemical reduction associated with anoxia and biological uptake increases S and N storage in wetland soils. Oxidation of stored S and N during drought periods, when water levels recede, makes these stored elements potentially mobile again, along with dissolved organic carbon. Flushing from the wetland to downstream surface water during subsequent storm flow can cause large pulses of mineral and organic acidity in downstream waters.

Carbon also has a large influence on Hg cycling in watersheds. Wetland- and soil-derived dissolved organic carbon acts as a vehicle for Hg transport downstream to receiving waters, thus facilitating bioaccumulation of Hg in fish and piscivorous wildlife. Export of dissolved organic carbon from wetlands and organic soil horizons to streams and lakes benefits producer communities by detoxifying Al.

4.3.3.   Surface Water Processes

The chronic concentration of SO_4^{2-} and associated acidity in surface waters downslope from a wetland area can be substantially lower than they would be in the absence of the wetland, largely as a consequence of S reduction in wetland soils. Wetlands can then contribute to wide fluctuations in downstream surface-water acid-base chemistry during rainfall and snowmelt hydrologic episodes when stored S is released. Such fluctuations in water chemistry can include pulses of acidity that may be toxic to aquatic biota. Thus, wetlands can provide important controls on downstream surface-water chemistry. These include decreases in chronic mineral acidity under low-flow conditions and increases in episodic mineral and organic acidity during rainstorms and snowmelt, especially when preceded by a long dry period. Wetlands also reduce the toxicity of dissolved Al by forming alumino-organic complexes, and they increase the potential for Hg methylation (see Section 3.3).

In addition to modifying the storage and release of S, thereby exerting strong control on mineral acidity, wetlands and other near-stream sources of organic C largely control the organic acidity of drainage water. Dissolved organic C contains a wide range of humic and fulvic acids with different acid dissociation constants. The cumulative effect of the presence of these acids is that pH and ANC are lower than they would be in the absence of the dissolved organic carbon.

4.4. BASE CATIONS AND ALUMINUM

4.4.1. Upland Processes

Aluminum is an abundant naturally occurring element in soils throughout acid-sensitive portions of New York. It is nearly insoluble in water if the pH is above about 6.0, but the solubility increases substantially at pH below about 5.5. Solubility is further increased by the formation of soluble organic complexes upon addition of dissolved organic carbon (Schnitzer and Skinner 1963; Lind and Hem 1975). One of the most important geochemical effects of acidic deposition is increased solubilization and subsequent transport of inorganic monomeric Al from soil to water (Cronan and Schofield 1979; Mason and Seip 1985). The concentration of Al_i in drainage water having pH below about 5.0 can be an order of magnitude or more higher than the concentration in water having pH above 6.

Mobilization of Al_i from soil to soil solution in response to soil acidification inhibits the uptake of Ca by tree roots (Shortle and Smith 1988). Evidence documenting the effects of Al_i on Ca uptake has been compiled through field studies (McLaughlin and Tjoelker 1992; Schlegel et al. 1992; Minocha et al. 1997; Shortle et al. 1997; Kobe et al. 2002) and laboratory studies (e.g., Sverdrup and Warfvinge 1993; Cronan and Grigal 1995). Efforts have focused on identification of a threshold value for the ratio of Ca to Al in soil solution that might be used to identify soil conditions that put trees under physiological stress. Alternatively, the ratio of the SBC to Al is also used, expressed either as the sum of the base cations ($Ca^{2+} + Mg^{2+} + K^+ + Na^+$) or as the sum of the nutrient base cations, omitting Na^+ (Bc; $Ca^{2+} + Mg^{2+} + K^+$) in soil solution. After an extensive literature review, Cronan and Grigal (1995) estimated that there was a 50% risk of adverse effects on tree growth if the molar ratio of Ca^{2+} to Al^{n+} in soil solution was less than 1.0 and that there was a 100% risk for adverse effects on tree growth at a molar ratio value of less than 0.2 in soil solution.

Estimates of vegetative health risk associated with the ratio of Ca^{2+} to Al^{n+} in soil solution such as those Cronan and Grigal summarized have mostly been based on laboratory experiments on seedlings grown hydroponically or in artificial soils. There are uncertainties associated with interpretation of such laboratory data. For example, addition of organic matter to the soil would be expected to partially ameliorate the harmful effects of Al by forming Al-organic complexes. Organically complexed Al, which predominates in upper, organic-rich soil horizons, is believed to be essentially nontoxic. Higher ratios of Ca^{2+} to Al^{n+} generally occur in solution of the organic horizons than in the upper mineral soil (Lawrence et al. 1995). Trees might adjust the distribution of fine roots among horizons to minimize chemical stress (Wargo et al. 2003). This may partly account for the observation that risk levels for the Ca^{2+}-to-Al^{n+} ratio defined in laboratory experiments have not been successfully applied to natural systems (Johnson et al. 1994a; de Witt et al. 2001; Kobe et al. 2002).

Despite the uncertainties of interpretation, a molar ratio of Ca^{2+} (or base cation sum) to Al^{n+} in soil solution has been used as a general index, suggesting an increasing probability of stress to forest ecosystems as the ratio decreases. However, the ratio value of 1.0 cannot be interpreted as a universally applicable stress threshold in natural systems (U.S. EPA 2008). Tree species vary in their sensitivity to Al stress. The form of Al present in solution also influences toxicity.

Effects of Al mobilization from soil to soil waters and surface waters include effects on nutrient cycling (Dickson 1978; Eriksson 1981), pH buffering (Driscoll and Bisogni 1984), toxicity to aquatic biota (Driscoll et al. 1980; Muniz and Levivestad 1980; Schofield and Trojnar 1980; Baker and Schofield 1982), and toxicity to terrestrial vegetation (Ulrich et al. 1980). High concentrations of Al^{n+} in soil solution reduce the growth of plants and their ability to take up water and nutrients, especially Ca^{2+} (Parker et al. 1989). Calcium ameliorates Al toxicity to both plant roots and fish. High concentrations of Al_i in surface water disrupt the function of gills and blood ionic regulation in fish (Baker et al. 1990a).

Dissolved organic acids derived from leaf litter and other decomposing organic materials in the organic horizons of the soil move with drainage water into the mineral soil, where they contribute to the weathering of soil particles and the release of Al into solution. As drainage water moves down into the soil profile, the natural acidity associated with these organic materials is neutralized and Al is deposited as a secondary mineral or as an organic Al (Al_o) complex (DeConinck 1980). Because the mobility of organic anions down through the soil profile is limited, most Al is retained in the mineral soil (often in the B_h horizon) in the absence of acidic deposition. Nevertheless, complexation of Al with dissolved organic matter increases the mobility of Al in the soil and can cause Al_o to be transported into surface waters from shallow soils that are high in organic matter (Lawrence et al. 1986).

Mineral acidity in surface water derived from acidic deposition is caused by the presence of anions, especially SO_4^{2-}, that are more mobile in the mineral soil than those derived from organic matter. As a consequence, acidic deposition transports

Al that had previously been deposited in the upper mineral soil to surface waters. This occurs mainly when base cation release is not sufficient to neutralize atmospheric inputs of acidity.

The study of Shortle and Smith (1988) was the first to attribute the decline of red spruce in the northeastern United States to an imbalance in the availability of Al (increased) and Ca (decreased) to fine roots. Soils in red spruce forests that had been acidified to varying degrees showed that Al dissolution in the mineral soil caused by leaching of SO_4^{2-} and NO_3^{-} decreased availability of Ca in the forest floor (Lawrence et al. 1995). This was because Al can move upward from the mineral soil to the forest floor through capillary water movement, a rise in soil water level during saturated conditions, and/or cycling by litterfall. The Al^{3+} has a higher affinity than Ca^{2+} for negatively charged soil exchange sites. Thus, movement of Al from the lower mineral soil into the overlying forest floor displaces Ca^{2+} from the cation exchange complex. As a consequence, Ca^{2+} is more easily leached into drainage water and is less available for root uptake from organic soil horizons (Lawrence et al. 1995; Lawrence and Huntington 1999).

Modeling future change in exchangeable Al concentrations in soils or dissolved Al in soil water is uncertain. We have only limited understanding of how organic matter influences dissolved Al concentrations in soil water. Also, changes in the formation, cycling, and breakdown of organic matter in the soil attributable to climate change will likely add further uncertainty to our ability to predict the future behavior of Al in soils.

4.4.2.   Wetland and Surface Water Processes

Wetlands are important in controlling the chemistry of Al in surface waters because wetlands contribute substantial organic matter to drainage waters. This organic matter binds with Al to produce relatively non-toxic Al_o. Pulses of high surface-water SO_4^{2-} concentration in surface waters that lie down-gradient from wetlands occur mainly during high-flow conditions after drought conditions.

Organically complexed Al occurs in surface waters as a result of natural soil and hydrologic processes and is generally not harmful to aquatic life (Gensemer and Playle 1999). However, at pH below about 5.5 to 6.0, Al_i is a health risk to some species of aquatic biota (Gensemer and Playle 1999). Reduced growth and survival of some species of fish have been found in streams and lakes with Al_i concentrations in the range of about 2 to 8 mmol/L (Muniz and Levivestad 1980; Baker and Schofield 1982; Baker et al. 1996; Baldigo et al. 2007). Baldigo and Murdoch (1997) found that variation in mortality of caged juvenile brook trout that were exposed to episodic changes in water quality in a Catskill Mountain stream varied with the concentration of Al_i and the duration of exposure. Lawrence et al. (2007) found that 25% (49 of 195) of western Adirondack streams surveyed during August base flow had Al_i concentration above 2 μmol/L, the level above which toxic effects on biota are more likely to occur (Driscoll et al. 2001a; Baldigo et al. 2007). Because brook trout is one of the most acid-tolerant fish species found in the lakes and streams of New York and some life stages are more acid-sensitive than is the YOY, it is possible that some effects of Al_i on fish communities may also occur at concentrations below 2 mmol/L (Baldigo et al. 2007).

The Neversink River basin in the Catskill Mountains contains tributaries that vary from well buffered to severely acidified. In the upper, most acidified stream reaches, fish populations are rare or absent, whereas lower reaches contain up to six fish species (Baldigo and Lawrence 2000). Species richness and distributions of aquatic biota are correlated with stream pH, ANC, and the concentrations of Al_i and Ca^{2+}. Baldigo and Lawrence (2001) surveyed stream-water chemistry, fish habitat conditions, and fisheries in 16 stream reaches in the Neversink River basin. Characteristics of brook trout and slimy sculpin (*Cottus cognatus*) populations were strongly correlated with the concentration of Al_i in stream water and other water-quality parameters.

Mixing zones in lakes and streams, where water of varying chemistry comes together, can contain water that is especially toxic to aquatic biota. When acidic water comes in contact with non-acidic water, Al hydroxides ($Al[OH]_2^+$ and $Al[OH]^{2+}$) are likely to form at the boundary. Aluminum hydroxides are relatively insoluble and can precipitate out of solution as Al hydroxide solid ($Al[OH]_3$) on to fish gills or egg membranes if the pH is suddenly increased in a mixing zone. These forms of Al are acutely toxic to fish.

4.5.   ACID-BASE INTERACTIONS

4.5.1.   Soil-Water Interactions

As acidified water moves through the soil profile, various biogeochemical processes influence the ANC of drainage water. These processes regulate the extent to which soils and drainage waters will be acidified by acidic deposition. The concentrations of acid anions from both human origins and naturally occurring organic acids are important in this regard. These include SO_4^{2-}, NO_3^-, and organic acid anions. The latter are produced in the upper soil horizons by biomass decomposition processes. They normally precipitate out of solution (Lundström et al. 2000) and adsorb to soil surfaces as drainage water moves into the lower mineral soil horizons. Partially as a consequence of this normal anion movement and because water flow at depth maintains greater contact time between solution and solids, drainage waters at greater depth generally have higher ANC. The addition of the strong acid anions SO_4^{2-} and NO_3^- from atmospheric deposition to soil solution partially short-circuits the natural neutralization processes, accelerates soil acidification, and allows drainage water rich in SO_4^{2-} and/or NO_3^- to flow from mineral soils into streams and lakes. To the extent that these anions are charge-balanced by H^+ and/or Al^{n+} cations, this water will have low pH and could be toxic to aquatic biota.

The major human-caused acidifier of soil and water in the Adirondack Mountains is SO_4^{2-}. This is reflected in the observation that concentrations of SO_4^{2-} greatly exceed concentrations of NO_3^- in most surface waters during most of the year (Stoddard et al. 2003). Prolonged SO_4^{2-} and base cation leaching depletes the exchangeable base supply of the soil, thereby impairing the soil's ability to neutralize acidic deposition in the future, especially if the natural supply of base cations through weathering is low. Decreased soil base saturation increases the sensitivity of the watershed to future acidification. Watersheds that were able to neutralize a particular level of acidic deposition in the past may no longer be capable of fully neutralizing that same level (or even lower levels) of acidic deposition today because of the cumulative effects of acidic deposition on soil base saturation (Likens et al. 1996; Lawrence and Huntington 1999).

In the western Adirondack and Catskill Mountains, soils have been acidified by acidic deposition over approximately the last 50 to 100 years. Exchangeable base cation reserves in these acid-sensitive soils were probably naturally low prior to the advent of acidic deposition. These reserves have been further depleted by acidic deposition in more recent decades. Where the soil is low in base cations, Al can be more readily transported to drainage water by acidic deposition. The soil can become further acidified when the stored base cations are depleted by mineral acid anion leaching, which can affect the ability of the soil to support plant growth.

4.5.2.   Upland Processes

Neutralization of drainage water acidity is largely achieved via release of base cations from the soil into soil water. This occurs through weathering, cation exchange on soil surfaces, and mineralization processes (van Breemen et al. 1983). If the acidity of soil water is caused by anions such as SO_4^{2-} and NO_3^- that are relatively mobile in the soil environment, cations (including base cations) will be leached out of the soil into surface water. Loss of base cations from soil to drainage water is a natural process, but the naturally occurring (largely organic acid and carbonic acid) anions typically have limited mobility, which restricts the rate of base cation leaching in the absence of acidic deposition. Because acidic deposition provides anions (SO_4^{2-}, NO_3^-) that are more mobile in the soil environment than anions of naturally derived acids, natural rates of base cation leaching are accelerated by acidic deposition (Cronan et al. 1978). This explains the finding of Lawrence et al. (1999) that pronounced differences in base cation leaching rates in New York occur in upper B-horizon soil along an elevational gradient that reflects changing acidic deposition levels.

Soil acidification from acidic deposition can involve a number of changes to soil properties, including a decrease in soil pH or percent base saturation, an increase in Al mobilization or exchangeable acidity, or a combination of such changes. Inputs to the soil of sulfuric and nitric acid can be partially neutralized by base cation release, which in turn is affected by weathering and base cation exchange. If atmospheric inputs of SO_4^{2-} and NO_3^- pass through the soil to drainage water without neutralization, the soil is not acidified but the drainage water is acidified. Neutralization of drainage water occurs at the expense of the pool of stored base cations in the soil.

Atmospherically deposited acidity is not the only driver that can reduce the soil pH. Net uptake of nutrient cations by vegetation can also acidify the soil. Considerable natural organic acidity is produced in the O_a horizon through the partial decomposition of organic matter and uptake of nutrient base cations by plant roots.

Johnson et al. (1994a, 1994b) found a statistically significant decrease in soil pH from 1930 to 1984 in Adirondack soils that had an initial pH of 4.0 to 5.5. No decrease in pH was found for more strongly acidic soils that had an initial pH of less than 4.0. The decrease in O_a-horizon soil pH was attributed to a combination of acidic deposition and aggrading forests. Bailey et al. (2005) also found a statistically significant decrease in pH (average 0.9 pH units) in combined O_a/A soil horizons in western Pennsylvania from 1967 to 1997. This acidification of the soil was attributed primarily to acidic deposition.

Atmospheric N deposition does not always cause soil or water acidification. One of the most significant effects of N deposition can be increased plant productivity (Kauppi et al. 1992). Nevertheless, some terrestrial ecosystems in New York have moved toward N saturation and showed increased NO_3^- leaching losses in drainage waters (Aber et al. 1989, Stoddard 1994, 1998). It is also possible that the level of N deposition limits increases in atmospheric CO_2 by constraining primary production.

Uptake of N into terrestrial biota is accomplished through complex interactions among plants and microbes (Schimel and Bennett 2004). Long-term N retention in the ecosystem is accomplished in large part by assimilation of N into soil organic matter (Aber et al. 1998) and to a lesser extent by abiotic processes (Dail et al. 2001). The forms of N that are assimilated are determined by N availability at the site (Schimel and Bennett 2004). Under N limitation, competition among plants and microbes to obtain N is high and N is assimilated primarily in depolymerized organic forms. This contributes to low rates of mineralization and minimal buildup of inorganic N in the soil. Where the availability of N in the soil is high, mineralization tends to also be high. This increases competition among plants and microbes for the available NH_4^+ that is produced by mineralization. Increased availability of N under high atmospheric N input reduces competition for NH_4^+ and contributes to increased production of NO_3^- by autotrophic nitrifying bacteria. Only a portion of this NO_3^- can be taken up by plants and microbes. Because much of the N demand is satisfied by NH_4^+ under these conditions and because NO_3^- is more mobile in soils than NH_4^+, a substantial amount of the NO_3^- that is produced by nitrification can leach to drainage water. According to the definitions of N saturation Aber et al. (1989, 1998) and Stoddard (1994) have proposed, the first stage of N saturation is reached when competition among plants and microbes for NH_4^+ has decreased to the point that net nitrification is stimulated.

4.5.3. Base Cation Depletion

Base cations occur in rocks and soils in structural forms in minerals that are largely unavailable to plants. However, a pool of bioavailable or exchangeable base cations is adsorbed to negatively charged surfaces of soil particles. They can enter soil solution by exchanging with other dissolved cations, including acidic cations such as H^+ and Al^{n+}. Base cations in the exchangeable soil pool are gradually leached from the soil with drainage water and are resupplied through weathering and atmospheric input. Weathering breaks down rocks and minerals, slowly releasing base cations to the available pool of adsorbed base cations on the soil. There is also evidence that fungal hyphae can extract Ca^{2+} and other base cations from mineral structures (Jongmans et al. 1997).

The balance between base cation supply and loss determines whether the pool of available base cations in the soil increases or decreases over time. Forest growth lowers exchangeable base cation concentrations in soil through uptake of nutrient base cations into tree biomass. Unless the trees are harvested and removed from the site, these base cations remain in the ecosystem and can become available to the soil again in the future via canopy leaching, litterfall, and plant death and subsequent decay and mineralization. Thus, base cations can be recycled in the ecosystem after uptake into plant biomass. This process does not constitute a loss from the ecosystem unless logs or other vegetative materials are removed from the site such as during tree harvest.

Acidic deposition can leach exchangeable bases from the mineral soil faster than they are resupplied by natural processes (Cowling and Dochinger 1980). For the most part, data that documented this process did not become available in the United States until approximately the 1990s, although decreases in soil exchangeable Ca between the periods 1947 to 1950 and 1987 to 1988 had been documented in Europe through repeated soil sampling (Billett et al. 1990; Falkengren-Grerup and Eriksson 1990).

Sullivan et al. (2006a) reported that soil base saturation and exchangeable Ca had decreased in the upper 10 cm of the B horizon in the Adirondack Mountains between the mid-1980s and 2003 in generally acid-sensitive watersheds (those having lakes with ANC ≤ 200 µeq/L). Soil chemistry in 36 lake watersheds in the mid-1980s was compared with soil chemistry in 32 lake watersheds in 2003. Although there was no repeated sampling of the same sites, the sampling locations were selected randomly in both soil surveys, thus allowing a statistically valid comparison.

The degree to which soil base cations have been depleted affects the chemical recovery potential of acidified soils. Replenishment of exchangeable base cations on depleted soils will require that base cation inputs from weathering and atmospheric deposition exceed losses from leaching and vegetative uptake.

Leaching of the mineral acid anions derived from acidic deposition must be balanced by leaching of an equivalent amount of cation charge. When SO_4^{2-} and NO_3^- anion leaching are charge-balanced by base cation leaching, soils acidify. However, this base cation leaching prevents or retards acidification of soil water and surface water, at least temporarily. Thus, there are both positive and negative effects on ecosystems from base cation leaching. In base-poor soils, exchangeable Ca, Mg, or K can become so depleted that vegetation nutrient deficiencies develop. If base cation leaching is not sufficient to neutralize the mineral acidity in acidic deposition, the pH in drainage water will decrease and Al_i concentrations will increase, possibly to toxic levels.

Sulfur and N emissions controls required by the CAAA have slowed and in some cases reversed acidification damage to sensitive aquatic and terrestrial ecosystems in New York by decreasing the leaching of mineral acid anions and base cations (Driscoll et al. 2001a). Limited chemical and biological recovery has occurred in some aquatic ecosystems. Nevertheless, additional emissions reductions might be necessary if sensitive aquatic ecosystems are to continue the recovery process and prevent renewed acidification in response to base cation loss from soils under continued (albeit substantially reduced) levels of acidic deposition. If the pool of stored base cations on the soil becomes severely depleted, it might take many more decades or longer for weathering processes to replenish the supply.

Thus, base cations play two important roles in the Adirondack and Catskill Mountains. First, they buffer some of the acidity contributed by acidic deposition, decreasing the degree to which lakes, streams, and soil water become acidified. Second, they serve as plant nutrients that promote vegetative growth and are cycled between soil and vegetation. The base cations that neutralize acidity from acidic deposition are lost from the soil and are no longer available to be recycled as plant nutrients. Furthermore, they cannot neutralize acidity in future acidic deposition.

In the absence of acidic deposition, the base cation cycle is highly efficient. Plants take up base cations from the soil to sustain plant growth and return them to the soil when leaves and wood decay. Only small amounts of new base cations are provided as watershed inputs from weathering and in atmospheric deposition, and only small amounts of base cations leach from the soil into drainage water and are lost from the watershed. Acidic deposition speeds up the cycle and causes more base cation loss from soil to drainage water. If the initial soil base cation pool is modest, high levels of acidic deposition can substantially deplete the storage reservoir in a matter of decades.

Soil resampling studies to investigate changes in soil chemistry over time in the northeastern United States have yielded mixed results. Some did not detect changes in soil acid-base chemistry over the past one or two decades (Yanai et al. 2000; Yanai et al. 2005; Hazlett et al. 2011). Other studies have found evidence of continued soil acidification as reflected in decreased exchangeable base cations (Bailey et al. 2005; Courchesne et al. 2005; Sullivan et al. 2006b; A. Johnson et al. 2008; D. Johnson et al. 2008). Other than the study of Lawrence et al. (2012), no studies have shown recent improvement in mineral soil acid-base chemistry, and the Lawrence et al. finding of some soil recovery from acidification was limited to sites in Maine, New Hampshire, and Vermont and to O_a-horizon samples. The one New York site included in that study (at Big Moose Lake watershed in the Adirondack Mountains) did not show improvement in the condition of soil acid-base chemistry in either the O_a or B horizons.

4.5.4.   Wetland and Surface Water Processes

Mountainous areas having high topographic relief and low soil base cation supply are often highly sensitive to surface-water acidification, as revealed by spatial patterns in surface-water ANC. In New York, such areas are found primarily in the Adirondack and Catskill Mountains (Plate 10). Rocks and thin soils that occur in some portions of these mountainous areas have naturally low availability of base cations that might neutralize added acidity. Water moves quickly down steep hillsides, providing limited opportunity for acid neutralization in the soil. Rain and snow volumes tend to be high in mountainous areas in response to orographic processes, causing more base cations to leach out of the soil with the increased water flux. As a consequence of these characteristics, the Adirondack and Catskill Mountains contain some of the most acid-sensitive lakes, streams, and forest soils in the United States.

Wetlands play multiple roles with respect to the acid-base chemistry of down-gradient surface waters. The concentration of natural organic acids is generally high in the surface waters that drain from wetlands. This organic acidity lowers the pH and ANC of drainage waters and provides substantial pH buffering of mineral acids contributed to the watershed from atmospheric deposition of S and N. Wetlands retain S and N and store them (at least temporarily), largely in reduced form. In addition, abundant organic acids in wetlands bind with Al_i to form Al_o, thereby reducing its toxicity.

Because surface-water chemistry integrates the results of all ecosystem processes that occur upstream in a watershed, that chemistry also reflects the results of watershed-scale terrestrial effects, such as N saturation, nutrient depletion, and soil acidification (Stoddard et al. 2003). Drainage-water chemistry provides an index of soil and vegetative properties and processes that predominate throughout the watershed.

Several chemical metrics can be used to assess the effects of acidic deposition on lake or stream acid-base chemistry. The most frequently used include surface-water pH and the concentrations of SO_4^{2-}, NO_3^-, Al_i, Ca^{2+}, SBC, and ANC. Also of value is the more recently developed base-cation surplus. Each can provide useful information regarding the acid-base status of surface-water chemistry. These variables reflect both the sensitivity of surface waters to acidification and the level of acidification that has occurred. Effects of water acidification on aquatic biota are most commonly evaluated using either Al or pH as the primary chemical indicator. ANC is also used because it integrates overall acid-base status and because surface-water acidification models are better able to simulate ANC than pH or Al_i concentration. The utility of the ANC criterion lies in the close relationship between ANC and the surface-water constituents that directly contribute to or ameliorate acid stress to biota, especially pH, Ca^{2+}, and Al_i. The most widely used water-chemistry indicator for both atmospheric deposition sensitivity and effects over the past three decades has been ANC.

Water acidification estimates that are derived using different approaches can be quantitatively compared by computing responses as fractions of the change in the mineral acid anion drivers, expressed as SO_4^{2-} or $[SO_4^{2-} + NO_3^-]$ concentration compared with the SBC concentration. This is commonly done using the F-factor (Henriksen 1984), which is defined as the fraction of the change in mineral acid anions that is neutralized by base cation release. The F-factor is based on the observation that the addition of mineral acid anions to drainage water through acidic deposition will cause the concentrations of some of the cations to increase and/or the concentrations of other anions to decrease in order to maintain electroneutrality. When surface water acidifies in response to acidic deposition, changes in ANC and/or Al_i concentration constitute an appreciable percentage of the overall response to increased $[SO_4^{2-} + NO_3^-]$ anions and therefore the F-factor is considerably less than 1.0 (Sullivan 1990). If, however, most of the $[SO_4^{2-} + NO_3^-]$ increase causes a corresponding increase in base cation concentrations, the F-factor will approach 1.0 and water acidification is minimized. Although F-factor calculations can be useful in the early stages of water acidification, this simple approach is less applicable in the latter stages of acidification or during recovery, when base cation depletion can play an important role. Negative values of the F-factor commonly occur during recovery from prior acidification because of base cation depletion (Rapp and Bishop 2009).

Assessment of changes in surface-water acid-base chemistry must distinguish between acidic waters and acidified waters. Acidic describes a condition that can be measured (i.e., Gran ANC or calculated ANC ≤ 0 µeq/L) and is due to either the effects of acidic deposition or other causes such as the presence of organic acidity or the weathering of S-containing minerals in the watershed. Acidified refers to the process of acidification (an increase in acidity or a decrease in ANC observed through time). It does not require that the water body be acidic and does not imply a particular cause for the change in chemistry. The term anthropogenically acidified implies that human activity was responsible for an increase in acidity that has occurred.

Surface-water ANC reflects the culmination of effects of all of the ionic interactions that occur as atmospheric deposition and precipitation move from the atmosphere into the soil and eventually to drainage water. The ANC reflects the relative balance between base cations and strong mineral acid anions in drainage water. When the sum of the equivalent concentrations of the base cations exceeds those of the strong mineral acid anions, the water will have positive ANC. Higher ANC is generally associated with higher pH and Ca^{2+} concentrations; lower ANC is generally associated with higher H^+ and Al^{n+} concentrations and a greater likelihood of toxicity to biota. In response to approximately a century of acidic deposition, many (but not all) lakes and streams in the Adirondack and Catskill Mountains have experienced decreases in ANC.

4.5.4.1. *Chronic Acidification Processes*

Acid Neutralizing Capacity. In the mid-1980s, the estimated percent of Adirondack lakes that had ANC ≤ 0 µeq/L ranged from about 14% to 26%, largely depending on the size range of lakes included in the calculation. The U.S. EPA's Eastern Lakes Survey (ELS) estimated that 14% were acidic, but it included only lakes larger than about 4 ha. The smaller lakes that were not considered in the estimate were more likely to be acidic in response to both natural and human-caused factors. From 1984 to 1987, the ALSC sampled 1,469 Adirondack lakes greater than 0.5 ha in size and estimated that 26% were acidic (Driscoll et al. 1991). The higher percentage of acidic lakes in the ALSC sample was mainly due to inclusion of lakes and ponds that were only 1 to 4 ha in area. Some of these small lakes were acidic as a consequence of high concentrations of naturally occurring organic acids (Sullivan et al. 1990). Many of the acidic and low ANC surface waters in New York are located in the southwestern and high peaks areas of the Adirondack Mountains.

Chronic acidification is most commonly documented by measuring changes in surface-water chemistry during relatively stable periods. These are generally summer or fall for lakes and spring or summer base flow for streams and avoid periods of rapid snowmelt and large rainstorms.

Much of the rainfall and snowmelt water in most acid-sensitive watersheds in New York passes through at least part of the soil profile before it enters a stream or lake. The typical soil profile in these areas has lowest pH in upper soil horizons, increasing down the profile. Drainage-water chemistry during base flow mostly reflects chemical conditions in lower soil horizons, where the acidity is more strongly buffered.

Summer or fall is a good time for surveying lake water in terms of minimizing variability in acid-base chemistry. However, lake-water samples collected during low-flow periods (which predominate during summer) do not directly reflect the chemistry that might occur during high-flow conditions and in particular do not reflect the dynamics or importance of N as an agent of acidification.

Concentrations of NO_3^- in lake water are chronically high during the summer and fall seasons only in lakes having watersheds that are at a fairly advanced stage of N saturation (Stoddard 1994). Driscoll et al. (1985) found that NO_3^- concentrations in 20 Adirondack lakes in the early 1980s averaged 12% of SO_4^{2-} concentrations. Nitrate was proportionately more important to base-flow chemistry in Catskill Mountain streams, where Lovett et al. (2000) found that base-flow NO_3^- concentrations in 1994 to 1997 were an average of 37% of SO_4^{2-} concentrations in 39 study streams.

Lake or stream ANC provides an index for assessing the biologically relevant status of surface waters. Baker et al. (1990c) used ANC cutoffs of 0, 50 and 200 µeq/L for reporting on national lake and stream population estimates. More recent assessments in the eastern United States commonly focused on cutoffs of 0, 50, and 100 µeq/L (U.S. EPA 2009). A value of ANC ≤ 0 µeq/L is a common benchmark because waters at or below this level have no capacity to neutralize further acidic inputs. Waters having negative ANC are defined as acidic. Waters having ANC ≤ 50 µeq/L have been termed "extremely acid sensitive" (Schindler 1988), are prone to episodic acidification in some regions (DeWalle et al. 1987; Eshleman 1988; Driscoll et al. 1991), and may be susceptible to future chronic acidification. The ANC value of 100 µeq/L is commonly used as a general benchmark for potential acid sensitivity.

The use of titrated ANC to assess the effects of acidic deposition on aquatic ecosystems is complicated by natural acidity derived from acidic organic acid functional groups and the influence of Al_i on titration results. The influence of organic acidity on titrated ANC varies with the concentration of dissolved organic carbon. Application of a triprotic model of organic acidity showed that an increase in the concentration of dissolved organic carbon of 183 mmol/L would decrease Gran ANC by about 10 meq/L in Adirondack lakes (Driscoll et al. 1994). The presence of dissolved Al_i increases titrated ANC. This is somewhat counterintuitive because Al_i can be harmful to both aquatic and terrestrial biota.

Change over time in the amount of organic acidity also complicates evaluation of the recovery of surface water from acidification. Concentrations of dissolved organic carbon have increased over the past 10 to 15 years in the Adirondack region and in other regions in the eastern United States. This may be a response to decreased acidic deposition, but a generally warming climate may also play a role in this increase in dissolved organic carbon. This effect has limited the ANC recovery of surface waters. Clair et al. (2011) postulated that the increase in dissolved organic carbon in lake water in Canada's Atlantic provinces has slowed the

ANC recovery but also decreased lakewater pH. They concluded that increasing lake temperature has most likely been the main factor driving the increase of dissolved organic carbon in their study region, where mean annual air temperature increased by about 1.5 to 2.5 °C over a period of about 20 years.

Concentrations of dissolved organic carbon in surface waters have increased over the last two decades or more throughout the northeastern United States and elsewhere in areas that previously received substantial acidic deposition (Driscoll et al. 2003a; Evans et al. 2006; Monteith et al. 2007). The cause of this increase in dissolved organic carbon is not known but may include one or more of the following (Burns et al. 2011):

- decreasing acidic deposition (Evans et al. 2006; Monteith et al. 2007)
- climate warming (Worall and Burt 2007; Clair et al. 2008)
- changes in precipitation (Hudson et al. 2003; Worall and Burt 2007)

These changes in dissolved organic carbon are important because it plays major roles in a variety of ecosystem processes, including episodic acidification, Al toxicity, Hg methylation and transport, light penetration into the water column, water temperature, and lake stratification (Snucins and Gunn 2000). In addition, increased dissolved organic carbon contributes to decreased ANC, thereby limiting the ANC recovery of surface waters in response to reductions in acidic deposition.

Waller et al. (2012) demonstrated that calculated ANC values increased twice as much as titrated Gran ANC in Adirondack lakes from 1991 to 2007 in association with partial chemical recovery from acidification. This is because calculated ANC does not include the influence of organic acid anions on the resulting ANC. Organic acid anion concentrations (and associated dissolved organic carbon) have increased during recovery from acidification. Some of those organic acid anions behave as strong acids in an ANC titration. Driscoll et al. (1994) estimated that percentage to be about one-third. Lawrence et al. (2013a) used the base cation surplus index and estimated concentrations of strong organic acid anions ($RCOO_s^-$) to fractionate the Al into organic and inorganic forms and evaluate the decrease in toxic Al_i that has occurred in Adirondack lakes. The observed higher dissolved organic carbon concentrations in more recent times increased complexation with Al and decreased the concentration of Al_i from 57% of the total Al in 1994 to 23% in 2011. Thus, the increase in dissolved organic carbon has accelerated chemical recovery from acidification by reducing Al_i toxicity. This finding is further reinforced by results of Adirondack lake fish surveys that found that lakes showing base cation surplus-to-$RCOO_s^-$ < 2 generally yielded no fish, whereas lakes showing base cation surplus-to-$RCOO_s^-$ > 2 generally did yield fish (Dukett et al. 2013).

Across the 42 Adirondack Long Term Monitoring Program (ALTM) drainage lakes examined, dissolved organic carbon increased on average by 78 µM over the

17-year study period while the base-cation surplus increased by 26 μeq/L. Even though the increase in RCOO^-_s constituted only 13% of the average annual increase in the base-cation surplus, the change in the relative importance of Al$_i$ as a fraction of total Al decreased by 40% from 1994 to 2011. This response effectively decreased toxic Al$_i$ levels beyond what would have occurred solely in response to reduced atmospheric deposition of strong mineral acid anions (Lawrence et al. 2013b).

Recent research results suggest that decreases in acidic deposition over the past few decades may have been the most important cause of increased dissolved organic carbon in previously acidified fresh waters (Pound et al. 2013). The widely observed pattern of a recent increase in dissolved organic carbon may be a return toward the natural state of these waters prior to the onset of acidic deposition (Krug and Frink 1983; Driscoll et al. 2003a; Evans et al. 2005; Pound et al. 2013). Furthermore, the presence of dissolved organic carbon in stream water may contribute to higher biodiversity (Passy 2010).

Dissolved organic C in the Buck Creek North tributary in the Adirondack Mountains increased by about 400 μM from 1991 to 2013. Over that time period, ANC increased but not by a statistically significant amount. The Al$_i$ concentration decreased but remained above the fish response threshold of 2 μm (Lawrence 2013).

Model projections of surface-water acidification usually quantify ANC as the SBC concentrations (Ca^{2+}, Mg^{2+}, Na$^+$, K$^+$, NH$_4^+$) minus the sum of inorganic strong-acid anion (SAA) concentrations (SO$_4^{2-}$, NO$_3^-$, Cl$^-$). This difference between SBC and SAA is often called calculated ANC (Cosby et al. 2001). The SBC measurement provides an index of the ability of a watershed to neutralize acid inputs through the release of base cations to soil solution. The SAA reflects the input of acidic deposition. Thus, the difference in these two quantities (SBC - SAA) provides a relative measure of both acidic deposition sensitivity and effect.

Surface Water pH. Surface-water pH is frequently used as an index of acidification. It provides an indication of the concentration of H$^+$ in solution. The pH also correlates with other biologically important components of surface-water acid-base chemistry, including ANC and concentrations of Al$_i$, Ca^{2+}, and organic acid anions. Low pH can be toxic to many aquatic species (Driscoll et al. 2001a). Common reference pH values for evaluating potential biological effects are 5.0, 5.5, and 6.0. Only the most acid tolerant fish species can survive below pH 5.0. About half of all fish species that occur in the Adirondack region are restricted to lakes with pH higher than 6.0 (Kretser et al. 1989). A pH value of 6.0 is sometimes considered the level below which biota are at increased risk from acidification (Driscoll et al. 2001a).

Natural acidity complicates the use of pH as an indicator of acidic deposition effects. Naturally acidic aquatic and wetland environments are common in some areas affected by acidic deposition (Dangles et al. 2004), including the Adirondack Mountains. Environments rich in organic acids contain plant and animal species that are adapted to such conditions. Thus, natural organic acidity is just that: it is natural. It does not reflect human-caused alteration of acid-base chemistry. However, if dissolved organic carbon increases in response to human-caused warming of the environment, then a portion of that response will be anthropogenic.

Base-Cation Surplus. Organic acidity is not included in determining the calculated ANC, and this representation of ANC does not reflect the mobilization of potentially toxic Al_i concentrations in a consistent manner if dissolved organic carbon concentrations vary (Lawrence et al. 2007). In contrast, the base-cation surplus distinguishes between the effects of organic acidity and acidic deposition (Lawrence et al. 2007). It is similar to calculated ANC, but it also explicitly accounts for strongly acidic organic acids. The base-cation surplus is defined as the summed concentrations of the four principal base cations (Ca^{2+}, Mg^{2+}, Na^+, and K^+) minus the summed concentrations of all strongly acidic anions (SO_4^{2-}, NO_3^-, Cl^-, and $RCOO_s^-$), where $RCOO_s^-$ represents the estimated concentration of strong organic acid anions. Thus, the base-cation surplus equals calculated ANC minus $RCOO_s^-$. If dissolved organic carbon is low, the base-cation surplus is approximately equal to calculated ANC. A plot of the base-cation surplus versus Al_i concentrations in Adirondack streams shows a distinct threshold for Al mobilization at base-cation surplus = 0 (Figure 4.3). This pattern is observed regardless of the dissolved organic carbon concentration (Lawrence et al. 2007). This threshold at base-cation surplus = 0 μeq/L provides an unambiguous reference point for evaluating the effects of human-caused acidic deposition on the mobilization of Al_i in surface water.

4.5.4.2.   *Episodic Acidification Processes*

During periods of snowmelt and rainfall, drainage water flow-routing favors flowpaths through upper soil horizons. During such high-flow conditions, drainage-water chemistry more closely reflects the lower pH and higher organic content of upper soil horizons (Sullivan 2000). Storm flow and snowmelt cause increased drainage water acidity because of an increase in the proportion of the total flow that is derived from water that has moved laterally through the upper soils.

Under high-flow conditions, surface waters tend to have a lower ANC, pH, and base-cation surplus and often higher concentrations of Al_i as compared with reference conditions under base flow. As a consequence, water chemistry can at some locations be suitable for supporting acid-sensitive aquatic biota under base flow, but water in that same lake or stream can be toxic under high-flow conditions. This difference is more pronounced in streams than lakes.

Acid-sensitive surface waters in New York commonly have ANC and pH values during the spring season that are lower than those found during the low-flow

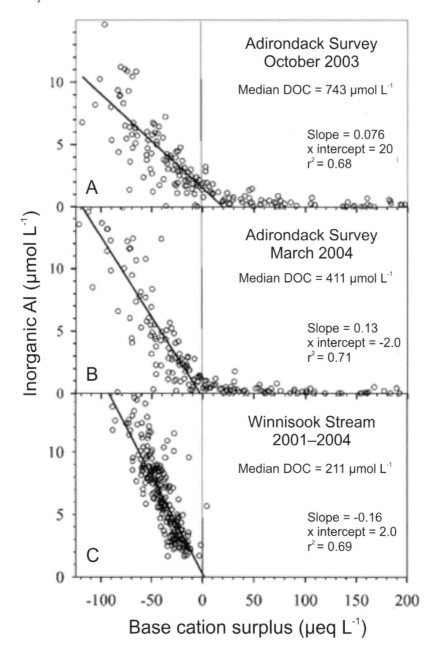

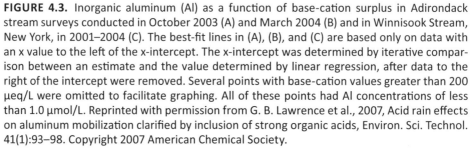

FIGURE 4.3. Inorganic aluminum (Al) as a function of base-cation surplus in Adirondack stream surveys conducted in October 2003 (A) and March 2004 (B) and in Winnisook Stream, New York, in 2001–2004 (C). The best-fit lines in (A), (B), and (C) are based only on data with an x value to the left of the x-intercept. The x-intercept was determined by iterative comparison between an estimate and the value determined by linear regression, after data to the right of the intercept were removed. Several points with base-cation values greater than 200 µeq/L were omitted to facilitate graphing. All of these points had Al concentrations of less than 1.0 µmol/L. Reprinted with permission from G. B. Lawrence et al., 2007, Acid rain effects on aluminum mobilization clarified by inclusion of strong organic acids, Environ. Sci. Technol. 41(1):93–98. Copyright 2007 American Chemical Society.

conditions that predominate during summer and fall. Sites that have ANC in the range of 0 to 50 µeq/L or even higher during the summer or fall can experience short-term episodic acidification to negative ANC values during spring snowmelt or intense rain events. For example, Lawrence (2002) found that 16% of total stream reaches in the West Branch Neversink River in the Catskill Mountains were chronically acidic, whereas four times as many (66%) of the stream reaches had a high likelihood of becoming temporarily acidic during high-flow periods.

The ANC and pH of surface waters usually decrease with increases in discharge (Wigington et al. 1990). These and other changes in surface-water chemistry during hydrologic episodes are controlled by both natural processes and interactions between acidic deposition and the watershed. The former include dilution of base cation concentrations with increasing discharge, nitrification of mineralized organic N, flushing of organic acids from terrestrial to aquatic systems, and the neutral salt effect, or the process whereby a neutral salt (i.e., NaCl) is added to the soil and some of the neutral salt cation (i.e., Na^+) is exchanged for H^+ on the soil ion exchanger. This results in acidification of drainage water. This effect tends to be more pronounced during hydrological episodes than during base-flow chemistry. It occurs commonly in near-coastal areas that receive atmospheric marine aerosol deposition of Na^+ and Cl^-.

Disturbance processes related to erosion and/or forest pests and pathogens can also be important in modifying the acid-base chemistry of surface water during periods of high discharge. Pulses of episodic acidification of stream water may last from hours to weeks. They can cause decreases in the ANC of acid-sensitive streams and lakes to low or negative values and increases in Al_i concentration to toxic levels. The toxicity of mobilized Al_i depends both on Al_i concentration and the duration of elevated levels of Al_i (Baldigo and Murdoch 1997).

Episodic acidification during rainstorms and snowmelt is largely a natural phenomenon caused by the mobilization of organic acids from upper soil horizons to drainage water and by dilution of base cation concentrations. Episodic decreases in pH and ANC in New York can also be caused, in part, by an increase in NO_3^- concentration in drainage water, some of which is due to acidic deposition (Murdoch and Stoddard 1992; Wigington et al. 1996b). Episodic acidification in the Adirondack and Catskill Mountains is most common in the spring as a result of snowmelt and rainstorms and is less common during summer. Water chemistry in late summer tends to be least acidic (Lawrence et al. 2007). Stoddard et al. (2003) compared seasonal data from New England lakes, Adirondack lakes, and northern Appalachian streams that were collected at intervals that ranged from monthly to quarterly to evaluate differences between the chemistry of surface waters in the summer and in the spring. Spring ANC values were on average about 30 µeq/L lower than summer ANC values (Figure 4.4).

The number of lakes and streams prone to episodic acidification can be estimated by combining detailed episodic data from a few sites with base-flow values of

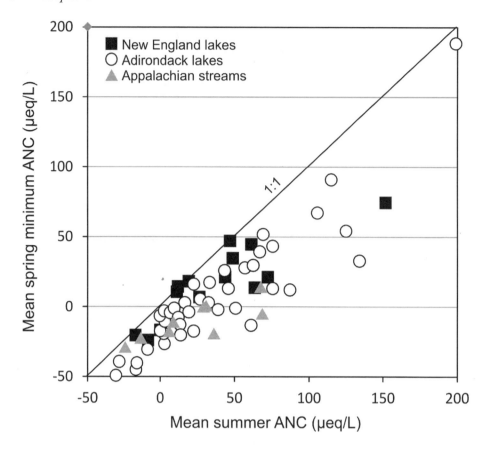

FIGURE 4.4. Relationship between summer and spring ANC values at LTM sites in New England, the Adirondacks, and the northern Appalachian Plateau. Data are plotted as mean summer values for each site during the period 1990–2000 (horizontal axis) and mean spring minima for each site for the same time period. On average, spring ANC values are 30 µeq/L lower than summer values. Source: Stoddard et al. 2003.

ANC that have been determined in surface-water surveys (Eshleman et al. 1995; Bulger et al. 2000; Driscoll et al. 2001a). If an assessment includes water bodies that might become episodically acidic, estimates of the extent of surface-water acidification will typically be higher. For example, base-flow lake-water samples collected from 1991 to 1994 through the Temporally Integrated Monitoring of Ecosystems (TIME) program indicated that 10% of the 1,812 lakes (> 1 ha surface area) in the Environmental Monitoring and Assessment Program (EMAP) sampling frame in the Adirondacks had ANC ≤ 0 µeq/L. An additional 31% of these lakes had base-flow ANC values less than 50 µeq/L and were estimated by Driscoll et al. (2001a) to be susceptible to short-term episodic acidification to ANC below 0 µeq/L.

Murdoch and Stoddard (1993) found high concentrations of NO_3^- in streams during high-flow conditions in Catskill streams. In fact, NO_3^- concentrations periodically equaled or exceeded SO_4^{2-} concentrations. This study also documented increasing NO_3^- concentrations over time during the period 1970–1990 in all 16 Catskill streams for which long-term data were available.

Lawrence (2002) estimated the extent of episodically acidified stream reaches in an 85-km^2 Catskill watershed. Sampling was conducted at 122 stream sites, representing a high density of small streams. Whereas approximately 16% of the total upstream reaches were chronically acidified to ANC below 10 µeq/L, about two-thirds of the stream reaches became episodically acidified to that same ANC level.

Episodic variation in stream chemistry in the eastern United States was characterized through the U.S. EPA's Episodic Response Project (ERP). In that study, 13 low-order streams with watershed areas of less than 24 km^2 were monitored during hydrologic episodes from 1988 to 1990 (Wigington et al. 1996b). Study streams were located in the Adirondack and Catskill Mountains and the Appalachian Plateau in Pennsylvania. Episodes of increased acidification involved decreases in ANC of up to 200 µeq/L, decreases in pH of up to one unit, and increases in concentrations of Al_i of up to 15 µmol/L (Wigington et al. 1996b).

Lawrence et al. (2008a) found that 124 out of 188 (66%) western Adirondack streams were prone to acidification during high-flow conditions to the level at which Al_i becomes mobilized from soil to stream water. A survey of western Adirondack streams in March 2004 was chosen to represent episodic conditions, and a survey conducted in August 2004 was chosen to represent base-flow conditions. These two surveys found that 35% of the streams were chronically acidified, 30% were episodically acidified, and 34% were not acidified (Lawrence et al. 2008a).

4.6. NUTRIENT INTERACTIONS

Nutrient enrichment refers to a variety of environmental changes that can occur when the availability of a key nutrient is increased as a consequence of air, soil, or water pollution. Nutrient enrichment effects can occur in terrestrial, wetland, and aquatic environments. In all cases, the addition of a key nutrient can contribute to changes in the species mix of the plant and algal communities. Some species will thrive at the expense of others. The mix of species present in the ecosystem can change as a consequence of atmospheric nutrient addition.

Fertilization with N can increase tree growth up to a point and cause changes to vegetation in the forest understory. Long-term effects on forest understory plants and/or differential growth effects on different tree species have not been as well studied in the United States as they have been in Europe.

In forested ecosystems, most of the nutrients required for tree growth are generally supplied through internal ecosystem cycling, including decomposition of dead and dying trees, limbs, and foliage. Inputs of nutrients are low to nutrient-poor soils unless they receive high levels of anthropogenic pollutants (Chapin et al. 2002). Leaching of nutrient base cations from the soil in response to SO_4^{2-} and/or NO_3^- leaching is a significant ecosystem stressor on sites that have naturally low nutrient base cation supply from weathering (Reuss and Johnson 1986).

Estuaries and near-coastal marine waters are highly sensitive to nutrient enrichment effects from human-caused atmospheric N deposition. In addition to atmospheric N inputs, however, a wide variety of point and nonpoint pollution sources often contribute N and other nutrients to these waters. Contributions from non-atmospheric sources are often difficult to quantify, and the relative importance of atmospheric versus non-atmospheric sources of N to marine waters is uncertain.

Many species are adapted to a particular supply and proportion of key nutrients. If one nutrient becomes more or less available, the result can be a nutrient imbalance, which in turn can affect a variety of ecosystem processes (Sterner and Elser 2002). Nutrient enrichment from N deposition at sufficiently high levels can reduce species richness and contribute to an overall decline in biodiversity. Such effects can occur in sensitive terrestrial, wetland, and aquatic ecosystems.

Some aquatic, wetland, and terrestrial ecosystems are highly sensitive to the effects of nutrient enrichment from atmospheric N deposition. These include remote lakes at high elevations, estuarine and near-coastal marine waters, and alpine, meadow, and nutrient-poor wetland ecosystems. Although nutrient enrichment effects are expected to be most pronounced in plant communities that lack trees, the same kinds of effects of N deposition can also occur in forested plant communities. However, long-term effects on forest understory plants and growth effects on different tree species are not well known in New York or elsewhere in this country.

4.6.1. Terrestrial Effects

In many terrestrial ecosystems in temperate zones, N is the most important nutrient that limits the growth of plants. If an appreciable amount of N is added to an N-limited ecosystem, plant growth can increase. Some species are better able to take advantage of added N than others. The end result can thus be an increase in the growth of some plant species at the expense of others.

Terrestrial ecosystems throughout New York have been exposed to a continuous supply of N from the atmosphere for many decades. The extent to which these ecosystems have responded to N addition has depended largely on the extent to which the growth of plants and algae was limited by N availability and the availability of P, light, water, or some other nutrient (Tessier and Raynal 2003). In terrestrial ecosystems that are at least partially N-limited, many plants will tend to grow faster if the availability of N is increased. Because some plant species are better able to take advantage of this addition of the key limiting nutrient than other plant species are, a shift in the competitive advantage happens that favors those plants. This can decrease species diversity, alter the species makeup of the plant community, and eliminate some of the rare or otherwise valued species. The species that benefit most from increased N supply are often nonnative opportunistic species; those that are suppressed often include rare native species.

The extent to which added N from atmospheric deposition increases primary production and tree growth in New York forests is not well known. Bedison and McNeil (2009) used tree diameter measurements made in 1984 and 2004, allometric equations, and estimates of wet N deposition at 32 permanent plots along a deposition gradient in the Adirondack Park to estimate the effects of atmospheric N deposition on the basal area increments and woody biomass increments of individual tree stems. They found that N deposition had either a neutral or a positive effect on basal area increments and woody biomass increments. Increased growth effects were most evident in the smaller size classes of particular tree species. Significant growth increases with increasing N input across the gradient were observed, especially for red maple, red spruce, and balsam fir. Bedison and McNeil (2009) hypothesized that differences in temperature, O_3 exposure, and base cation depletion in soils may have largely counteracted the N fertilization response.

Plant communities dominated by herbaceous species are believed to be especially sensitive to N enrichment. This may be partly because it takes a long time for scientists to document effects on trees, which can grow for decades or centuries and are affected by a wide range of potential stressors. Nutrient enrichment effects on plant communities in the United States have been most convincingly demonstrated for herbaceous arctic and alpine plant communities, meadows, wetlands, and arid or semi-arid lands (U.S. EPA 2009). Hurd et al. (1998) reported the results of an experimental N addition to the understory at three study sites in the Adirondack Mountains. Nitrogen was added as HNO_3 (2x ambient) and $(NH_4)_2SO_4$ (2x and 4x ambient). The cover of common herbaceous species (*Oxalis acetosella*, *Maianthemum canadense*, and *Huperzia lucidula*) declined substantially after only three years of treatment. The observed response was attributed to shading from the increased growth of ferns, *Dryopteris intermedia* and *Dennstaedtia punctilobula*.

Gilliam (2006) reviewed available data regarding the response of the herbaceous layer of forest communities of the eastern United States to increased N input. This focus was driven in part by the observation that most of the species diversity in eastern forests occurs in the herb layer (defined as including all vascular plants ≤ 1 m tall). He identified six issues that affect herbaceous plants and that can be altered by N addition: competition, herbivory, mycorrhizal infection, disease, species invasion, and the activity of exotic earthworms. Gilliam concluded that each of these processes contributes to changes in species composition and decreases in biodiversity in response to atmospheric N deposition.

Alpine plant communities are found at scattered high-elevation locations in the Adirondack Mountains. Alpine communities are known for their high diversity of vascular plant species (Bobbink et al. 2010) and are especially sensitive to N enrichment. Alpine sedges appear to benefit more from N addition than do grasses or forbs (Bowman et al. 2006; Bobbink et al. 2010). Changes in the abundance of a species of alpine sedge (*Carex rupestris*) have been shown at N deposition levels near 4 kg N/ha/yr in Colorado, although research of this type has not been conducted in New York. Effects on general vascular plant community composition in a relatively short term of a few years probably require higher N deposition levels (Bowman et al. 2006). Model simulations suggest that effects on alpine plant communities can develop over centuries in response to even lower levels of N input (Sverdrup et al. 2012; McDonnell et al. 2014). Various alpine plant community characteristics govern sensitivity to N input, including low temperature, restricted growing season, low primary production, and wide variation in moisture regime (Bowman et al. 1993; Bowman 1994; Bowman and Fisk 2001). Soil formation is a slow process in the harsh alpine environment and as a consequence alpine plants have evolved under conditions of low nutrient supply (U.S. EPA 2008). Ground-layer bryophytes and lichens in alpine scrub habitat are especially sensitive to N addition (Fremstad et al. 2005; Britton and Fisher 2007) but have not been studied in New York.

Increased N input reduces plant biodiversity in grasslands in both Europe and North America (Bobbink et al. 2003; Stevens et al. 2004; Clark and Tilman 2008) across a range of soil conditions. Changes in grassland species composition occur at N deposition levels as low as about 10 kg N/ha/yr (Bobbink et al. 2003). Nonnative nitrophilous (N-loving) grasses have displaced native plant species in response to greater N supply from the atmosphere combined with cessation of livestock grazing at some locations in California (Fenn et al. 2003b; U.S. EPA 2008). A survey across acidic grasslands in the United Kingdom by Stevens et al. (2004) estimated a decrease of one plant species for every 2.5 kg N/ha/yr of atmospheric N deposition. Effects on meadow communities in New York are not known.

Lichens are types of fungi that grow in association with algae or cyanobacteria. They are among the most N-sensitive terrestrial receptors (Bobbink et al. 2003; Fenn et al. 2003a; Vitt et al. 2003). The algae or cyanobacteria provide photosynthetic capability to augment the moisture-holding and attachment capabilities of the fungus. This symbiotic association of two different life forms is important to nutrient cycling processes in a variety of ecosystem types. Although some lichen taxa exhibit broad tolerance to N supply, most can be classified into groups as oligotrophic (restricted to N-poor environments), mesotrophic (requiring moderate N supply), and eutrophic (preferring N-rich environments). Atmospheric N deposition causes a shift from oligotrophic toward eutrophic conditions, and this shift is accompanied by a change in the lichen community that favors the mesotrophic and eutrophic forms and produces a decrease in the oligotrophic species (Geiser et al. 2010). This change in lichen species in response to a change in atmospheric N deposition can have multiple ecosystem effects. For example, many oligotrophic lichen species are important as forage, for nutrient cycling, and as nesting materials for other species, including birds and mammals. Documenting the presence or abundance of lichen indicator taxonomic groups is difficult in New York, however, because N and S deposition over the past century have been so high that many of the oligotrophic lichen species were probably eliminated a long time ago (Pardo et al. 2011).

4.6.2. Wetland Effects

Nitrogen limitation is widespread in North American marsh lands (Bedford et al. 1999). This might be caused by differences in nutrient cycling in wetlands as compared to open-water ecosystems (Morrice et al. 2004). Wetland sediments tend to be O_2-poor. Oxidation of organic C by decomposers in anoxic environments can be limited by the availability of NO_3^- for use as an electron acceptor (Sundareshwar et al. 2003).

Wetlands are common throughout northern New York. In the Oswegatchie-Black River watershed of the southwestern Adirondack Mountains, about 15% of the surface area is occupied by wetlands (Roy et al. 1996). Wetlands considered sensitive to N deposition often contain plant species that have evolved under N-limited conditions. The competitive balance among plant species in sensitive wetland ecosystems can be altered by N addition; some species will be displaced by those that can more efficiently use the additional N (U.S. EPA 1993). This effect has been well documented in heathlands in The Netherlands, where common heather (*Calluna vulgaris*) has been replaced by grass species (Heil and Bruggink 1987; Tomassen et al. 2003). However, these heathlands have received very high levels of atmospheric N deposition, several times higher than levels found in New York.

Wetland types differ in how they respond to N addition and the response varies with hydrological conditions. Wetlands such as bogs that receive much of their water input from precipitation are most sensitive to the effects of N input (Morris 1991). Ombrotrophic bogs are often acidic and dominated by mosses. They are especially common in northern boreal forests, where precipitation is higher than evapotranspiration and where there is an impediment to downward drainage in the soil (Mitsch and Gosselink 2000). Peatland bog ecosystems are highly sensitive to nutrient enrichment (Krupa 2003) and are common in areas that were glaciated, especially in portions of the Northeast and upper Great Lakes region (U.S. EPA 1993). Bogs are considered to be vulnerable to ecological damage in part because they can host several federally listed rare and endangered plant species, including quillworts (*Isoetes* spp.), some sphagnum mosses (*Sphagnum* spp.), and the green pitcher plant (Greaver et al. 2011). The sensitivity of *Sphagnum* species in peatlands to elevated atmospheric N deposition has been well documented in Europe (Berendse et al. 2001; Tomassen et al. 2004). *Sphagnum squarrosum* and *S. fallax* have been observed to be negatively affected by experimentally elevated atmospheric N and S inputs (Kooijman and Bakker 1994). Roundleaf sundew is also highly sensitive to atmospheric N deposition (Redbo-Torstensson 1994).

Shrub-dominated wetlands in the Adirondacks generally contain extensive stands of speckled alder (*Alnus incana* ssp. *rugosa*; Mitchell and Tucker 1997), an N-fixing species in symbiotic association with an actinomycete (*Frankia* spp.). These riparian wetlands cover substantial areas, especially near lake inlets (Roy et al. 1996). Alder fixes about 40 kg N/ha/yr in Adirondack wetlands (Hurd et al. 2001). Kiernan et al. (2003) found that Adirondack wetlands that had a high density of alder had about six times higher accumulation of NO_3^- in exchange resins than wetlands lacking alder. Substantial amounts of N are added by N fixation to Adirondack lake inlet ecosystems dominated by alder (Hurd et al. 2005).

Nitrogen and P loading to coastal wetlands is a particular concern throughout the lower Great Lakes (Lakes Erie and Ontario and the southern part of Lake Michigan; Hill et al. 2006). Both agricultural and atmospheric sources of nutrients contribute to this stress. Atmospheric N deposition increases from Lake Superior in the north to Lake Ontario in the south (NADP 2004; Hill et al. 2006). Agricultural inputs of nutrients to wetlands also follow this north-south pattern, with the highest agricultural loadings among the Great Lakes to Lake Erie and the lowest to Lake Superior. Total N dissolved in surface waters of Great Lakes wetlands is directly correlated with both atmospheric N deposition and the degree of nearby agricultural activity (Hill et al. 2006). Hill et al. 2006 also reported lower N concentration in wetland sediments in the upper Great Lakes compared with those of the lower Great Lakes. Increased N loading to the lower Great Lakes can contribute to eutrophication, turbidity, sedimentation, and algal blooms.

Wetland waters associated with the Great Lakes often have nutrient concentrations that suggest N limitation. Hill et al. (2006) found that more wetlands were N- than P-limited at each of the Great Lakes, and Morrice et al. (2004) measured a ratio of N to P in Lake Superior coastal wetlands that suggested N limitation. Trebitz et al. (2007) reported data from 58 coastal wetlands around the Great Lakes that had been selected for study across gradients in hydromophic type, water quality, and agricultural influence. They showed that Great Lakes coastal wetland sites exhibited nutrient (N, P) levels and water clarity that were strongly associated with agricultural intensity in the contributing watershed. Lake Erie wetlands had the highest nutrient levels and lowest water clarity.

Hill et al. (2006) found evidence of N limitation in coastal wetlands of the Great Lakes, based on surface water and sediment chemical stoichiometry and microbial enzyme activity ratios. This result contrasts with results for open Great Lakes waters, which are generally considered to be P-limited (Schelske 1991;

Downing and McCauley 1992; Rose and Axler 1998). This difference is likely caused by differing nutrient dynamics in wetlands and open waters (Morrice et al. 2004).

Most wetlands that occur along the margins of the southern Great Lakes have been substantially altered by intensive agriculture and urban development along the shoreline. In addition, manipulation of the water level has been a major stressor to shoreline plant communities around Lake Ontario (Environment Canada and U.S. EPA 2005). Because of widespread habitat perturbations, there are few relatively undisturbed Great Lakes wetlands to use as a baseline for evaluation of potential degradation in response to atmospheric nutrient addition. Several Great Lakes–wide studies have been conducted over the past decade to quantify the ecological condition of these wetlands, identify potential landscape stressors, develop cost-effective indicators of both stressors and biological communities, and better link diagnostic indicators to actual stressors (e.g., Niemi et al. 2004; Danz et al. 2007; Niemi et al. 2007).

Nitrogen deposition causes nutrient imbalances in wetland vegetation, including an increased shoot-to-root ratio of plants. It increases the sensitivity of plants to drought stress, frost stress, and attack by insect pests. Plant dieback following these disturbances can cause gaps in the canopy of the wetland shrub layer, followed by invasion by grasses that are typically more efficient in using the additional N and therefore gain a competitive advantage (Krupa 2003).

Data are not available for evaluating the extent to which wetlands in New York have been affected by nutrient enrichment from N deposition. Wetlands are widely distributed throughout the state, including in some areas that receive relatively high levels of atmospheric N deposition. These levels of N deposition may or may not be sufficiently high to cause species shifts in wetland plants. If such effects do occur, they are most likely in wetlands such as bogs and poor fens that normally receive most of their nutrients from atmospheric inputs. These wetlands have been shown in Europe to experience substantial changes in plant species composition in response to high levels of atmospheric N deposition.

Based on consideration of the results of the studies of Aldous (2002), Moore et al. (2004), Rochefort et al. (1990), and Vitt et al. (2003), Greaver et al. (2011) estimated that the critical load of atmospheric N deposition to protect peatlands in the United States from increased plant productivity was in the range of 2.7 to 13 kg N/ha/yr. Ambient deposition falls in this range throughout most of New York.

4.6.3.   Surface Water Effects
4.6.3.1.   *High Elevation Lakes*

Surface waters that are often N-limited or N and P co-limited are responsive to additions of N. In contrast, adding N to a P-limited water body would not be expected to result in a change in primary productivity. In high-elevation areas that have received relatively high levels of atmospheric N deposition, past deposition may have shifted the nutrient supply from a relatively balanced but primarily N-deficient condition in the past to a more P-limited condition today (Elser et al. 2009). Changes in nutrient availability and ratios may have contributed to changes in plankton community structure, species diversity, and trophic interactions (Elser et al. 2009). In general, lakes are expected to retain more N than streams because of in-lake processes, including sedimentation and denitrification. Nevertheless, N retention may be less efficient in small Adirondack lakes with short retention times (Canham et al. 2012).

Sensitivity to aquatic N enrichment effects can be pronounced in remote high-elevation lakes, which are common in the Adirondack Mountains. Small N additions (less than 2 kg N/ha/yr) have been shown to alter the species composition of diatom communities in Rocky Mountain lakes (Wolfe et al. 2001; Baron 2006). It is not known what level of N input might have similar effects at high elevation in New York.

High-elevation lakes are a major focus among those who study aquatic ecosystems because of potential impacts from atmospheric N enrichment. There are multiple reasons for this. Many high-elevation lakes are dilute and nutrient-poor. This makes them more susceptible to biological change in response to addition of relatively low quantities of a limiting nutrient. It is likely that many high-elevation lakes in New York are N-limited. Because soils in the watersheds of high-elevation lakes are typically shallow and poorly developed, the transport of atmospherically deposited N to lakes can be increased, making them more responsive to atmospheric inputs.

4.6.3.2.   *Great Lakes*

The Great Lakes constitute the largest freshwater system in the world; they cover nearly a quarter of a million km². Two of the Great Lakes (Erie and Ontario) border New York State. Among the Great Lakes, Lake Superior is the largest, coldest, and most oligotrophic. The other four lakes are more heavily impacted by human activities and have had higher nutrient concentrations and productivity. Lake Erie has the smallest volume and is relatively shallow (average depth about 19 m). It has an intensively farmed watershed that contains numerous large urban areas. Lake Ontario is much deeper (86 m, on average) and has a longer hydraulic retention time (about 6 versus 2.6 years), which affects its response to N loading.

Lake Ontario and its watershed have been subjected to multiple stressors, including atmospheric deposition; agricultural, urban, and industrial development; introduction of nonnative species; nutrient and toxic contaminant contributions; and regulation of water level (Environment Canada and U.S. EPA 2005). More than eight million people live in the Lake Ontario watershed, and that number is expected to continue to increase. Forested areas occur primarily in the northern (Ontario, Canada) and eastern (New York) portions of the watershed.

Many of the bays and river mouths of Lake Ontario and the south shore coastal zone of the lake have experienced nutrient enrichment and associated algal blooms that have commonly been associated with agricultural land use (Makarewicz 2009). Concentrations of P in lake water have often exceeded the New York State Department of Environmental Conservation's Ambient Water Quality Guidelines for P of 20 µg/L. Water quality of the offshore zone has generally been better than that of the coastal zone, suggesting a significant connection between watershed sources of nutrients and the observed eutrophication of the lake.

In the early 1970s, P loads to the Great Lakes were curtailed by international treaty (Dolan 1993). By 1990, total P loads to Lake Erie had decreased to about one-third of peak loads observed in the early 1970s (Dolan 1993). Nitrate concentrations appear to have increased in Lake Ontario (Dove 2009) and Lake Superior (Finlay et al. 2007), likely due partly to nitrification of N derived from external sources such as agriculture, human waste, and atmospheric deposition (Bennett 1986; Finlay et al. 2007).

Based on data collected in the Canadian Great Lakes Surveillance Program monitoring effort, the trophic status of Lake Ontario has changed significantly over the last four decades (Dove 2009). Concentrations of P were high in the late 1960s and early 1970s. Management changes associated with the Great Lakes Water Quality Agreement were successful in reducing the observed eutrophication of the lake. For example, secchi depth, a measure of water transparency, increased substantially since the late 1980s. Despite reductions in total P concentration from levels of 20 to 25 µg/L prior to 1977 to levels consistently below 8 µg/L between 1998 and 2008, the concentration of inorganic N (nitrate plus nitrite) increased steadily in Lake Ontario from 1969 (0.256 mg/L) to 2008 (0.447 mg/L), likely due to both watershed and atmospheric sources (Dove 2009).

Several studies have been conducted on nutrient limitation in the Great Lakes, which are generally believed to be P-limited (Schelske 1991; Downing and Mc-Cauley 1992; Rose and Axler 1998). Water quality in the open waters of these lakes has been improving in response to controls on point sources of P (Nicholls et al. 2001), although increasing nonpoint sources of nutrients from agricultural and urban storm-water runoff and food web impacts from invasive species (e.g., zebra and quagga mussels) have been difficult to separate and likely have had synergistic effects in the Great Lakes.

Levine et al. (1997) demonstrated a complicated pattern of response to nutrient addition for Lake Champlain. They added nutrients to in situ enclosures and measured indicators of P status, including alkaline phosphatase activity and orthophosphate turnover time. Although P appeared to be the principal limiting nutrient during summer, N addition also resulted in algal growth stimulation during that season. During spring, phytoplankton growth was not limited by P, N, or silicon (Si), but perhaps it was limited by light or temperature. Phosphorus sufficiency appeared to be as common as P deficiency. In Lake Erie, N input was also found to be limiting to phytoplankton on a seasonal basis (Moon and Carrick 2007; North et al. 2007).

Silica depletion in response to nutrient enrichment has also been reported for the Great Lakes (Conley et al. 1993). Increased growth of silicate diatoms as a result of eutrophication induced by NO_3^- and phosphate (PO_4^-) followed by removal from the water column of fixed biogenic Si through sedimentation can change the ratios among nutrients such as Si, N, and P. Such changes might cause shifts from diatoms to nonsiliceous phytoplankton (Ittekot 2003).

The water quality and fish and wildlife habitat quality of Lake Erie have been substantially impacted by changes in land use in the watershed, nutrient loading, chemical contamination, the introduction of nonnative species, alterations to the shoreline, and other human activities. Lake Erie is both the smallest (by volume) and shallowest of the Great Lakes. It warms quickly in spring and is the most biologically productive of the Great Lakes (Environment Canada and U.S. EPA 2005). More than 11 million people live in the watershed, mostly in the United States.

The western basin of Lake Erie is shallow (average depth 7.4 m), the eastern basin is substantially deeper (average 25 m), and the central basin is intermediate (18.3 m). Waters of the central and eastern basins thermally stratify, contributing to the development of seasonally anoxic bottom waters. The size of the anoxic zone in the central basin has increased in recent years (Environment Canada and U.S. EPA 2005). Oxygen depletion in this basin is likely driven by the thickness of the hypolimnion (influenced by temperature) more than by nutrient loads or the expansion of invasive species (Charlton and Milne 2004). Development of a thin hypolimnion by solar heating or reduced wind mixing can contribute to strong thermal stratification of the waters of the basin, favoring development of anoxia. Lake Erie also has 34 nonnative invasive fish species (Environment Canada and U.S. EPA 2005) and these can have various impacts on native lake biota.

Changes in water chemistry in the Great Lakes are due, at least in part, to the 1989 and 1991 introductions of two exotic dreissenid mussel species. Zebra and quagga mussels have substantially disrupted the food web, especially of Lake Ontario. Their high filtration capacity can drastically deplete nutrient levels and reduce lake production (Mills et al. 1993). Mills et al. (2006) reported declines in open-water P concentrations between 1995 and 2005 in Lake Ontario. This result is consistent with observed increases in water transparency (Barbiero et al. 2006) and declining phytoplankton biomass (Munawar et al. 2006) in the lake. Diatom abundance in Lake Ontario has decreased since the invasion of dreissenid mussels. Silicon concentrations in lake water have increased in response to reduced Si uptake by diatoms (Millard et al. 2003). While some species of phytoplankton have decreased in abundance, others have increased, altering the species composition of the phytoplankton (Auer et al. 2010).

Filtering of lake water by these nonnative mussels has reduced particulate matter in the water column, thereby increasing light penetration and contributing to growth of extensive beds of macrophytes in the littoral zone (Environment Canada and U.S. EPA 2005). Mussels have altered Lake Ontario's physical habitat and its chemistry and biology. Filtration by the introduced mussel species has redirected nutrients from pelagic to benthic production (Johannsson et al. 2000).

The EPA Great Lakes National Program set trophic state goals for each of the Laurentian Great Lakes. Most recent data show that Lakes Superior, Huron, and Michigan are meeting their oligotrophy goals. Surface-water quality in these lakes is considered "excellent," and water quality in Lakes Huron and Michigan shows improvement since monitoring began in the 1980s (U.S. EPA 2012a; see also U.S. EPA 2012b). Lake Ontario is also meeting its trophic state goal of oligomesotrophy. However, Lake Erie shows signs that are cause for concern. It was the first lake to show evidence of nuisance algal blooms and O_2-depletion problems associated with eutrophication. Lake Erie is also most affected by pollutants from nearby agricultural and urban areas. Surface waters in Lake Erie's eastern, central, and western basins are currently meeting their EPA trophic state goals of oligotrophy, oligomesotrophy, and mesotrophy, respectively. However, the central basin has experienced elevated P concentrations since the 1990s and exhibits summer O_2 depletion in deeper waters, threatening the health of fish and other aquatic biota (U.S. EPA 2012a; see also U.S. EPA 2012b).

However, nutrient contaminant levels have declined in Great Lakes ecosystems over the last several decades. In particular, there has been a substantial reduction in P loading since the 1970s and 1980s. These loading reductions were achieved by limiting wastewater discharge, restricting the P content of laundry detergents, and controlling nonpoint agricultural sources. In response, total P concentrations have decreased or remained constant in all of the Great Lakes except Lake Erie.

Other drivers have and will influence the response of the Great Lakes to nutrient addition. In addition to the introduction of nonnative invasive species, these include changing climate, decreased ice cover, earlier summer stratification, warmer and thicker upper water column layers, and faster C cycling. Dreissenid mussels (zebra and quagga) have been introduced to all of the Great Lakes since about the 1980s. The mussels contributed to greater nutrient retention in near-shore areas and increased benthic productivity (Hecky et al. 2004).

4.6.3.3.   Coastal Waters

Because N tends to be the main growth-limiting nutrient in estuaries and coastal marine waters, these environments are highly susceptible to nutrient enrichment effects from atmospheric N deposition. Many coastal waters in New York are located in proximity to urban development and receive substantial contributions of N from a wide range of point and nonpoint sources of both atmospheric and non-atmospheric N. It is difficult to isolate the atmospheric N contributions from other sources, thereby complicating assessment of the influence of atmospheric N deposition on the condition of sensitive marine resources.

4.7.   MERCURY INTERACTIONS

The greatest threat from atmospherically deposited Hg to the environment and humans is exposure to organic MeHg, among the most toxic and widespread environmental contaminants that affect natural ecosystems (cf. Brumbaugh et al. 2001). Methylmercury is a persistent bioaccumulative toxin. It is formed when bacteria convert inorganic Hg into MeHg, which then can enter through the digestive system in animals and bind to proteins. It bioaccumulates as each successive predator in the aquatic food web consumes more MeHg. Low concentrations of MeHg in surface water (less than about 1 part per trillion) can bioaccumulate over a millionfold and reach harmful levels in fish (Driscoll et al. 2007b). Predatory fish, piscivorous birds and mammals, and humans that eat contaminated fish can accumulate levels of MeHg that cause adverse health effects.

Much of the research conducted on Hg methylation, the influence of SO_4^{2-} on methylation rates, and controls on Hg transport in watersheds has focused on the northeastern and upper Great Lakes regions of the United States (see Chen et al. 2005; Evers 2005; Evers et al. 2005; Kamman et al. 2005; Driscoll et al. 2007b; Evers et al. 2007; Evers et al. 2008). These regions experience moderate levels of atmospheric Hg deposition. More important, watershed characteristics are frequently conducive to Hg biomagnification. These characteristics promote Hg transport, methylation, and bioaccumulation (Evers et al. 2011a). Myers et al. (2007) proposed a national map showing the relative sensitivity of aquatic ecosystems to Hg contamination. Areas showing the highest estimated sensitivity included the Adirondack Mountains region, which has been the focus of substantial research on Hg biogeochemistry and Hg contamination of fish and wildlife (Evers et al. 2007). A study of 44 Adirondack lakes by Yu et al. (2011) found many that had average Hg concentrations in fish higher than the established fish whole-body health criterion for protecting piscivorous wildlife (0.27 ppm). One-fifth of the common loons (*Gavia immer*) sampled had blood Hg concentration above 3 ppm, a level at which adverse effects on reproduction have been documented. Mercury methylation and bioaccumulation are influenced by both Hg input and S input. Possible synergistic effects of atmospherically deposited S and Hg have been the focus of research interest in the Adirondack Mountains for several decades (cf. Driscoll et al. 2007b).

Although the Catskill Mountain region lacks the extensive wetlands found in the Adirondacks, large Catskill lakes and reservoirs host large piscivorous fish that have bioaccumulated large amounts of Hg. Other parts of New York also have fish consumption advisories for Hg. Thus, the Hg problem is statewide.

4.7.1. Upland Processes

The scientific research community has not yet focused much research on the study of Hg cycling in terrestrial ecosystems. Most work on Hg bioaccumulation has focused on aquatic and wetland communities. In addition, terrestrial species have generally not been considered to be at heightened risk of Hg pollution. However, recent research has suggested that some upland ecosystems may also be affected by Hg bioaccumulation (Rimmer et al. 2005; Rimmer et al. 2009). Mercury binds to soil particles and especially to organic matter (Obrist et al. 2011). As a consequence, upland forest soils can serve as Hg sinks (Juillerat et al. 2012). In one study in Vermont, differences among Hg pools in northern hardwood and mixed hardwood/conifer stands were positively correlated with C pools that, in turn, were affected by historical land use (Juillerat et al. 2012). Isolated studies have been conducted in the Adirondacks (see Blackwell and Driscoll 2011). Bradley et al. (2011) reported seasonal and spatial variations of Hg in different landscapes of the Fishing Brook Basin in the Adirondacks.

Most Hg found in vegetation is derived from the atmosphere (Grigal 2003). Mercury uptake from soil to vegetation is limited, largely because Hg transport from roots to foliage is minimal (Grigal 2002). Plant foliage accumulates elemental Hg largely in response to atmospheric deposition (Ericksen et al. 2003; Frescholtz et al. 2003), although foliar/air Hg exchange is bidirectional (Millhollen et al. 2006).

Millhollen et al. (2006) compared foliar Hg accumulation in three tree species with fluxes measured using a plant gas-exchange system after soil amendment with mercury chloride ($HgCl_2$). Measured foliar Hg fluxes indicated that deposition of atmospheric Hg constituted the dominant flux of Hg to the leaf surface. Litterfall was the dominant cause of Hg deposition to the soil in forested ecosystems. Soil uptake was of secondary importance. Nevertheless, root Hg concentrations were strongly correlated with soil Hg concentrations.

Laboratory studies showed that Hg concentration in foliage increases throughout the growing season for multiple tree species (Ericksen et al. 2003; Millhollen et al. 2006). Bushey et al. (2008) investigated Hg accumulation in foliage and the flux of Hg to the forest floor with litterfall. Leaf samples were collected monthly during two growing seasons from an upland forest stand in the Adirondack Mountains to determine Hg accumulation rates, assess the Hg pool in the canopy, and measure the Hg flux to the ground surface via litterfall. The average total Hg pool increased tenfold over the growing season. Linear accumulation rates documented for yellow birch (0.23 ng/g/day), sugar maple (0.22 ng/g/day), and American beech (0.35 ng/g/day) suggested that Hg accumulation is regulated by mass transfer processes between the atmosphere and the leaf surface (Bushey et al.

2008). Litterfall Hg fluxes to the forest floor (~ 16 mg/m^2) for both years of the study constituted the largest ecosystem Hg input, substantially larger than wet deposition.

Accumulation of Hg in forest and wetland soils, including legacy Hg deposition that has occurred over many decades, may moderate the recovery of connected aquatic ecosystems in response to ongoing decreases in Hg deposition (Demers et al. 2013). Large repositories of previously deposited Hg in forest soils will gradually be transported to down-gradient surface waters (cf. Bookman et al. 2008). Thus, the hydrogeologic setting of wetlands links landscape position to the hydrologic regime and ultimately to surface-water chemistry and wetland soil chemistry (Demers et al. 2013). Because organic matter is so central to Hg retention and transport (Grigal 2003), a host of hydrologically linked factors control the dynamics of Hg cycling in wetlands of a variety of types. Demers et al. (2013) examined the pool sizes of Hg and Hg retention in forest soils and wetlands in varying hydrogeologic settings. They concluded that differences in Hg stoichiometry with C, N, and S in soils across an upland/wetland interface reflected differences in mechanisms of Hg retention. In upland soils, aerobic decomposition of organic matter is followed by transport of Hg into mineral soils. As a consequence, Hg is correlated with C and N. Headwater wetland sediments provide different environmental conditions (often anaerobic) that lead to strong correlations between Hg and S, as described below.

4.7.2.   Wetland Processes

Increased S deposition increases rates of Hg methylation in freshwater wetlands (Galloway and Branfireun 2004; ICF International 2006; Jeremiason et al. 2006), and wetlands act as important sources of MeHg to freshwater ecosystems. This is because wetlands have high availability of dissolved organic carbon and often have anaerobic conditions in sediments, both of which are associated with increased methylation. Because Hg binds to organic matter (Grigal 2002), dissolved organic carbon increases MeHg transport to downstream waters. Many lake watersheds in the Adirondack Mountains contain extensive wetlands and many wetland-influenced Adirondack lakes exhibit relatively high concentrations of dissolved organic carbon (Kretser et al. 1989). Because wetlands increase Hg methylation and transport (Aiken et al. 2003), the percentage of wetland land cover is commonly correlated with MeHg levels in lakes (Grigal 2002).

Wetland sediments in the Adirondacks tend to be rich in organic matter and S, anoxic, relatively nutrient rich, and warm during the summer season. Thus, they provide favorable conditions for MeHg production (Hammerschmidt and Fitzgerald 2004). In the sediment of an anaerobic wetland, lake, or stream, methylation is largely carried out by sulfate-reducing bacteria. Because these bacteria are responsible for production of MeHg from inorganic Hg (Compeau and Bartha 1985; Gilmour et al. 1992), atmospheric S deposition is an important driver of Hg methylation and biomagnification. The MeHg produced in wetlands can bind with organic matter and be transported to hydrologically connected lakes and streams (St. Louis et al. 1996; Sellers et al. 2001; Grigal 2002; Bradley et al. 2011). In chains of streams and lakes, such as are common in the Adirondacks, lakes are generally net sinks for MeHg due to in-lake loss processes (Bradley et al. 2011; Burns et al. 2012). Methylation is further correlated with water acidity and SO_4 concentrations (Wiener et al. 2006; Driscoll et al. 2007b). Methylating sulfate-reducing bacteria require SO_4^{2-} to carry on their metabolic activities, and much of that SO_4^{2-} in New York is provided by atmospheric S deposition. Low concentrations of inorganic Hg in surface water can bioaccumulate to very high levels when and where environmental conditions favor methylation. This pattern is evident in the Neversink Reservoir in the Catskill Mountains, where large piscivorous fish accumulate relatively large body burdens of Hg.

Although Hg methylation in natural environments is generally driven by dissimilatory sulfate-reducing bacteria, research by Warner et al. (2003), Fleming et al. (2006), and Kerin et al. (2006) suggests that dissimilatory Fe-reducing bacteria might also play a role. Kerin et al. (2006) tested multiple species of bacteria in the genera *Geobacter, Desulfuromonas* and *Shewanella*. All of the strains in *Geobacter* and *Desulfuromonas* tested showed an ability to methylate Hg; both genera are closely related to known sulfate-reducing bacteria in the *Deltaproteobacteria*.

Selvendiran et al. (2008a, 2008b) measured fluxes and pools of total and methyl Hg in two wetlands in the Arbutus Lake watershed in the Adirondack Mountains. They found that the wetlands were net sources of Hg to drainage water, but the magnitude of the source varied according to wetland connectivity to the stream. Groundwater in a riparian peatland in the Arbutus Lake watershed in the Adirondack Mountains showed little interaction with water in the adjacent stream because of the low hydraulic conductivity of the peat (Selvendiran et al. 2008a). The proportion of total Hg that was present in the methylated form increased from 2% at the inlets to the wetland to 6% at the wetland outlet (Selvendiran et al. 2008a).

In an experimental SO_4^{2-} addition field experiment, Jeremiason et al. (2006) demonstrated increased Hg methylation and increased MeHg concentrations in a Minnesota wetland and in surface-water outflow from that wetland. Concentrations of MeHg increased 2.4 times in response to a fourfold increase in SO_4^{2-} loading. These field study results suggested that reduced SO_4^{2-} deposition would lower MeHg export from wetlands to connected down-gradient receiving waters.

In SO_4^{2-}-limited fresh waters, reduced levels of S deposition may partly or wholly mitigate Hg biomagnifications in fish and wildlife. Reduced atmospheric S inputs have been identified as a major factor in the observed recent decline in fish Hg to levels that are now safe for human consumption in Isle Royale, an island in Lake Superior (Drevnick et al. 2007). Similarly, Coleman-Wasik et al. (2012) experimentally manipulated the S loading to a small boreal peatland and measured resulting changes in MeHg production and concentrations in pore waters. Peat acted as a sink for the newly produced MeHg. Wetland pore waters achieved peak MeHg levels within one week of the addition of SO_4^{2-}. Five years after initiation of the experimental treatment, SO_4^{2-} addition was discontinued to part of the wetland. Four years after the addition of SO_4^{2-} was stopped, MeHg concentration in pore waters and peat had decreased significantly, but not to the level of the reference site. Effects were also noted on animal life. Mosquito larvae collected at the end of the experiment had total Hg levels reflective of MeHg content in the peat and pore waters where the mosquitoes were collected (Coleman-Wasik et al. 2012).

Because large amounts of Hg are stored in wetland sediments, wetlands might constitute a long-term total Hg source to surface waters even if industrial Hg emissions decline substantially (Selvendiran et al. 2009). However, MeHg exhibits a short residence time in the active upper zone of peat. As a consequence, MeHg production, re-emission, and transport might be highly responsive to changes in inputs of Hg or SO_4^{2-} (Selvendiran et al. 2008b).

4.7.3. Surface Water Processes

Atmospheric deposition is the main source of Hg to freshwater ecosystems in remote areas of New York, such as in the Adirondack and Catskill Mountains (Fitzgerald et al. 1998). Surface waters in these mountains appear to have higher Hg concentrations than most fresh waters in the rest of New York (Simonin et al. 2008). However, levels of Hg in aquatic ecosystems are relatively high at locations throughout New York. The distribution and magnitude of the impacts of Hg contamination of natural ecosystems in the Great Lakes region and throughout the northeastern United States are substantial. Concentrations of Hg in fresh waters are high enough to cause Hg body burdens in fish that exceed risk thresholds for many species of fish and wildlife across the region (Evers et al. 2011a).

Many studies of Hg in surface waters of New York have been conducted on Adirondack lakes, where wetland influence is an important driver of Hg concentration due to increased Hg methylation and transport in and down-gradient from wetlands. Once Hg is transported to a lake, it can be removed by incorporation into lake sediments, volatilization to the atmosphere, or export with water flow from the lake outlet. Volatilization loss of elemental Hg from the lake surface is an important part of the Hg cycle in New York watersheds. Selvendiran et al. (2009) measured volatilization loss from Arbutus Lake at levels comparable to atmospheric deposition input to the lake surface.

Hydraulic residence time is an additional important variable that controls Hg cycling in lake watersheds. Selvendiran et al. (2009) measured negligible in-lake Hg loss at Sunday Lake, which has a very short hydraulic residence time of 0.02 yr. In contrast, Arbutus Lake, located in the same general region, had substantially higher (~ 6%) Hg loss from the lake, in large part due to the fact that its hydraulic residence time is 0.6 yr (30x higher than Sunday Lake).

CHAPTER 5

Biotic Effects of
Atmospheric Deposition

In response to an environmental stress, the more sensitive species in an ecosystem often decrease in abundance and can be eliminated from the ecosystem. The more tolerant species tend to increase in abundance (Woodwell 1970). Removal of the more sensitive species in response to exposure to the stress can impair ecosystem function, change community structure, and decrease taxonomic richness and diversity. Such changes can occur in response to atmospheric inputs of N, S, and Hg to terrestrial, aquatic, and wetland ecosystems in New York.

5.1. TERRESTRIAL RESOURCE RESPONSE TO ACIDIFICATION, EUTROPHICATION AND MERCURY INPUT

Changes in plant species occurrence have been documented or inferred in N-limited terrestrial ecosystems throughout much of the United States in response to increased N deposition (Pardo et al. 2011). Effects of atmospherically deposited N on terrestrial vegetation are often most pronounced in the soil that surrounds plant roots (Wall and Moore 1999). Relationships among plant roots, mycorrhizal fungi, and microbes constitute integral components of the N cycle and contribute to overall plant growth and ecosystem health. The plant provides structure and a source of C; the fungi and bacteria provide the plant with access to N and other nutrients. The diversity of mycorrhizal fungi can be decreased by increased N availability (Egerton-Warburton and Allen 2000), which alters aboveground biodiversity and ecosystem productivity (Wall and Moore 1999). Added N might also affect the ratio of bacteria to fungi in the soil (Kopáček et al. 2013).

Because of the importance of N as a limiting nutrient in terrestrial ecosystems, addition of N from atmospheric deposition can have significant biological consequences. The most immediate impact of N addition in terrestrial ecosystems has probably been increased growth of trees (Bedison and McNeil 2009; Thomas et al. 2010). It is likely that some changes in biodiversity have also occurred. Although atmospheric N deposition in many parts of New York may have been high enough to cause some plant species to outcompete others, this effect has not been well documented.

McNeil et al. (2012) concluded that spatial patterns of foliar N in Adirondack forests respond to factors related to the composition of tree species, gradients of resource availability, and human impacts on resource condition, including land use and atmospheric deposition. Tree species adjust foliar N content according to species-specific resource strategies and in response to human impacts on forest environments (McNeil et al. 2012).

Nutrient limitation is key to understanding how an ecosystem responds to atmospheric deposition of N or any other nutrient. This concept is complicated, however, by the diversity of species and age classes that occur in forest ecosystems. Species differ in their nutrient requirements. What is limiting for one species may not be limiting for others. What limits a particular species may change over time. In addition, it is likely that relief of the primary limitation would result in establishment of the next limitation rather quickly (Davidson and Howarth 2007; Vadeboncoeur 2010). Ecosystems that do not exhibit pronounced shortages of N or P (for example) tend to move toward co-limitation, whereby the ratios of the two nutrients more or less match biotic demand (Davidson and Howarth 2007; Vadeboncoeur 2010). Temperate-zone terrestrial ecosystems typically show increased primary productivity after adding N. If N deposition causes N-saturation, P or Ca may become limiting. Vadeboncoeur (2010) conducted a meta-analysis and synthesized results from 35 fertilization experiments to determine the nutrient limitation status of hardwood forests in the northeastern United States. Although evidence was strong for widespread N-limitation, forest productivity also increased with addition of P and Ca. In addition, multiple element additions had larger effects than single additions.

Species shifts and ecosystem changes can occur even if the ecosystem does not exhibit signs of N saturation. Plant species also differ in their response to NO_3^- versus NH_4^+ inputs. Interactions with other air pollutants such as O_3 are poorly understood. Fast-growing annual species, including many agricultural crops, and fast-growing pioneer trees such as birch (*Betula* spp.) seem to prefer NO_3^- (Pearson and Stewart 1993). Slow-growing perennial plant species generally prefer NH_4^+.

Some plant species readily take up both NO_3^- and NH_4^+ (Krupa 2003). These include members of the family Ericaceae (e.g., *Calluna, Erica, Vaccinium*), coniferous trees, and climax tree genera such as *Quercus* and *Fagus* (Krupa 2003).

Experimental N addition to forest ecosystems has caused increased growth of some tree species (Emmett 1999; Elvir et al. 2003; DeWalle et al. 2006; Högberg et al. 2006). Increased tree growth, to the extent that it occurs, can exacerbate other nutrient deficiencies, such as Ca, Mg, or K. This can affect the health of the forest. Some long-term experiments have found initial and relatively short-term growth increases followed by increased mortality, especially at higher rates of N fertilization (Elvir et al. 2003; Magill et al. 2004; McNulty et al. 2005; Högberg et al. 2006). Note, however, that these experimental application rates were invariably higher than atmospheric N deposition in New York.

Fertilization with N can cause trees to allocate less photosynthate to roots and mycorrhizal fungi, decreasing the capacity to take up water and potentially limiting nutrients. This can increase trees' susceptibility to drought. Thus, evaluation of forest ecosystem response to N fertilization must also consider the confounding effects of climate.

Increased growth from N fertilization occurs mainly in parts of the plant that are above ground level (Dueck et al. 1991). This changes the shoot-to-root ratio and decreases plant resistance to drought and frost. As the plant grows faster, the demand for additional C to balance assimilated N increases CO_2 uptake through plant leaves. This causes increased stomatal opening and can increase water loss. Also, because shoot growth is increased more than root growth, the water uptake by the roots needed to balance transpirational water loss can become insufficient during drought (Fangmeier et al. 1994; Krupa 2003). Addition of N also can prolong the growth phase of plants during autumn, delaying development of winter hardiness. This can damage plants if frost occurs early in the autumn (Cape et al. 1991).

Forest responses to added N are complex and variable. Improved understanding of below-ground processes and responses to N input is needed. Root responses affect the functioning of an entire ecosystem, but they have not been well studied. Models that incorporate such below-ground processes might help explain the heterogeneous patterns in tree productivity noted in gradient studies (c.f., Thomas et al. 2010; Smithwick et al. 2013). Smithwick et al. (2013) examined the role of roots and associated feedbacks in regulating growth and above-ground function. They suggested five mechanisms by which added N might affect fine root production and mortality:

1. Increased nutrient supply increases above-ground primary productivity and changes below-ground allocation
2. High N availability shifts C allocation from below ground to above ground, thereby affecting fine root production

3. Increased root stress is caused by increased reactive N species in fine roots
4. Increased Al toxicity is caused by decreases in base cations in the soil
5. Decreased rooting depth causes increased susceptibility to drought stress

Developing a better understanding of the effects of N deposition on root formation may also help improve our understanding of other forest perturbations, including fire, insects, and disease (Smithwick et al. 2013).

Soil acidification contributes to deficiencies in plant nutrients. In particular, Ca is essential for many aspects of plant physiology, including root development, wood formation, cell membrane integrity, and cell wall structure. Mobilized Al binds to fine root tips, limiting Ca and Mg uptake (Shortle and Smith 1988; Shortle et al. 1997).

Most effects of acidic deposition on terrestrial vegetation appear to be mediated primarily through the soil (National Science and Technology Council 1998), partly through depletion of nutrient cations and partly through Al toxicity. Base cations leach from both foliage and soils in response to the high mobility of mineral acid anions. Under low pH and low Ca conditions, Al mobilization from soils to soil solution and drainage water in order to maintain the charge balance, can impede Ca and Mg uptake by plant roots and induce deficiencies in these nutrient base cations. Because of these processes, foliar Ca levels and soil and root Ca-to-Al ratios are considered low to deficient over large portions of the spruce-fir region in the eastern United States (Joslin et al. 1992; Cronan and Grigal 1995; NAPAP 1998; U.S. EPA 2008).

Aluminum in soil solution reduces Ca uptake by plant roots because it competes for binding sites in the cortex of the fine roots. Plants can be damaged if high Al-to-nutrient cation ratios prevent the uptake of sufficient quantities of Ca and Mg (Shortle and Smith 1988; Garner 1994). As a consequence of high Al-to-nutrient cation ratios, cambial growth, rate of wood formation, and the amount of functional sapwood and live crown decrease.

Some tree species, especially red spruce and sugar maple, have been damaged by soil acidification in some areas in New York. The degree of damage to terrestrial ecosystems and the extent of recent recovery, if any, in response to recent reductions in pollutant emissions are not well known. Different trees and plant species in the forest understory probably exhibit differences in sensitivity to acid and Al stress. A recent study showed that air pollutant deposition and soil acidification broadly affect tree mortality across the eastern United States, including in many species that had not previously been identified as being susceptible to acidification (Dietze and Moorcroft 2011). However, in situ experimental data for quantifying sensitivities and levels of effect are generally not available, especially for plant species of the forest understory.

It is unlikely that plants have recovered from previous soil acidification to a significant extent in the Adirondack or Catskill Mountains in response to recent decreases in the levels of air pollution. It is more likely that soil conditions and plant health are continuing to decline (Sullivan et al. 2006a). This is because even with reduced levels of acidic deposition, it appears that the base cation supply of the soil in the most acid-sensitive watersheds continues to be depleted by ambient acidic deposition levels.

Quantifying the effects of acidification on plants is especially challenging because plants are simultaneously affected by multiple stressors in addition to soil acidification. These include changing climatic conditions, insect pests, disease, and competition from introduced species. However, studies have revealed that red spruce, sugar maple, and probably other plant species throughout the western Adirondack and Catskill Mountains have been damaged by acidic deposition. Adverse impacts have been caused by acidity deposited directly on plant foliage, Al toxicity to plant roots, and insufficient Ca and perhaps other base cation nutrients in the soil.

5.1.1. Red Spruce Response to Acidification

Red spruce forests are generally considered to be highly susceptible to adverse impacts of acidification. However, other factors, including topography, can also influence the response of a particular stand to acidification from acidic deposition (Aber et al. 1998; Webster et al. 2004). Red spruce is common in New York at both high and low elevations. At low elevations, red spruce often occurs as pure stands or in association with hardwoods or black spruce (*Picea mariana*). At high elevations, this species more typically occurs in association with balsam fir. Red spruce in New York is considered to be highly sensitive to acidification partly as a consequence of its inherent sensitivity and also as a consequence of its common distribution at high elevation, where precipitation and acidic deposition levels tend to be relatively high.

Red spruce forests in New York are often underlain by unreactive bedrock, and base cation production via weathering tends to be limited (Turner et al. 1990). Soils often have thick organic horizons, high organic matter content in the mineral horizons, and low pH (Joslin et al. 1992; Sullivan et al. 2006b). Because of unreactive bedrock, base-poor litter, high precipitation, and high leaching rates, soil base saturation in these forests often tends to be below 10% and the soil cation exchange complex is generally dominated by Al (Johnson and Fernandez 1992).

Red spruce trees died at a rapid pace at some locations in the Adirondacks and northern New England in the 1970s and 1980s (LeBlanc 1992). More than 50% of the canopy red spruce trees died at high elevations in the Adirondack Mountains and the Green Mountains of Vermont. In the White Mountains in New Hampshire, about 25% of the canopy spruce died during that same period (Craig and Friedland 1991). Dieback of red spruce trees was also observed in mixed hardwood-conifer stands at relatively low elevations in the western Adirondack Mountains, which received high inputs of acidic deposition (Shortle et al. 1997). This effect was linked to exposure of spruce foliage to acidic cloud water (which can be more acidic than wet deposition) at high elevations and an increase in the amount of dissolved Al^{n+} relative to the amount of dissolved Ca^{2+} in soil water. NAPAP (1998) concluded that some of the best evidence for significant forest damage in the United States attributable to acidic deposition was the observed increased mortality and decline of red spruce in portions of the Northeast, including New York. Red spruce decline in this region has been mainly attributed to reduced cold tolerance and increased general tree stress (Hawley et al. 2006; Halman et al. 2008).

The membrane-associated pool of Ca influences the response of plant cells to changing environmental conditions. The plant plasma membrane plays an important role in mediating cold acclimation and low-temperature injuries (U.S. EPA 2004). Findings of DeHayes et al. (1999) suggest that direct acidic deposition on red spruce needles preferentially removed membrane-associated Ca. In addition, decreased availability of soil Ca has been linked with injuries to needles in the wintertime (Hawley et al. 2006).

Results of controlled exposure studies showed that acidic mist or cloud water reduced the cold tolerance of current-year red spruce needles by 3 to 10 °C (De-Hayes et al. 1999), and this may have been at least partially responsible for the observed dieback. The frequency of freezing injuries in red spruce increased over the 1980s and 1990s (DeHayes et al. 1999), and there was a significant positive association between cold tolerance and foliar Ca in trees that exhibited deficiency in foliar Ca. Although continued red spruce mortality has not been reported in more recent years, winter injury to red spruce foliage has been documented as recently as 2003 (Lazarus et al. 2004; Lawrence et al. 2012).

5.1.2. Sugar Maple Response to Acidification

The hardwood tree species in New York that is most commonly associated with acidification effects caused by acidic deposition is sugar maple. This species is distributed throughout the northeastern United States as a major component of the northern hardwood forest. Sugar maple is found throughout much of upstate New York. The health of sugar maple trees appears to be highly influenced by the availability of Ca in the soil (Horsley et al. 2000). Trees growing on soil with a low supply of Ca are stressed and can become more susceptible to damage from defoliating insects, drought, and other forms of extreme weather. Effects can include reduced growth of mature trees, poor canopy condition, and poor regeneration of seedlings.

Acidic deposition may be causing or contributing to episodic dieback of sugar maple in the northeastern United States through depletion of nutrient cations from marginal soils. Horsley et al. (1999) documented dieback at 19 sites in northwestern and north-central Pennsylvania and southwestern New York. The observed dieback was correlated with combined stress from defoliation and soil deficiencies of Mg and Ca. Dieback occurred predominately on ridgetops and upper slopes, where soil base availability was much lower than at middle and lower slopes (Bailey et al. 1999). Hallett et al. (2006) further substantiated the importance of soil cation nutrition to sugar maple health. However, multiple factors, including soil mineralogy and landscape position, can affect soil base status (U.S. EPA 2008).

Lovett and Rueth (1999) found a twofold increase in mineralization rate in soils of sugar maple stands along an N deposition gradient in the Northeast from 4.2 to 11.1 kg N/ha/yr but no significant relationship between N deposition and mineralization in American beech stands. This difference might be attributable in part to the lower litter quality typically observed under beech stands. Sugar maple may be more susceptible to effects of N deposition and associated soil acidification from either direct leaching of NO_3^- or increased nitrification. Aber et al. (2003) found a decrease in the soil C-to-N ratio in northern hardwood forests from 24 to 17 over a deposition gradient of 3 to 12 kg N/ha/yr. This decrease was similar to but less steep than the decrease seen in conifers.

Sugar maple is the dominant canopy tree species throughout much of the northern hardwood forest in New York State. A recent research project (Sullivan et al. 2013) focused on quantifying the extent of damage to sugar maple trees in the Adirondack Mountains in response to acidic deposition. This study documented the effects of acidic deposition and soil acid-base chemistry on the growth, regeneration, and canopy condition of sugar maple trees. The field study, conducted in 2009, sampled 50 study plots in 20 small Adirondack watersheds. Soil acid-base chemistry and sugar maple growth, canopy condition, and regeneration were evaluated. Atmospheric S and N deposition were estimated for each plot. Trees growing on soils with poor acid-base chemistry (low exchangeable Ca and low % base saturation) that received relatively high levels of atmospheric S and N deposition exhibited little to no sugar maple seedling regeneration, decreased canopy condition, and short- to long-term growth declines compared to study plots that had better soil conditions and lower levels of acidic deposition. These results suggested that the aesthetic, cultural, and monetary value of the ecosystem services provided by sugar maple in the western and central Adirondack Mountain region are at risk from ongoing soil acidification caused largely by acidic deposition.

Sugar maple has been shown to respond positively to Ca addition. For example, Long et al. (2011) found that sugar maple basal area increments increased significantly in limed plots on the Allegheny Plateau in northern Pennsylvania from 1995 to 2008, whereas American beech was unaffected by the liming.

Less-sensitive tree species found in hardwood forests throughout New York are probably also experiencing gradual losses of base cation nutrients in response to acidic deposition. This may reduce the quality of forest nutrition over the long term (National Science and Technology Council 1998). However, the extent to which acidic deposition at ambient levels actually affects the growth or health of forest tree species other than red spruce and sugar maple in New York and elsewhere in the eastern United States is less certain.

5.1.3. Vegetation Response to Nitrogen Supply

Atmospheric N inputs to natural systems are not necessarily harmful. For each species and ecosystem, there is an optimum N level that will maximize productivity without causing significant changes in species distribution or abundance. Above the optimum level, soil and water acidification, nutrient enrichment, and other harmful effects can occur in both aquatic and terrestrial ecosystems (Gunderson 1992; Aber et al. 1998). High levels of N deposition might increase the susceptibility of forests to other stressors, including reducing the resistance of some tree species to frost, insect damage, or drought. These include effects that occur solely in response to acidic deposition or in response to any one of the other stressors individually or in combination (U.S. EPA 2008).

Addition of N to an ecosystem can disrupt nutrient cycling, change intra- and inter-species competitive interactions of plants, and alter soil processes (Aber et al. 1989). In particular, nutrient uptake and allocation in plants, litter production, N immobilization, nitrification, NO_3^- leaching, and trace gas emissions can be altered (Aber et al. 1989; Garner 1994). Nitrogen availability can also modify the effects of increased atmospheric CO_2 on plant growth at locations where ecosystems are limited by N availability.

Although the primary route of plant uptake of N is generally through roots, some uptake also occurs through the foliage. Differences between the amount of N deposition measured in throughfall, and the amount measured in adjacent forest clearings suggest that the canopy can take up an average of 16% of total atmospheric N input in the northeastern United States (Lovett 1992). Canopy uptake can be much higher (up to about 90%) in some N-limited forests with large epiphyte loads (e.g., Klopatek et al. 2006).

Differing responses of forest stands to experimental N additions may reflect differences in N status and the N cycling history of the treatment sites. Such differences have commonly been attributed to disturbance history dating back a century or more (Goodale and Aber 2001). Disturbances that caused loss of soil N, such as logging, fires, and agriculture, would be expected to be effective in increasing N retention. Retention capability often decreases with the age of a stand, and thus older forests are more susceptible than younger forests to becoming N saturated (Hedin et al. 1995). Aber et al. (1998) speculated that land-use history may be more important than cumulative atmospheric deposition of N in determining the N status of a forest.

Peatlands and bogs are especially vulnerable to the adverse nutrient-enrichment effects of N deposition (Krupa 2003). The sensitivity of peatland *Sphagnum* species to elevated atmospheric N deposition is well documented (Berendse et al. 2001; Tomassen et al. 2004). Roundleaf sundew is also highly susceptible to effects from atmospheric N deposition (Redbo-Torstensson 1994). This plant is native to and is broadly distributed across the United States and is listed as vulnerable in New York (USDA Natural Resources Conservation Service 2014).

Gotelli and Ellison (2002) calculated model estimates of extinction risk for the northern pitcher plant in two ombrotrophic bogs in the northeastern United States under varying N deposition scenarios. Simulated extinction risk within the next 100 years was relatively low under ambient N deposition loading, although model estimates suggested increased risk under increased N deposition scenarios.

5.1.4.   Avian Response to Acidification

A number of studies have focused on the likelihood that S, N, and/or Hg deposition have influenced bird populations in New York and elsewhere in the eastern United States. Such studies have considered effects on foraging and diet, breeding, distribution, and reproduction. The available evidence does suggest effects on birds, especially in response to Hg methylation and biomagnification, and such effects appear to primarily have been detrimental (Table 5.1; Longcore et al. 1993). Effects of surface-water acidification on birds have been less clear, although some studies have found linkages between acidic deposition, Ca availability in soils, and the reproductive success of birds (Graveland 1998; Hames et al. 2002).

The common loon, a species of special concern in New York, breeds in the Adirondack Park, which is near the southern extent of its breeding range (Schoch 2006). Loon decline was earlier linked to lake acidification in the Adirondack Mountains, perhaps in response to reduction of the food supply in acidified lakes (Alvo et al. 1988). However, a study of 24 Adirondack lakes by Parker (1988) did not find a significant ($p > 0.1$) relationship between lake acidity status and the reproductive success of loons. In addition, the response of loons and other birds to acidic deposition are complicated by the fact that S deposition contributes to increased Hg methylation and therefore to potential Hg toxicity, especially to piscivorous species. There also may be synergistic effects associated with climate change and increased shoreline development and other habitat disturbances.

5.1.5.   Mercury Methylation

Simonin et al. (2008) developed models to predict concentrations of Hg in fish tissue in lakes throughout New York. Lake variables most strongly correlated with high levels of Hg in fish included pH, conductivity, ANC, and the concentrations of base cations. The concentrations of Hg and chlorophyll *a* in lake water, the presence of an outlet dam, and the amount of contiguous wetlands were also important predictor variables.

TABLE 5.1. Studies that either did (yes) or did not (no) yield evidence that acidic deposition affected certain species of birds

Species	Diet/foraging		Breeding distribution		Reproductive measures		Reference[a]
	Yes	No	Yes	No	Yes	No	
Common loon	x		x	x	x	x	1–3, 17, 19
Common merganser			x		x		19
Belted kingfisher			x				4
Osprey	x		x		x		5, 6
Black duck	x		x		x		7–9
Common goldeneye			x[b]				8
Ring-necked duck	x					x	10, 11
Eurasian dipper	x		x		x		12–14
Eastern kingbird				x	x		15
Tree swallow	x			x	x		16–18

Source: Longcore et al. 1993. Copyright 1993 by The Wildlife Society, Bethesda, MD. Used with permission.

[a] 1 = Alvo et al. 1988; 2 = Parker 1988; 3 = Wayland and McNicol 1990; 4 = Goriup 1989; 5 = Eriksson 1983; 6 = Eriksson 1986; 7 = Hunter et al. 1986; 8 = DesGranges and Darveau 1985; 9 = Rattner et al. 1987; 10 = McAuley and Longcore 1988a; 11 = McAuley and Longcore 1988b; 12 = Ormerod et al. 1985; 13 = Ormerod et al. 1986; 14 = Ormerod and Tyler 1987; 15 = Glooschenko et al. 1986; 16 = Blancher and McNichol 1988; 17 = Blancher and McNichol 1991; 18 = St. Louis et al. 1990; 19 = Blair 1990.

[b] The effect on common goldeneye was beneficial.

Local environmental factors can be important regulators of Hg bioaccumulation in topographically heterogeneous landscapes. Riva-Murray et al. (2011) demonstrated positive correlations between dissolved MeHg and organic C in stream water in the Fishing Brook basin in the Adirondack Mountains. Hydrologic transport distance was negatively correlated with Hg concentration in macroinvertebrate feeding groups and was a stronger predictor of MeHg concentration than wetland density.

Riva-Murray et al. (2013a) studied macroinvertebrates, both primary consumers and predators, and selected forage fish from three locations in the Adirondack Mountains. Scraper and filterer primary consumers had higher levels of MeHg and more depleted ^{13}C than shredders collected at the same location. Consumers collected from more shaded locations had more highly enriched ^{13}C. This study provided evidence that the source of dietary C is an important control on Hg bioaccumulation in stream biota. Therefore, factors that influence primary production, including riparian canopy cover, could account for some of the spatial variability in Hg bioaccumulation.

There is evidence that the observed decrease in acidic deposition during the last four decades might actually have contributed to increased Hg methylation and bioaccumulation in fish. A nationwide survey in Norway found increased Hg concentrations in brown trout (*Salmo trutta*) and European perch (*Perca fluviatilis*) in lakes in southern Norway after pronounced decreases in acid and Hg deposition (Hongve et al. 2012). The lake showing the highest percent increase in fish Hg also had a high percentage increase in total organic C, from 4 to 9 mg/L, in lake water. The lake showing no increase in fish Hg exhibited a smaller percentage increase in total organic C (from 12 to 15 mg/L). Thus, it is possible that interactions between Hg and C that promote Hg transport might facilitate methylation. To the extent that total organic C (and dissolved organic carbon) increases in the future in response to reduced acidification, climate change, and/or other factors, the amount of Hg methylation may be impacted.

Bioaccumulation factors for Hg in game fish are used for purposes of effects assessment, contaminant regulation, and the calculation of total maximum daily loads of Hg (U.S. EPA 2010). The bioaccumulation factors are calculated as

$$\text{bioaccumulation factor(Hg)} = \text{Hg}_{fish} \div \text{Hg}_{water} \qquad (2)$$

where Hg_{fish} and Hg_{water} represent the concentrations of Hg in fish and water, respectively. The calculation of bioaccumulation factors is sensitive to sampling issues for both fish and water. Riva-Murray et al. (2013b) evaluated the influence of the timing of collections of water samples, filtration of the samples, and choice of Hg species (total or methyl) on the modeled bioaccumulation factor in 11 streams and rivers located in five states. Higher bioaccumulation factors occurred most often during the growing season and were sensitive to discharge. Thus, analysis of levels of Hg in stream water can be optimized by considering sampling conditions and specifications (Riva-Murray et al. 2013b). In particular, consideration of the flow regime at the time of sampling can help target the higher Hg concentrations in fish, although the concentration-discharge relationship varies from site to site (Riva-Murray et al. 2013b).

Atmospherically deposited Hg must be methylated in order to bioaccumulate in the food web. Methylation of Hg is a biological process. Thus, efforts have been made to develop an understanding of the bacteria responsible for methylation. Sulfate-reducing bacteria appear to be the main agents of Hg methylation in sediments and wetlands (St. Louis et al. 1994; Branfireun et al. 1999; Jeremiason et al. 2006; Kolka et al. 2011; Coleman-Wasik et al. 2012; Drevnick et al. 2012). If the SO_4^{2-} supply is low, the total amount of biologically available S controls bacterial activity and the rate of methylation. Atmospheric SO_4^{2-} deposition causes the stimulation of sulfate-reducing bacteria and increases Hg methylation. Because coal combustion releases substantial amounts of both S and Hg, these stressors often covary. This may have caused or contributed to the observed postindustrial amplification of MeHg concentrations in fish (Jeremiason et al. 2006). Studies have demonstrated increased levels of MeHg production with experimental additions of SO_4^{2-} (Gilmour and Henry 1991; Gilmour et al. 1992; Branfireun et al. 1999; Jeremiason et al. 2006). Fish have shown increased MeHg levels in lakes acidified by S deposition compared to levels in non-acidified lakes that received similar atmospheric Hg deposition (Gilmour and Henry 1991). Reduced SO_4^{2-} emissions and deposition may be expected to cause a reduction in Hg methylation and bioavailability (Jeremiason et al. 2006; Drevnick et al. 2007). Whether this occurs will depend on the availability of SO_4^{2-} in soils and sediments, changes in dissolved organic carbon, and perhaps the extent to which Fe-reducing bacteria contribute to Hg methylation.

Although studies of the transfer of Hg in food webs have focused mainly on aquatic ecosystems, which are considered to be at greatest risk of Hg biomagnification, some studies have also been conducted on terrestrial upland ecosystems, including studies of Hg bioaccumulation in passerine birds. Rimmer et al. (2005) documented MeHg in the blood of insectivorous passerine birds at 21 mountaintop locations in the northeastern United States. Mean blood Hg concentration at breeding sites varied from 0.08 to 0.38 µg/g (wet weight) and were highest in the southern portions of the study area.

Evers et al. (2009) sampled invertebrates and breeding songbirds in 2006 at 20 locations in New York and Pennsylvania. Forty-seven bird species and 359 individuals were sampled to measure blood levels of Hg. Songbirds with the highest blood Hg levels were generally bog-obligate species such as the palm warbler (*Dendroica palmarum*), upper canopy foragers such as the red-eyed vireo (*Vireo olivaceus*), and species commonly associated with riparian habitats (e.g., the Louisiana waterthrush [*Seiurus motacilla*]). Blood Hg concentration varied with elevation and probably also trophic level. Evers et al. (2009) suggested that the wood thrush (*Hylocichla mustelina*) is a good indicator of Hg impacts on passerine birds. This species generally tended to have high blood Hg levels, and populations have experienced substantial declines in recent years in distribution and abundance throughout New York and the Appalachian Mountains (Evers et al. 2009). Breeding populations of wood thrush appeared to be diminished in areas with low soil pH, high elevation, and forest habitat fragmentation (Hames et al. 2002). Evers et al. (2009) hypothesized that Hg is incorporated from leaf litter into invertebrates that feed on leaf tissue and by insect predators that are in turn preyed upon by songbirds.

5.1.6. Effects of Mercury on Humans

Humans can accumulate high levels of MeHg from eating contaminated fish. Avoiding exposure to Hg is especially important for children, pregnant women, and nursing mothers because Hg poisoning can cause neurological impairment in children. This can lead to effects on memory, visual and spatial ability, information processing, and general intelligence (Mahaffey 2005).

The U.S. EPA (2001) recommended a human health standard of 0.3 ppm for Hg in fish and shellfish tissue to protect the general population. More stringent restrictions may be appropriate for women of child-bearing age and for children. Some states, including Maine and Minnesota, have adopted a more restrictive human health standard of 0.2 ppm. All 50 states issued Hg advisories for human fish consumption in 2008 (Fenn et al. 2011), leading to an increase from 2006 in lake areas under advisory of 19% and an increase in river lengths under advisory of 42%. Mercury advisories for coastal waters increased by 25% from 2004 to 2008. These increases were likely due mainly to increased availability of measured data rather than to increases in Hg deposition, methylation, or bioaccumulation (Fenn et al. 2011).

Both the Adirondack and Catskill Parks have fish consumption advisories due to high concentrations of Hg in fish. Detailed consumption advisories are available on the Web site of the New York State Department of Health (New York State Department of Health 2014).

5.2. EFFECTS ON THE BIOLOGY OF FRESHWATER ECOSYSTEMS

The most important chemical parameters associated with biological effects of acidification and eutrophication on fresh surface waters are generally pH, ANC, Al_i, N, P, chlorophyll *a*, and Ca^{2+}. Complex interactions among these chemical parameters govern the effects of atmospheric S and N deposition on aquatic biota (e.g., Mount et al. 1988; Ingersoll et al. 1990; Wood et al. 1990).

Acidification of lakes and streams in response to acidic deposition can cause effects on organisms at all trophic levels. Many studies have focused on the loss of fish populations, especially salmonids, largely because of the perceived value of these sport fish to humans. Other studies reported that many species of phytoplankton, zooplankton, insect larvae, crayfish, snails, and freshwater mussels are less common or absent from acidified lakes and streams (Havas 1986; Baker et al. 1990a).

Effects of acidification on aquatic biota have been demonstrated in laboratory and field bioassays (e.g., Baker et al. 1996), whole-ecosystem acidification experiments (e.g., Schindler et al. 1985), and field surveys (e.g., Baker and Schofield 1982; Gallagher and Baker 1990). Many of the species that commonly occur in acid-sensitive surface waters susceptible to acidic deposition cannot reproduce or survive if the water has ANC near or below 0 μeq/L. Some sensitive species of fish, invertebrates, and algae cannot survive even at moderate levels of acidity. For example, some zooplankton predators, sensitive mayfly species, and sensitive fish species are affected at pH values below the range of about 5.6 to 6.0 (Baker and Christensen 1991), which generally corresponds to ANC of about 25 to 50 μeq/L in Adirondack and Catskill waters. In general, however, aquatic life is considered to be at risk if the pH falls below about 6.0, ANC is less than 50 to 100 μeq/L, or the concentration of Al_i is greater than 2 μmol/L (U.S. EPA 2008).

5.2.1. Phytoplankton

Effects of atmospheric N deposition on estuaries and coastal marine waters differ somewhat from effects on fresh waters, soils, and forests. Estuaries and coastal marine waters are especially susceptible to effects from N enrichment because N tends to be the main growth-limiting nutrient in such waters. As a consequence, addition of N from outside the watershed might be expected to increase algal growth, whereas addition of a nutrient other than N may not necessarily be expected to increase growth. Also, estuarine and coastal marine waters in New York typically receive rather substantial non-atmospheric contributions of N in addition to the atmospheric contributions.

Excess algal growth can cause a variety of environmental effects, including decreased dissolved oxygen in the water, fish kills, bad odors, and loss of shellfish. Dense populations of algae in the water column can shade out the sea grasses that normally grow on some portions of the estuary bottom. This loss of submerged aquatic vegetation can destroy important habitat for fish and other aquatic life. When algae reproduce and then die in large quantities, the microorganisms that decompose them consume such large quantities of O_2 from the water that dissolved oxygen becomes depleted and fish and shellfish can suffocate from insufficient O_2. The shifts in nutrient availability in the estuary caused by human inputs of N can also promote the growth of cyanobacteria that release toxins into the water or that are aesthetically unpleasant.

Eutrophication effects on fresh waters have been most extensively studied in lakes. Nevertheless, effects in response to excessive nutrient loading to rivers and streams are also of concern, although other factors often limit the extent to which nutrient additions cause increased algal growth in flowing waters. In particular, light availability is an important limiting factor in small, low-order streams where tall trees in the riparian zone provide shade. Restriction of light penetration into

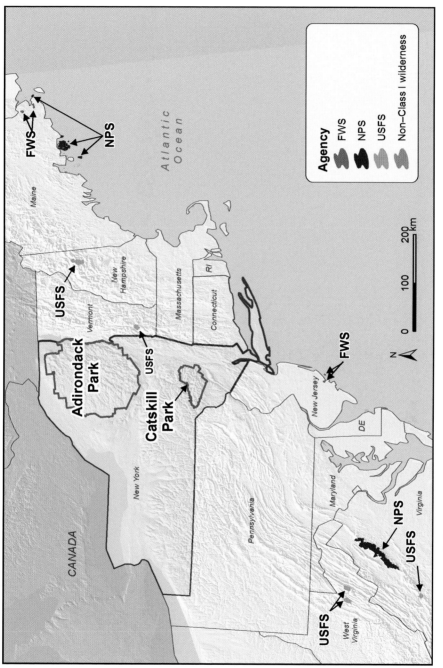

PLATE 1. Locations of Class I areas around New York that receive maximum protection by the Clean Air Act from effects of air pollution. Agencies: U.S. Fish and Wildlife Service (FWS), National Park Service (NPS), U.S. Forest Service (USFS).

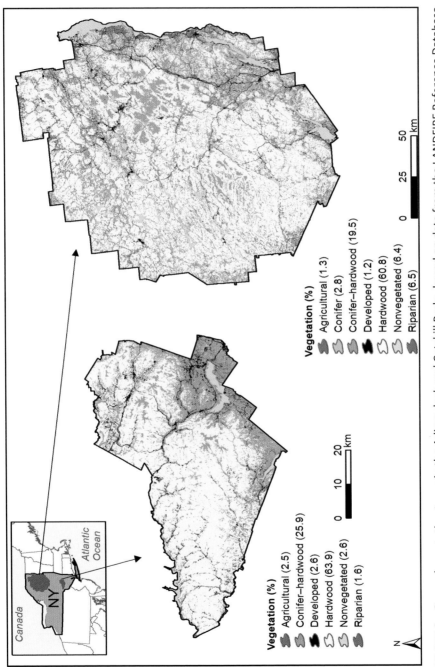

PLATE 2. General vegetation types in the Adirondack and Catskill Parks, based on data from the LANDFIRE Reference Database of the USDA Forest Service (http://www.landfire.gov/). Areas classified as nonvegetated are mainly open water. Vegetation with less than 1% coverage is not included.

Vegetation (%)

Agricultural (2.5)
Conifer–hardwood (25.9)
Developed (2.6)
Hardwood (63.9)
Nonvegetated (2.6)
Riparian (1.6)

Vegetation (%)

Agricultural (1.3)
Conifer (2.8)
Conifer–hardwood (19.5)
Developed (1.2)
Hardwood (60.8)
Nonvegetated (6.4)
Riparian (6.5)

Canada
NY
Atlantic Ocean

0   10   20
km

0   25   50
km

N

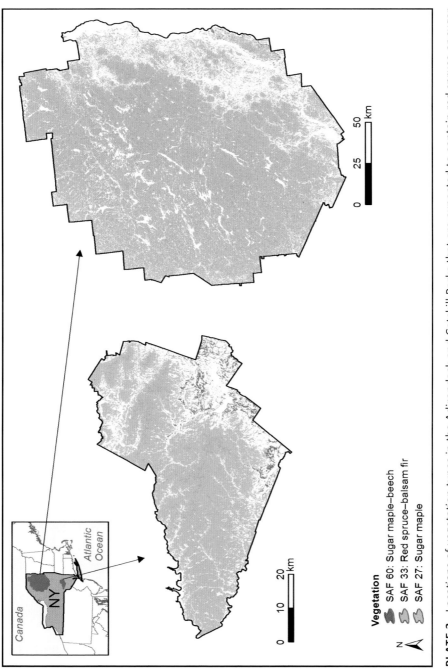

Vegetation

SAF 60: Sugar maple–beech

SAF 33: Red spruce–balsam fir

SAF 27: Sugar maple

PLATE 3. Locations of vegetation types in the Adirondack and Catskill Parks that are expected to contain red spruce or sugar maple trees, based on data from the LANDFIRE Reference Database of the USDA Forest Service.

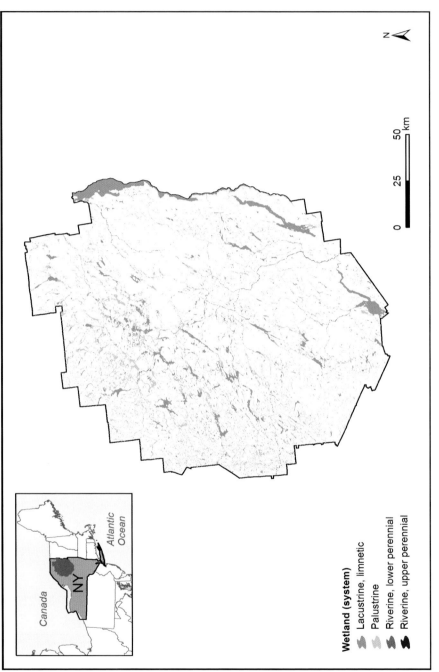

PLATE 4. Locations of wetlands in the Adirondack Park, based on data collected by the Adirondack Park Agency. Note that riverine wetlands are typically too small to see at the scale of this map.

Wetland (system)
Lacustrine, limnetic
Palustrine
Riverine, lower perennial
Riverine, upper perennial

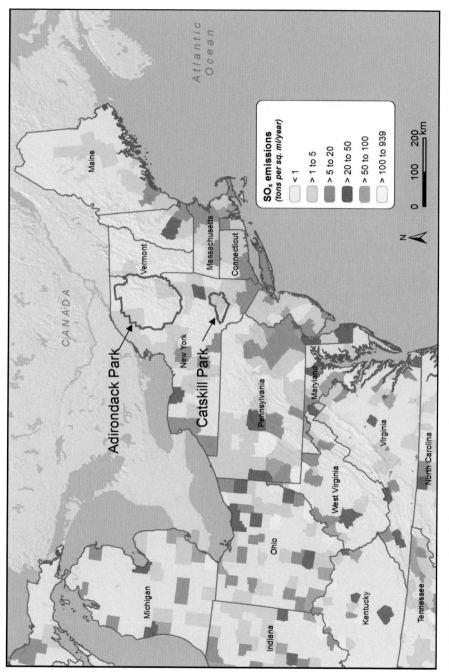

PLATE 5. Total annual emissions per square mile of sulfur dioxide (SO_2) by county for the year 2008 in and in reasonable proximity to New York. Data provided by the Western Regional Air Partnership.

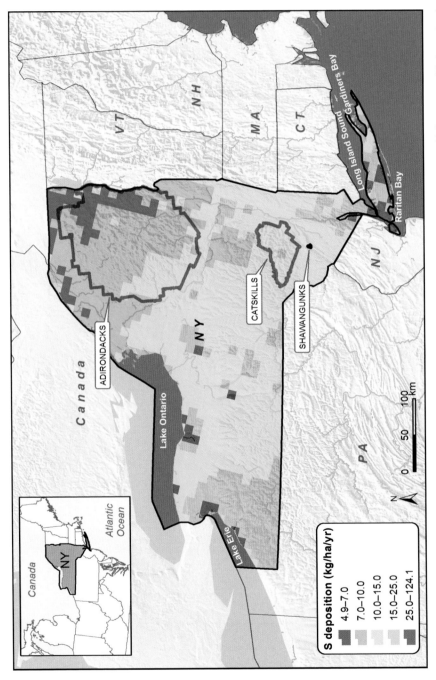

PLATE 6. Total wet plus dry sulfur (S) deposition in New York for the year 2006, expressed in units of kilograms of S deposited from the atmosphere to the earth's surface per hectare per year. Data were derived using the CMAQ model. Data provided by R. Dennis, U.S. EPA.

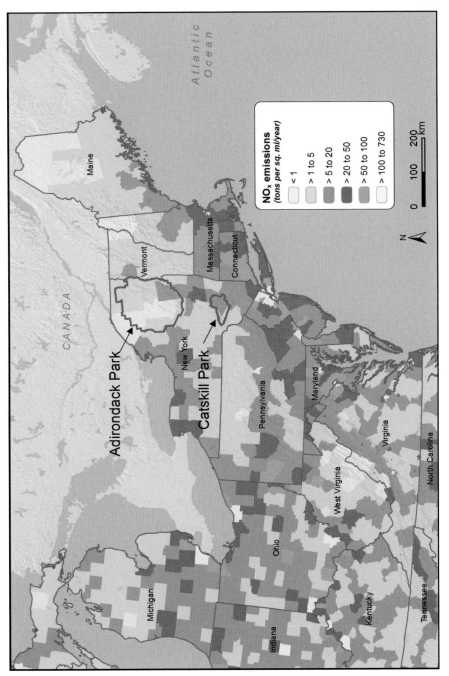

PLATE 7. Total annual emissions per square mile of nitrogen oxides (NO_x), by county for the year 2008 in and in reasonable proximity to New York. Data provided by the Western Regional Air Partnership.

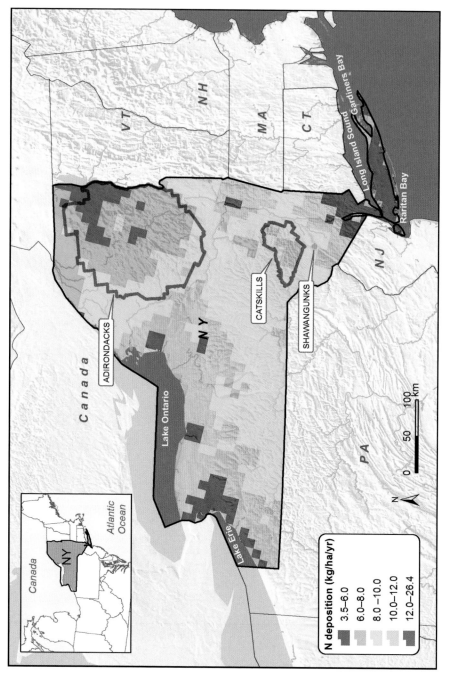

PLATE 8. Total wet plus dry deposition of total (oxidized plus reduced) nitrogen (N) in New York for the year 2006. Data were derived using the CMAQ model. Data provided by R. Dennis, U.S. EPA.

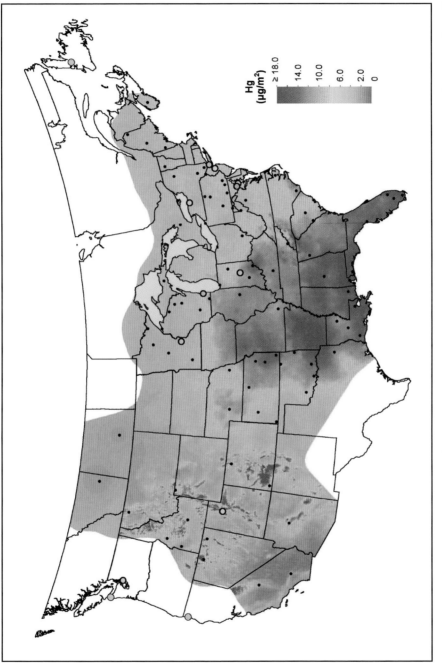

PLATE 9. Interpolated wet mercury (Hg) deposition for 2009 determined by the Mercury Deposition Network. Source: http://nadp.sws.uiuc.edu/mdn/.

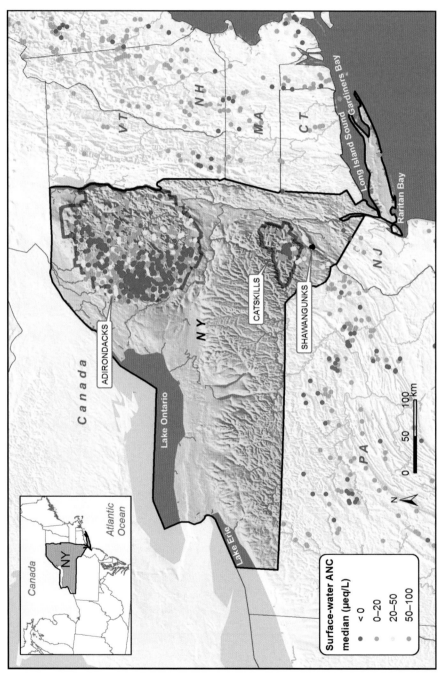

PLATE 10. Spatial distribution of lakes and streams in and around New York with measured surface-water ANC ≤ 100 μeq/L, based on available data compiled by Sullivan (in revision) from multiple sources.

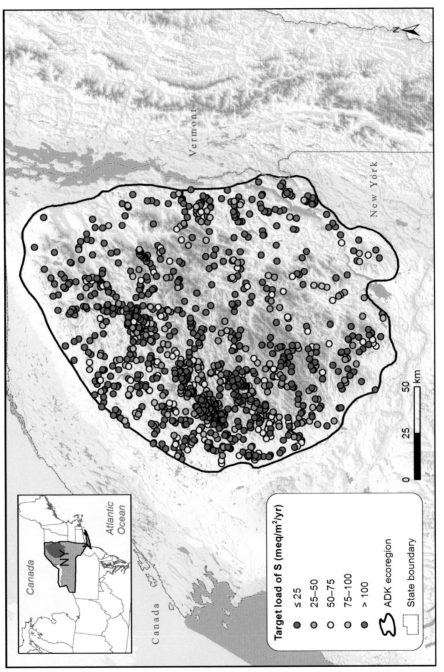

PLATE 11. Estimated target load of sulfur deposition to protect lake ANC to 50 μeq/L in the year 2100, based on extrapolation of MAGIC modeling results to the population of lakes surveyed by the Adirondack Lakes Survey. Source: Sullivan et al. 2012.

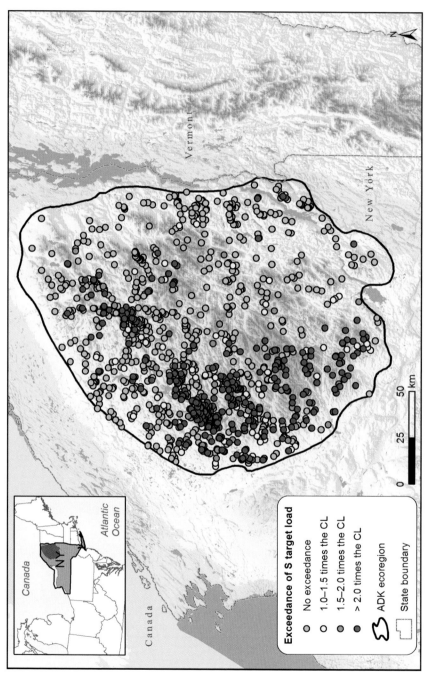

PLATE 12. Exceedance classes for 1,136 Adirondack lakes based on extrapolation of MAGIC model results of sulfur (S) target loads to the ALS surveyed lakes. Exceedances were calculated for the year 2100 using a critical threshold for ANC of 50 µeq/L. Source: Sullivan et al. 2012.

the water column in response to suspended sediments, especially during high-flow conditions, commonly limits algal growth in larger rivers (Hoyer and Jones 1983).

Experimental work by Makarewicz et al. (2007) illustrated that watershed sources of nutrients were more strongly associated with primary productivity in Conesus Lake, one of the Finger Lakes in New York, than with atmospheric nutrient sources. They performed experiments in the littoral zone of Conesus Lake to test the hypothesis that nutrient-rich tributary inflow from watersheds draining agricultural (60 to 80%) land use had stimulatory effects on littoral metaphyton growth, whereas inflow from a watershed draining largely forested land use (12% agricultural) did not. A statistically significant stimulatory effect was documented for each of the six experimental agricultural tributaries, including increased metaphyton cover in watersheds with increased agricultural land use. Inflow from the forested tributary watershed did not exhibit a stimulatory effect on metaphyton growth. The forested watershed, in contrast, received atmospheric N inputs but limited agricultural inputs of N and P.

Extreme cases of the reduction of dissolved oxygen can cause hypoxia or anoxia, fish mortality, odor problems, and other types of degradation of water quality. There has not been any documentation that atmospheric N deposition levels in the United States are high enough on their own to cause such extreme effects on fresh or marine waters.

Results from surface-water surveys, paleolimnological reconstructions of past lake chemistry, experimental nutrient enrichment studies, and meta-analyses of hundreds of studies have concluded that N limitation is common in fresh waters. This appears to especially be the case in remote areas. There is a nearly universal eutrophication response to N enrichment in lakes and streams that are N-limited (U.S. EPA 2008).

During the spring survey of western Adirondack streams conducted by the U.S. Geological Survey (Lawrence et al. 2008a), algal communities were judged to be severely affected by acidification in more than half of the study streams, and moderately to severely affected in 80% of the streams, based on changes in species composition. Passy (2006) documented the effects of episodic acidification on diatom communities in a tributary to Buck Creek in the Adirondack Mountains. She found that the origin of acidity (organic versus inorganic and SO_4^{2-} versus NO_3^-) can be as important as the pH depression in determining diatom response. The nature of the acidity in the stream can determine the species composition of the diatom flora, especially the role played by *Eunotia exigua*, a common species in acidified streams.

The Great Lakes are known to be sensitive to nutrient additions, but atmospheric inputs of N are just one part of a complicated story. Substantial reductions in point-source P loads to Lake Erie during the latter part of the twentieth century contributed to substantial reductions in phytoplankton and nuisance algae in the lake (Makarewicz and Bertram 1991). More recent changes in the water quality

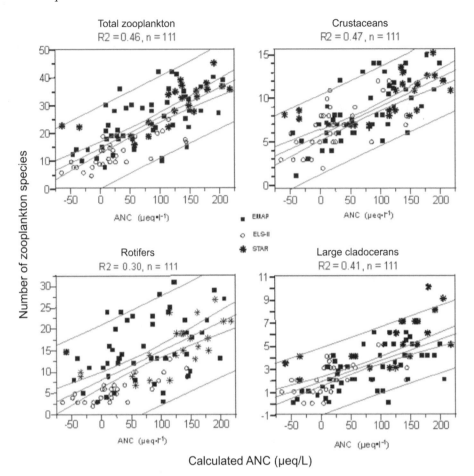

FIGURE 5.1. Relationship between zooplankton taxonomic richness and ANC levels for a combined Adirondack data set, based on 111 lake visits to 97 lakes in the EMAP, ELS, and the U.S. EPA's Science to Achieve Results (STAR) program zooplankton surveys for total zooplankton, crustaceans, rotifers, and large cladocerans. R2 is the square of the correlation coefficient. Source: Sullivan et al. 2006a.

of Lake Ontario have been strongly influenced by the introduction and expansion of invasive mussels that first appeared in the lake in about 1989. Total P concentrations have continued to decline, coincident with an increase in the proportion of P that is soluble. Silica concentrations during spring have increased, suggesting declining diatom populations.

5.2.2. Zooplankton

Acidification reduces the species richness of lake zooplankton (Nierzwicki-Bauer et al. 2010). Thresholds for zooplankton community alteration have been identified in the range of about pH 5 to 6 (Locke and Sprules 1994; Holt et al. 2003). Holt and Yan (2003) reported recovery of zooplankton community composition in lakes in Killarney Park, Ontario, Canada in which the pH increased from below to above 6.0 in response to reduced acidic deposition during the period 1971–2000.

Lake characterization studies focused on the response of zooplankton to acidification generally examine zooplankton data across ANC gradients to determine any relationships that might exist between zooplankton richness and lake ANC. These analyses have been conducted for all zooplankton groups combined (total zooplankton) and for discrete groups of zooplankton. The discrete groups have included crustaceans, rotifers, and large cladocerans. An example for Adirondack lakes in New York is shown in Figure 5.1, which depicts the number of zooplankton species in terms of ANC levels at the time of the zooplankton survey. In each case, the taxonomic richness of zooplankton increased with increases in lake ANC.

5.2.3.   Benthic Macroinvertebrates

Macroinvertebrates are relatively easy and inexpensive to sample in the field (Plafkin et al. 1989; Resh et al. 1995; Karr and Chu 1999; Potyondy et al. 2006). Stream macroinvertebrates are sensitive to various kinds of disturbance, including acidification from atmospheric deposition (Cairns and Pratt 1993). Acidification causes a loss of acid-sensitive benthic macroinvertebrate taxa, expressed at either the species or genus level. Taxa that are most sensitive to acidification include mayflies (order Ephemeroptera), amphipods, snails, and clams. Caddisflies (order Tricoptera) are also relatively sensitive. If stream pH decreases to the range of 5.5 to 6.0 or below, acid-sensitive species are replaced by more acid-tolerant species. This typically does not cause large changes in the richness, diversity, density, or biomass of species in the total community. If pH decreases to values below about 5.5, it is likely that more species will be lost without replacement, causing decreased richness and diversity (U.S. EPA 2008). More extreme acidification to pH < 5.0 has been found to virtually eliminate mayflies, crustaceans, and mollusks (Guerold et al. 2000).

Baldigo et al. (2009) conducted a study of the effects of the acid-base chemistry of stream water on resident macroinvertebrates in streams of the western Adirondack Mountains in conjunction with the Western Adirondack Stream Survey sampling effort of Lawrence et al. (2008b). Macroinvertebrate communities were sampled from a subset of 36 streams in the Oswegatchie and Black River basins from 2003 to 2005. The researchers documented effects of chronic and episodic acidification on macroinvertebrates across the western Adirondack Mountains. They also assessed indicators and thresholds of biological effects. Baldigo et al. concluded that macroinvertebrate communities were moderately to severely impacted by acidification in about half of the study streams.

Baldigo et al. (2009) applied community multimetric indices to reflect acidification impacts on stream macroinvertebrates in New York. They calculated the

New York State Biological Assessment Profile (NYSBAP) score, which is used for stream assessment statewide, for 36 western Adirondack streams (Riva-Murray et al. 2002; Smith et al. 2007). The NYSBAP was designed mainly for assessing the influence of wastewater treatment effluent and industrial pollutants and was not found to be especially sensitive to acidification impacts. They also calculated the acid biological assessment profile (AcidBAP) index (Burns et al. 2008b), which was developed from benthic macroinvertebrate taxa that are commonly abundant in acidic streams of the Adirondack and Catskill Mountains but uncommon in 20 reference streams surveyed across New York. The AcidBAP index is based on percent mayfly richness and an Acid Tolerance Index that reflects the percent of individuals from 10 genera known to contain one or more acidophilous species. Baldigo et al. (2009) assessed the relationships between AcidBAP index scores and aspects of acid-base water chemistry. The AcidBAP index reflected acid deposition effects in the study streams fairly well. It was correlated (r^2 values 0.58 to 0.76) with concentrations of Al_i, pH, ANC, and base-cation surplus in stream water.

5.2.4.   Fish

5.2.4.1.   *Effects of Acidification on Fish*

Acidification can affect fish populations in a number of ways, including increased mortality, emigration to refugia to escape adverse chemical conditions, and decreased availability of food (Baker et al. 1990a). Population decline is often due to failure to successfully recruit YOY fish (Mills et al. 1987; Brezonik et al. 1993). The response of fish populations to acidification typically becomes apparent first as changes in age distribution and decreased growth and condition of individual fish, followed by decreased biomass and density of acid-sensitive species, and finally as elimination of sensitive species (Baker et al. 1990a).

Effects of watershed disturbance can interact with acidification from atmospheric deposition, leading to more pronounced effects on biota than a single stress on its own would have. Baldigo et al. (2005) compared effects of clear-cut and timber-stand-improvement tree harvests on water chemistry and the mortality of caged brook trout in three Catskill Mountain streams. Effects of tree harvests on fish communities are potentially important because they could interact with acidic deposition and produce more substantial effects on biota than either stressor alone. Timber harvest removed 73% of the tree basal area from a clear-cut sub-basin, 5% basal area from a timber-stand-improvement sub-basin, and 14% basal area at a site below the confluence of both streams, selected to represent the combined effect of the two harvest methods. Stream acidity, NO_3^-, and Al_i concentration all increased in the clear-cut stream during high-flow conditions after the first growing season. Episodic pulses of H^+ and Al_i were severe during the first year and then decreased in magnitude and duration. All trout at the clear-cut site died within seven days during the spring of the second year, and 85% died during the spring of the third year. Only background mortality was observed in other years at this site and during all years at the reference sites that had not been clear-cut.

The Ca^{2+} concentration in surface water affects the ability of fish to survive at low pH. Survival is reduced at lower Ca^{2+} concentration (Baker et al. 1990a). Effects of Al_i and H^+ interact both synergistically and antagonistically, depending on conditions (Muniz and Levivestad 1980; Havas 1985; Rosseland and Staurnes 1994). The Al hydroxide forms, $Al(OH)_2^+$ and $Al(OH)^{2+}$, and free Al (Al^{3+}) appear to be most toxic; naturally occurring organic acids can reduce or eliminate that toxicity by forming nontoxic alumino-organic complexes.

Of the 53 fish species the ALS recorded in Adirondack lakes, about half were absent from lakes with pH below 6.0, which roughly corresponds with ANC below 50 µeq/L in this region. Those 26 species included important recreational species, such as Atlantic salmon (*Salmo salar*), tiger trout (*Salmo trutta* X *Salvelinus fontinalis*), redbreast sunfish (*Lepomis auritus*), bluegill (*Lepomis macrochirus*), tiger musky (*Esox masquinongy* X *E. lucius*), walleye, alewife (*Alosa pseudoharengus*), and kokanee (*Oncorhynchus nerka*) (Kretser et al. 1989), plus ecologically important minnows that serve as forage for sport fish. Nearly one-fourth of the 1,469 lakes surveyed in the ALS contained no fish at the time of the survey. The lakes that lacked fish were significantly lower in pH, dissolved Ca^{2+}, and ANC and had higher concentrations of Al_i than lakes that contained fish (Gallagher and Baker 1990). Among the lakes with fish, the number of fish species was correlated with lake pH (Kretser et al. 1989; Driscoll et al. 2001b) and lake ANC (Sullivan et al. 2006a). Similar results were found for lakes in Ontario, Canada, where the deviation between predicted and observed fish species varied with lake pH (Figure 5.2).

There are many possible causes for the absence of fish in some Adirondack lakes. In addition to acidification, these causes can include lack of suitable habitat, poor spawning conditions, winter kill due to O_2 depletion under ice and snow, and blocked access. Small high-elevation Adirondack lakes are more likely to be fishless than larger lakes at low elevations (Gallagher and Baker 1990; Driscoll et al. 2001a). Small, high-elevation Adirondack lakes with fish had significantly higher pH than those that lacked fish. The primary cause of the difference in fish presence was likely acidification in response to acidic deposition (Driscoll et al. 2001a).

Threshold pH values for adverse effects caused by acidification have been documented for many species of fish (Haines and Baker 1986; Baker et al. 1990a). Baker and Christensen (1991) published pH thresholds for significant adverse effects on populations of 25 fish species that occur in New York. Blacknose dace (*Rhinichthys atratulus*) was found to be relatively sensitive to acidification; population loss was documented at pH values as high as 6.1. Results of surveys of water chemistry showed decreased species diversity and absence of some species in the pH range of 5.0 to 5.5 (Haines and Baker 1986). Smallmouth bass are usually not found at pH values less than 5.2 to 5.5 (Haines and Baker 1986). Lake or stream

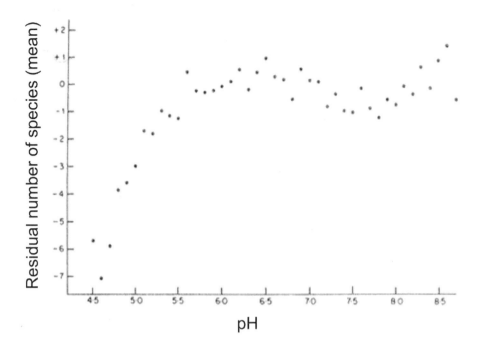

FIGURE 5.2. Mean residual number of species per lake for lakes in Ontario, by pH interval. The residual number of species for a lake is the deviation of the observed number from the number predicted by lake area. Source: J. E. Matuszek and G. L. Beggs, 1988. Fish species richness in relation to lake area, pH, and other abiotic factors in Ontario lakes, Can. J. Fish. Aquat. Sci. 45(11):1931–1941. Copyright 2008 Canadian Science Publishing or its licensors. Reproduced with permission.

pH values of less than 6.0 to 6.5 have been associated with adverse effects on populations of dace, minnows, and shiners (family Cyprinidae; Hall 1987; Klauda et al. 1987). Some life stages of some species (e.g., striped bass [*Morone saxatilis*] and fathead minnow [*Pimephales promelas*]) have experienced significant mortality in bioassays even at pH in the range of 6.0 to 6.5 (Buckler et al. 1987; McCormick et al. 1989). Mudminnow (*Umbra* spp.) is very acid tolerant; it survived at pH near or below 4.0 (Dederen et al. 1987). Some minnows and dace have shown adverse effects at pH in the range of 5.5 to 6.0, but brook trout, largemouth bass, and smallmouth bass are less sensitive, having threshold effects at pH in the range of about 5.0 to 5.5.

The U.S. EPA's Episodic Response Project conducted in situ bioassays to document effects of episodic acidification on fish in 13 small streams in the Adirondack and Catskill Mountains and the northern Appalachian Plateau in Pennsylvania (Baker et al. 1996). Toxicity was quantified for brook trout, blacknose dace, and mottled sculpin (*Cottus bairdi*) or slimy sculpin. Fish community status, fish movements, and the abundance and biomass of brook trout were documented in each stream during hydrologic episodes. Streams with suitable water chemistry during low flow but episodic acidification to low pH and high Al_i during high flow had higher fish mortality, greater downstream movement of brook trout during hydrological events, and lower brook trout abundance and biomass than streams that did not experience episodic acidification. The more acid-sensitive fish species (blacknose dace and sculpin) were not found in the episodically acidic streams. Brook trout movement into refugia with higher pH and lower Al_i during episodes partially mitigated the effects of episodic stream acidity (Baker et al. 1996).

Acidic episodes reduced fish numbers and eliminated the more acid-sensitive species in streams having median high-flow pH lower than 5.2 and Al_i concentrations higher than 3.7 μmol/L (Baker et al. 1996). Brook trout abundance was lower in Episodic Response Project streams with median high-flow pH < 5.0 and Al_i > 3.7 to 7.4 μmol/L. Acid-sensitive fish species were not found in streams with median high-flow pH < 5.2 and Al_i > 3.7 μmol/L. The concentration of Al_i was the single best predictor of fish mortality in the bioassays (van Sickle et al. 1996). High Al_i concentration during high-flow periods are most likely the main cause of impacts on fish during low pH episodes in the Adirondack and Catskill Mountains (U.S. EPA 2008). In general, larval stages and young fish tend to be more acid-sensitive and Al-sensitive than adults. Subsequent research by Baldigo et al. (2007) found that mortality of YOY brook trout occurred at Al_i concentrations as low as 54 μg/L (2 μmol/L). The effects of Al_i can be partially to completely ameliorated by high concentrations of dissolved organic carbon and/or Ca^{2+}. Thus, multiple factors, including chronic pH, Al_i, dissolved organic carbon, and Ca^{2+} concentrations and the timing and magnitude of episodic variations in water chemistry influence the effects of acidification on fish mortality (Baker et al. 1990a; Gagen et al. 1993; Simonin et al. 1993; van Sickle et al. 1996; Baldigo and Murdoch 1997).

Brook trout is often used as an indicator of acidification impacts on aquatic biota in the eastern United States, although it is relatively acid tolerant compared to many other fish species. The strong focus on this species is attributable to the fact that brook trout constitute a popular sport fishery that is native to many eastern streams and lakes. People place great recreational and aesthetic value on this native species. In acidified Adirondack lakes, brook trout is commonly lost if the ANC decreases below about 0 μeq/L (MacAvoy and Bulger 1995).

Although acidification and high concentrations of Al_i in lakes and streams have adversely affected fish populations in surface waters in the Adirondack and Catskill Mountains (cf. Baker and Schofield 1982; Colquhoun et al. 1984; Johnson et al. 1987; Schofield and Driscoll 1987; Kretser et al. 1989; Stoddard and Murdoch 1991; Simonin et al. 1993), the absence of fish from a given lake or stream does not necessarily indicate that acidification is the cause. Many Adirondack lakes have always had marginal spawning habitat for brook trout (Schofield 1993), and some of the currently fishless acidic lakes probably never supported fish.

Baker et al. (1990a) concluded that of the Adirondack lakes without fish, 42% had high organic acid content that may have caused the observed low pH; 13% were bog lakes of high acidity and naturally poor fish habitat; 9% had pH > 5.5, suggesting that other factors were likely responsible for the lack of fish; and 3% were small high-elevation lakes that were unlikely to have fish regardless of acid-base chemistry. Of the lakes Baker et al. (1990a) surveyed that had no fish at the time of sampling, 34% also had low pH that the researchers attributed to acidic deposition. In addition, for every lake that is fishless due to acidic deposition, there may be several others that still have some fish but that have lost the more sensitive species, have lost life forms other than fish such as acid-sensitive phytoplankton or zooplankton, or now contain fewer fish as a consequence of acidification. Model simulations suggested that some Adirondack lakes have probably lost up to four fish species in response to acidification caused by acidic deposition (Sullivan et al. 2006a).

Only a few fish species are commonly found in Adirondack lakes with pH levels below 5. These include central mudminnow, brown bullhead (*Ameiurus nebulosus*), and yellow perch. More than three-fourths of the Adirondack lakes with a pH of less than 5.0 are fishless. Most of the fishless lakes are located in the southwestern Adirondacks, where both acid deposition and watershed acid sensitivity are high.

Christensen et al. (1990) determined that lake pH was the best predictor of brook trout presence in Adirondack lakes. Other explanatory factors included silica, ANC, dissolved organic carbon, type of lake, substrate, and distance to the nearest road. The number of fish species in Adirondack lakes declines as ANC declines. Lakes that have ANC $\leq$ 0 μeq/L typically support only one or no fish species. At ANC above about 50 to 100 μeq/L, Adirondack lakes generally support four to six species of fish (Figure 5.3), and some lakes have 10 or more species (Sullivan et al. 2006a).

Research by Baldigo et al. (2007) has confirmed that brook trout mortality in Adirondack streams (as had been found for lakes) was most closely related to Al_i exposure. They evaluated the water quality and mortality of caged YOY brook trout during 30-day exposure periods. Mortality varied among years and study sites, depending on the timing of snowmelt and watershed neutralization of acidity, which limited the occurrence of acutely toxic levels of Al_i during high discharge periods. Two to four days of exposure to concentrations of Al_i greater than 4 μM caused 50% to 100% YOY brook trout mortality. Concentration of Al_i less than 2 μM was associated with low mortality (Baldigo et al. 2007). The researchers surmised that because observed stream Al_i concentrations in the Western Adirondack Stream Survey often exceeded these brook trout thresholds, entire fish assemblages throughout the region might be negatively impacted by episodes of high Al_i concentration. This is because other commonly occurring fish species are generally more acid- and Al_i-sensitive than brook trout.

Changes in fish populations in Adirondack lakes between the periods 1984–1987 and 1994–2005 were documented as part of the Adirondack Long Term Monitoring Program and reported by Roy et al. (2013). There were signs of fish recovery from previous acidification damage over the average 14-year period that

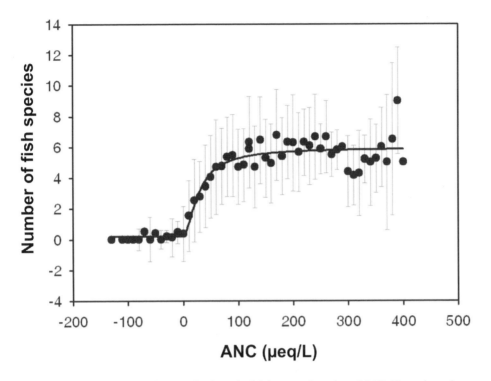

FIGURE 5.3. Fish species richness of Adirondack lakes as a function of ANC. The values shown represent the mean (filled circles) and standard deviation (bars) of 10 µeq/L ANC classes. Also shown as a solid line is the application of the logistic model. Source: Sullivan et al. 2006a. Reproduced with permission.

was examined, but they were modest. The largest gains in fish species richness occurred in moderately sized lakes that had pH in the range of 5.5 to 6.0. These gains generally occurred in lakes that also showed improvements in pH, ANC, and Al$_i$ (Roy et al. 2013). Lakes in that pH range (n = 15) gained, on average, about two fish species. As an example of substantial fish recovery, four species were caught in Cascade Lake in the 1984 survey (brown bullhead, yellow perch, brook trout, and white sucker [*Catostomus commersonii*]) and the pH was 6.5. By 1999, the pH had increased to 6.7 and the same four fish species were caught plus four additional species (pumpkinseed sunfish [*Lepomis gibbosus*], golden shiner, creek chub, and common shiner; Roy et al. 2013).

5.2.4.2. *Effects of Mercury on Fish*

Increased Hg concentrations in fish can decrease reproductive success and cause changes in fish behavior. Fish with increased MeHg exposure display less efficient predator-avoidance behaviors. Webber and Haines (2003) concluded that Hg exposure at levels found in lakes in northern New York can increase the vulnerability of fish to predation. Reproductive effects on fish-eating birds have been reported at fish Hg levels as low as 0.16 ppm (Fenn et al. 2011). Yellow perch and the piscivorous common loon are often studied as indicator species for Hg contamination in freshwater ecosystems.

Fish monitoring typically targets mid-trophic-level omnivorous prey fish and higher-trophic-level predatory fish. This is done in an effort to capture the full range of fish Hg conditions (Kamman et al. 2005). The former are expected to respond quickly to changes in Hg exposure (Wiener et al. 2007). The latter are more strongly influenced in their response to Hg exposure by such factors as the age and size of fish, nutrient input, and interspecies competition (Mason et al. 2005).

Histopathological biomarkers can provide evidence of effects of toxins on individual fish. The macrophage aggregate is a focal accumulation of macrophages in the spleen, kidney, or liver of fish that indicates an immune system response to adverse environmental conditions. Microphage aggregates are formed in response to tissue damage by toxic agents (Schwindt et al. 2008). Significant correlations have been reported between increased microphage aggregates in fish and exposure to Hg or other metals (Handy and Penrice 1993; Meinelt et al. 1997; Manera et al. 2000; Fournie et al. 2001; Khan 2003; Capps et al. 2004). Schwindt et al. (2008) showed increased microphage aggregate occurrence in fish exposed to high Hg and those that had kidney and spleen tissue damage. This research was conducted on four species of trout collected from 14 lakes in national parks or preserves in the western United States, suggesting that Hg and/or other contaminants might adversely impact fish that inhabit remote and protected lakes.

Wiener et al. (2012) summarized and assessed data on 6,400 yellow perch collected from the Great Lakes and associated inland lakes and reservoirs in the Great Lakes region. Perch from 6.5% of the sampled waters contained average whole-fish Hg concentrations high enough to cause adverse impacts on fish. Concentrations were higher in fish at higher trophic levels.

Mercury loads in fish collected in the Great Lakes region tended to be higher in inland waters than in the Great Lakes themselves. The median Hg concentration in yellow perch fillets collected from inland waters (0.14 ppm [wet weight]) was more than 50% higher than the median concentration in yellow perch collected from the Great Lakes (0.09 ppm [wet weight]; Wiener et al. 2012). Similarly, Hg concentrations in fish from the Great Lakes were also about 55% lower for walleye and 25% lower for largemouth bass (Monson et al. 2011) than concentrations in fish collected from inland waters in the same region.

Monitoring data from Great Lakes fish suggested a substantial decrease in Hg contamination from the late 1970s to the early 1980s. Subsequent to a period of substantial decline from about 1977 to 1983, Hg concentrations in Lake Erie walleye remained relatively constant through 2003. Similarly, no clear trend has been found for Hg in rainbow smelt from Lake Erie, where Hg levels in fish remained relatively constant from 1985 through 2003. Fish consumption advisories remain in place for Hg in all five Great Lakes (Environment Canada and U.S. EPA 2005).

5.2.4.3. *Effects of Environmental Factors on Mercury Bioaccumulation in Fish*

Lakes with low and high fish Hg content can be situated within a few kilometers of each other. For example, fish collected from two lakes in Voyageurs National Park had some of the highest and lowest Hg concentrations recorded in Minnesota for northern pike (Rolfhus et al. 2011). These two lakes receive nearly identical Hg inputs in atmospheric deposition (Wiener et al. 2006; Rolfhus et al. 2011). Thus, lake and watershed factors and trophic factors can have a large influence on Hg bioaccumulation in fish.

The concentration of Hg in fish tissue in the northeastern United States has often been shown to be positively correlated with lake area and watershed area and negatively correlated with lake pH, ANC, and zooplankton density (Chen et al. 2005; Driscoll et al. 2007b). Lake types that are generally associated with the most Hg bioaccumulation are poorly buffered, are low in pH and productivity, and have forested watersheds with limited human development (Chen et al. 2005). Hg-sensitive surface waters have also been described as those having high SO_4^{2-} concentrations, low pH and ANC, extensive wetlands, large watershed area relative to lake area, fluctuating water level, and low nutrient concentrations (Evers 2005).

High Hg concentrations in yellow perch and common loon in the northeastern United States are significantly correlated with elements of water chemistry: total P < 30 µg/L, dissolved organic carbon > 4 mg/L, pH < 6.0, and ANC ≤ 100 µeq/L (Chen et al. 2005; Driscoll et al. 2007b). Other studies have shown that high concentrations of Hg in biota are also correlated with wetland abundance, low lake-to-watershed area ratio, and high percent forest cover (Roué-LeGall et al. 2005; Driscoll et al. 2007b). These characteristics are related to the transport, methylation, and trophic transfer of Hg (St. Louis et al. 1996; Wiener et al. 2006; Driscoll et al. 2007b; Turnquist et al. 2011).

The review of Evers (2005) classified Hg-sensitive surface waters as those having

- high SO_4^{2-} concentrations
- low pH and ANC
- extensive wetlands
- large watershed area relative to lake area
- fluctuating water levels, and
- low nutrient concentration

Such lakes and watersheds are common in the Adirondack Mountains.

In addition to watershed characteristics, the level of atmospheric Hg deposition is an important influence on Hg concentration in fish. For example, Hrabik and Watras (2002) found an approximately 30% reduction in fish Hg concentrations from 1994 to 2000 in a Wisconsin seepage lake in response to decreased atmospheric Hg deposition of about 10% per year from 1995 to 1999 (Watras et al.

2000). However, seepage lakes such as the lake Watras et al. studied have a minimal contributing area and might be expected to respond more to decreased atmospheric Hg loading (Harris et al. 2007) than drainage lakes that typically receive much of their Hg input from the watershed.

Mercury transformation and transport in the terrestrial environment exert strong controls on Hg cycling in fresh waters. Gabriel and Williamson (2004) reviewed the major biogeochemical reactions involved in terrestrial Hg cycling. Key processes include Hg ligand formation, adsorption/desorption, and the reduction and subsequent volatilization of elemental Hg. Ligand formation differs significantly depending on oxidation state and the presence of organic matter. Adsorption of Hg to soil, sediment, and organic material is largely governed by pH and the identity and quantity of dissolved ions. For example, Hg adsorption decreases with increasing H^+ and/or Cl^- concentration (Gabriel and Williamson 2004).

In 2005 and 2006, Dittman and Driscoll (2009) resurveyed a group of 25 Adirondack lakes that had been surveyed in 1992 to 1993 in order to determine intervening changes in lake chemistry and Hg content of yellow perch. The concentration of Hg in fish increased in 6 lakes, decreased in 12 lakes, and stayed the same in 7 lakes, and these differences appeared to be related to watershed area, elevation, changes in pH, and changes in fish body condition. Dittman and Driscoll hypothesized that as lake acidity is reduced in response to reduced levels of acidic deposition, the lakes increase in productivity and/or the water chemistry becomes less stressful. In response, fish exhibit growth dilution of tissue contaminants.

Yu et al. (2011) studied food-web aspects of Hg cycling in 44 Adirondack lakes. Concentrations of Hg in fish and loons exceeded health criteria in many of the lakes. Bioaccumulation factors for MeHg increased from crayfish (mean $\log_{10}$ BAF = 5.7) to zooplankton (5.9) to prey fish (6.2) to larger fish (6.3) to common loon (7.2).

The annual fluctuation of the water level in a lake is another factor that can have a significant influence on Hg bioaccumulation in fish. This provides a potential management opportunity to control Hg bioaccumulation in reservoirs and other lakes that have stage control structures. Sorensen et al. (2005) reported results of a monitoring study of Hg concentration in YOY yellow perch and water-level fluctuations in 14 lakes in northeastern Minnesota. Over a three-year period of record across all lakes, mean fish Hg concentrations varied in each lake by nearly a factor of 2, on average. For the 12-year monitoring period at Sand Point Lake, which illustrated rather extreme variability, values ranged from 38 ng/g (ww) in 1998 to 200 ng/g (ww) in 2001. Annual water-level fluctuations were correlated with Hg concentrations in YOY perch.

5.2.5. Fish-Eating Birds and Mammals

Mercury in fish can affect the health of wildlife that consume large quantities of fish. Evers (2005) reported the results of a compilation of data on Hg effects on wildlife across the northeastern United States. Data were compiled on Hg levels in five general indicator types of Hg bioaccumulation: fish (multiple species), common loon, bald eagle, mink, and river otter. Biological hotspots of high Hg bioaccumulation were identified that covered relatively large areas, contained two or more indicator types having Hg concentration consistently above thresholds for adverse impacts (Table 5.2), had extensive data on Hg in biota, and exhibited large deviations in Hg levels in biota compared with the surrounding landscape. Nine biological hotspots were identified between New York and Nova Scotia, seven of which were not attributable to a single large Hg point source. One biological Hg hotspot, the western Adirondack Mountains, was identified in New York. This hotspot exhibited high Hg levels in fish, common loon, river otter, and mink. Evers offered a variety of reasons to help explain the occurrence of these biological hotspots, including local sources of Hg emissions, hydrological conditions in reservoirs, and (especially in the case of the western Adirondacks) water chemistry conducive to Hg methylation.

TABLE 5.2. Threshold mercury (Hg) levels for identification of biological Hg hotspots in the northeastern United States

Concentration	Medium	Description	Reference
0.16	Fish whole body	Poses risk at population level for fish-eating birds	Evers et al. 2004
3.0	Adult loon blood	Effects on physiology, behavior, reproduction, and survival	Evers et al. 2004
1.3	Loon egg	Effects on physiology, behavior, reproduction, and survival	Burger and Gochfeld 1997 Evers et al. 2003 Evers 2005
0.7	Young bald eagle blood	Level documented at locations where common loon blood Hg levels exceed 3.0 ppm	Evers 2005
20	Mink and otter fur	Sublethal toxicity	Mierle et al. 2000

Source: Evers 2005.
Note: Values expressed as parts per million Hg based on wet weight of medium.

Rolfhus et al. (2011) synthesized the results of 10 studies on trophic transfer efficiency of MeHg in the pelagic food webs of lakes in the western Great Lakes region. The largest biomagnification occurred at the base of the food web, between water and suspended particles. The similarity in the efficiency of trophic transfer suggested that the aqueous supply of MeHg in the water largely controls bioaccumulation in pelagic food webs in this region.

5.2.5.1.   *Fish-Eating Birds*

Fish-eating birds can be adversely affected by both acidification and Hg methylation. Acidification effects on birds may be indirect, caused by changes in the quantity and quality of food. Other potential effects of surface-water acidification and/or Hg methylation might include delayed egg laying, lighter/thinner egg shells, behavioral changes, and reduced chick growth (Tyler and Ormerod 1992). Reproduction is especially sensitive to chronic low-level MeHg exposure (Wolfe et al. 1998). Reduced clutch size, increased number of eggs laid outside the nest, eggshell thinning, and increased embryo mortality have all been shown to be associated with exposure of piscivorous birds to MeHg (Wolfe et al. 1998).

Methylmercury causes damage to the central nervous system of birds and other vertebrates (Scheuhammer 1991; Clarkson 1992). Low-level dietary Hg exposures that cause no observable effects on adult birds can nevertheless impair fertility, hatchling survival, and reproductive success (Scheuhammer 1991). In reproducing female fish, birds, and piscivorous mammals, MeHg passes directly to developing eggs or embryos (Evers et al. 2003; Hammerschmidt and Sandheinrich 2005; Heinz et al. 2010). Early life stages are more sensitive than adults to the adverse effects of MeHg (Evers et al. 2003; Wiener et al. 2003; Scheuhammer et al. 2007). In loons, bald eagles, and (presumably) other piscivorous birds, Hg poisoning can lead to brain lesions, reduced reproductive success, increased chick mortality, spinal cord collapse, and neuromuscular problems.

Decreased diversity and quantity of prey appear to affect the feeding behavior of nesting common loons on low-pH lakes in the Adirondacks (Parker 1988). Other fish-eating bird species may be similarly impacted. Scarcity of prey resources, decreased food quality, and/or elevated concentrations of MeHg in lake water might negatively impact foraging, breeding, and reproduction for such species as the common loon, the common merganser (*Mergus merganser*), the belted kingfisher (*Ceryle alcyon*), and the osprey (*Pandion haliatus*; Table 5.2).

The New York State Department of Environmental Conservation conducted population surveys of loons breeding in the Adirondack Park in the 1970s and 1980s, and the Adirondack Cooperative Loon Program initiated an annual loon survey beginning in 2001. The loon population in the park nearly doubled between the 1980s and 2001–2005 (Schoch 2006). The extent to which this increase in the loon population might be related to recovery from lake acidification is not known.

Recovery of fish-eating bird populations from acidification has not been clearly documented for lakes in the Adirondacks that are showing signs of pH and ANC recovery in response to recent S and N emissions controls. However, improvements have been shown in the Sudbury region of Ontario, where a statistically significant increase in fish-eating birds has been observed since the mid-1980s (McNicol 2002). This recovery was attributed to an increase in prey resources for fish-eating birds as a result of stricter S emissions controls in the United States and Canada.

Logistic regression modeling with measured pH and species occurrence data for acid-sensitive lakes in the Algoma region of Ontario found the occurrence of fish, common loons, and common mergansers to be positively related to lake water pH (McNicol 2002). Model predictions of common loon and merganser recovery were developed using the Waterfowl Acidification Response Modeling System. The number of lakes projected to be suitable for supporting breeding pairs and broods of common loons and common mergansers increased with simulated lake pH and assumed emissions controls. Improvements to fish-eating bird habitat were predicted to occur under hypothetical S emissions reductions of 50% and 75% for lakes that had pH below 6.5 (McNicol 2002).

The common loon is a well-studied example of Hg bioaccumulation in wildlife. However, this species is relatively insensitive (LC 50 [the concentration of a contaminant that is lethal to half of the exposed individuals] of MeHg injected into eggs > 1 ppm [ww]) to adverse effects of Hg compared with other bird species. For example, American kestrel (*Falco sparverius*), white ibis (*Eudocimus albus*), snowy egret (*Egretta thula*), osprey, and tricolored heron (*Egretta tricolor*) are all more sensitive to Hg toxicity (LC 50 of MeHg injected into eggs < 0.25 ppm [ww]), according to studies by Heinz et al. (2009) and Kenow et al. (2011).

Accumulation of MeHg in fish-eating birds in the northeastern United States is associated with lake acidification (Evers et al. 2007). Acidic deposition contributes to Hg toxicity in fish-eating birds in part because SO_4^{2-} addition to wetland environments acidifies downstream waters and stimulates the production of MeHg (Jeremiason et al. 2006). Methylation is a key process that regulates the effects of Hg on aquatic biota. Methylmercury is neurotoxic and bioavailable and accumulates in top predators to levels of concern for both human health and the environment (Table 5.3; Evers et al. 2007). Kramar et al. (2005) found that the extent of wetland in close proximity (less than 150 m) to loon territory was positively correlated with Hg concentration in loon blood.

The lowest observed adverse effect level benchmark quantifies potential injury to wildlife from Hg exposure. The lowest observed adverse effect level for the common loon has been reported to be 3.0 μg/g of Hg in adult loon blood (Evers et al. 2007, 2008), a level associated with reduced fledgling success (Burgess and Meyer 2008). Loon Hg exposure data can include either sex and different ages, locations, and time periods and can be derived from analysis of blood, tissue, or eggs. It has thus been difficult to standardize the data regarding MeHg exposure. Evers et al.

TABLE 5.3. Summary statistics of biological data layers for mercury (Hg) concentrations in fish and wildlife (μg per g) in the northeastern United States and southeastern Canada

Category/ species	Sample size	Hg Concentration Mean ± standard deviation	Range	Hg level of concern (tissue type)	Percentage of samples with concentration > level of concern
Human health					
Yellow perch[a]	4,089	0.39 ± 0.49	< 0.05–5.24	0.30 (fillet)	50
Largemouth bass[b]	934	0.54 ± 0.35	< 0.05–2.66	0.30 (fillet)	75
Ecological health					
Brook trout	319	0.31 ±0.28	< 0.05–2.07	0.16 (whole fish)	75
Yellow perch[c]	(841)[d]	0.23 ± 0.35	< 0.05–3.18	0.16 (whole fish)	48
Common loon[e]	1,546	1.74 ± 1.20	0.11–14.20	3.0 (blood)	11
Bald eagle	217	0.52 ± 0.20	0.08–1.27	1.0 (blood)	6
Mink	126	19.50 ± 12.1	2.80–68.50	30.0 (fur)	11
River otter	80	20.20 ± 9.30	1.14–37.80	30.0 (fur)	15

Source: Evers et al. 2007, by permission of the American Institute of Biological Sciences.
Note: All data are in wet weight except for fur, which is on a fresh-weight basis.
[a] Fillet Hg in yellow perch is based on individuals with a standardized length of 20 cm.
[b] Fillet Hg in largemouth bass is based on individuals with a standardized length of 36 cm.
[c] Whole-fish Hg in yellow perch is based on individuals with a standardized length of 13 cm. Whole-fish Hg for yellow perch was converted to fillet Hg.
[d] The sample population of 841 yellow perch examined for whole-fish Hg is included with the 4,089 fillets (i.e., the total number of all biotic data layers does not double-count yellow perch).
[e] Egg Hg for the common loon was converted to the adult blood equivalent.

(2011b) developed linkages among loon Hg measurements in eggs, blood, and fish prey in the Great Lakes region, which they normalized into standard loon tissue units. Use of a standard unit of measure that combines multiple tissues facilitates examination of spatial gradients in pollution effects. Based on analysis of over 8,000 male loon units, seven biological Hg hotspots were identified in the Great Lakes region. The average male loon unit concentration across the region was 1.8 µg/g; 82% were above 1 µg/g and 9.8% were above the lowest observed adverse effect level of 3 µg/g. Evers et al. explored Hg data for four focal regions where these biological hotspots were best understood. The focal area with the highest average male loon unit was northern New York (2.38 µg/g, approaching the lowest observed adverse effect level of 3 µg/L).

Male loon units tended to have higher Hg concentrations than female loon units. This is believed to be because male loons are typically 21% larger than female loons (Evers et al. 2010) and therefore eat larger fish (Barr 1996), which tend to have higher Hg concentrations (Sandheinrich and Wiener 2011).

The biological Hg hotspot in northern New York Evers et al. (2011b) reported contains largely mixed deciduous and coniferous forest (73%), scrub-shrub wetlands, sphagnum bogs, and lakes with low pH and ANC (Yu et al. 2011). Over 20% of the loon population in this region has been estimated to be at potential risk of adverse effects of Hg on reproduction (Evers et al. 2011b). There may be adverse effects on loon populations at existing levels of exposure (Evers et al. 2004; Burgess and Meyer 2008; Evers et al. 2008).

Several piscivorous bird and mammal species have been suggested as biomonitors of Hg bioaccumulation in the northeastern United States (Wolfe et al. 2007). High Hg concentrations have been associated with behavioral, physiological, and reproductive effects on common loons and bald eagles (Burgess and Meyer 2008; Evers et al. 2008). Evers et al. (2005) compiled a database of over 4,700 records of Hg levels in birds in the northeastern United States and eastern Canada, using the belted kingfisher and bald eagle as indicators. They reported increased Hg bioavailability from marine to estuarine to riverine systems, but bioavailability was highest in lakes. Differences in Hg body burden among species were most strongly correlated with trophic position and exposure to MeHg (Evers et al. 2005).

Bald eagles in the Great Lakes region have accumulated Hg levels that suggest subclinical neurological damage. An estimated 14% to 27% of the eagles Rutkiewicz et al. (2011) and Zillioux et al. (1993) studied had Hg tissue concentrations above the proposed risk threshold for liver toxicity (16.7 ppm). Bald eagles in the United States have been found to contain substantial amounts of Hg in feathers, eggs, livers, and brains (cf. Wood et al. 1996; Bechard et al. 2007; Scheuhammer et al. 2008). Mercury toxicity can affect survival and reproduction of these birds at both the individual and population levels.

In recent years, high Hg concentrations have been documented in additional species of wildlife across the Great Lakes region and the northeastern United

States. Estimated effects levels have also been reduced, suggesting sublethal effects on wildlife at whole-fish concentrations of only 0.2 to 0.3 ppm (Beckvar et al. 2005; Dillon et al. 2010; Sandheinrich and Wiener 2011).

The common loon is a good Hg bioaccumulation indicator of risk to piscivorous birds (Evers et al. 2008; Evers et al. 2011b). It is also listed as a species of special concern in New York. Because loons feed almost exclusively on fish and crayfish and are relatively long lived, they can bioaccumulate substantial Hg (Evers et al. 2011a). Even though common loons are considered less sensitive to adverse effects from Hg exposure than some other piscivorous birds, they do concentrate Hg in their blood to levels that can be high enough to impair reproduction (Burgess and Meyer 2008; Evers et al. 2011b).

Kejimkujik National Park, Nova Scotia, Canada, provides an excellent example of the importance of receptor-site condition in determining spatial patterns of biological response to Hg. Evers (2005) reported that 92% of adult loons analyzed from this park had blood Hg levels above 4 ppm, and impacts on loon reproduction have been observed since the early 1990s (Kerekes et al. 1994; Burgess et al. 1998). However, estimated total atmospheric Hg deposition at this park is relatively low compared with other locations in southeastern Canada. Watershed and lake chemistry conditions are ideal for Hg methylation (high dissolved organic carbon, low pH), and this explains the high Hg levels observed in wildlife.

Schoch et al. (2011) assessed Hg exposure and risk to the common loon in the Adirondack Park. Mercury levels were measured in water, lake sediment, zooplankton, crayfish, fish loon prey upon, and loon tissue (blood, feathers, eggs) collected from 44 lakes. Concentrations of Hg in the food web increased from lower to higher trophic levels. Concentrations of Hg in loon blood were strongly correlated with fish Hg. Lake acidity was correlated with Hg in both fish and loons. The birds were judged to be at high risk of reproductive and behavioral impacts based on blood Hg levels (in 21% of males and 8% of females) and feather Hg levels (in 37% of males and 7% of females). Loons in the highest Hg exposure category showed large reductions in number of chicks fledged per year (32% for females; 56% for males) compared with birds in the lowest exposure category. Loon productivity was never high at locations where Hg exposure was high. Schoch et al. (2011) concluded that Hg exposure causes a long-term impact on the population growth and size of the segment of the loon population that breeds on acidic lakes in the Adirondack Park.

5.2.5.2. *Fish-Eating Mammals*

Mink is an appropriate mammalian indicator species of toxic contaminant exposure because this species is widely distributed and abundant and occupies a high trophic position (Basu et al. 2007a; Martin et al. 2011). Tissue samples are available from trappers, and mink is thus a good candidate for biomonitoring (Mason and Wren 2001). Symptoms of Hg toxicity in mink include decreased coordination, loss of weight, and splaying of hind legs (Wobeser et al. 1976; Wren et al. 1987). Total measured Hg concentrations in mink in the Great Lakes region have been high enough to suggest subclinical effects (Basu et al. 2007b). In one survey, Hg concentrations in mink were highest in wetlands along impounded rivers that had fluctuating water levels and that were downstream of large historical Hg point sources (Hamilton et al. 2011). River otter also provide a good link to metal contamination of the food web, in part because they are a relatively sedentary species that bioaccumulates pollutants from small areas.

5.2.6. Other Life Forms

The snapping turtle (*Chelydra serpentina*) has been suggested as a good indicator of Hg contamination because of its omnivorous diet and long life span (Gibbs et al. 2007). It efficiently accumulates toxics, including Hg (Golet and Haines 2001; Bergeron et al. 2007). Turnquist et al. (2011) evaluated Hg concentration in 58 snapping turtles sampled at 10 lakes and wetlands across New York. Total Hg concentration in muscle tissue was significantly correlated with SO_4^{2-} concentration in lakes and maximum watershed elevation. Total Hg concentration in the shell was significantly correlated with lake ANC, maximum watershed elevation, percent open water, lake-to-watershed area ratio, and atmospheric Hg deposition (Turnquist et al. 2011).

Amphibians are considered good indicators of environmental contaminants, and many are sensitive to adverse impacts attributed to water acidification or atmospheric deposition of pesticides. Their sensitivity to Hg deposition in New York is not known.

5.2.7. Community Metrics

Ecological effects of surface-water acidification and eutrophication can occur at multiple levels of biological organization, including (1) the individual level; (2) the population level, which is made up of many individuals of the same species; (3) the biological community level, which is made up of many species (Billings 1978); and (4) the ecosystem level. Metrics have been developed to describe the effects of chemical stressors at each of these levels of organization. The most common involve assessment at the community level, using metrics that reflect species richness or community structure.

Acidification alters aquatic species composition in fresh waters. Many fish and other aquatic species are sensitive to acidification and cannot survive, compete, or reproduce in acidic or low-ANC waters. In response to changes in surface-water acidity, acid-sensitive species are typically replaced by more acid-tolerant species, resulting in changes in community composition. More species are affected at lower pH and ANC. A consistent pattern has been observed of lower community richness and diversity with acidification (U.S. EPA 2008).

Aquatic biological assemblages can be analyzed by calculating metrics from the list of the species, genera, or families identified and their abundances. For example, the number of different mayfly genera in the sample can be tallied and this number becomes the mayfly genus richness metric. Richness metrics can be calculated for any defined taxonomic group (e.g., mayflies, rotifers, insects). Richness can also be calculated for other autecological attributes such as functional feeding groups (shredder richness), habitat preference (swimmer richness), or tolerance to various pollutants. In addition, the same type of metrics can be calculated based on percent of individuals in the sample (e.g., percent mayfly individuals or percent shredder individuals). There are also species diversity metrics based on equations that express diversity as a combination of both overall sample richness (number of different taxa) and evenness (distribution of number of individuals across taxa). The most robust measures of biological condition typically require modeling or combining various metrics into one overall multi-metric index.

5.2.7.1. *Taxonomic Richness*

One commonly used biological community metric for quantifying the biological effects of an environmental stress such as acidification or eutrophication is taxonomic richness. The richness metric can be applied at multiple taxonomic levels. For example, the number of fish species found in a stream or lake is used as an index of acidification (see Bulger et al. 1999). Effects of acidification on zooplankton in lakes can be evaluated on the basis of richness of crustacean or rotifer zooplankton (Sullivan et al. 2006a). Acidification effects on benthic aquatic insects in streams can be evaluated based on the number of families or genera of mayflies or caddisflies (Sullivan et al. 2003; Cosby et al. 2006). During the process of acidification, the more acid-sensitive species of fish, crustaceans, rotifers, insects, and other life forms are eliminated. More species are lost with increased acidification (Nierzwicki-Bauer et al. 2010). As a consequence, acidification causes a decline in the total number of taxa in a stream or lake. The pattern and rate of loss of taxonomic richness can vary from watershed to watershed.

Taxonomic richness is often positively correlated with pH and ANC in surface water (Rago and Wiener 1986; Kretser et al. 1989). This is attributable to the elimination of acid-sensitive taxa during acidification (Schindler et al. 1985). Highly mobile organisms can sometimes move to refugia that have more favorable water chemistry in order to escape toxic conditions during acidic episodes. In addition, some, but relatively few, species are favored by increased acidity. However, the overall species richness typically decreases with decreasing ANC and pH.

Decreases in taxonomic richness in response to water acidification have been well documented for all major trophic groups of aquatic organisms (Baker et al. 1990a; Nierzwicki-Bauer et al. 2010). Baker et al. (1990a) summarized the results of 10 studies that documented this phenomenon, drawing on sample sizes that ranged from 12 to nearly 3,000 lakes and streams per study.

Lake or stream size can be a significant confounding factor in interpreting taxonomic richness data. Larger lakes and streams in larger watersheds typically contain more species than smaller lakes or streams in smaller watersheds, irrespective of acid-base chemistry. This pattern may be due to a decrease in the number of available niches as stream or lake size decreases. However, when adjusted for lake size, lakes with pH levels of less than approximately 6.0 contain significantly fewer species than lakes with pH levels greater than 6.0 (Figure 5.1; Harvey and Lee 1982; Frenette et al. 1986; Rago and Wiener 1986; Schofield and Driscoll 1987; Matuszek and Beggs 1988; U.S. Environmental Protection Agency 2008).

Fish species richness is a good indicator of acidification response for Adirondack lakes. Lakes with pH below about 5.0 or ANC below about 0 µeq/L typically do not support any fish (U.S. EPA 2008). Lakes that have pH above about 6.5 and ANC above about 50 to 100 µeq/L generally support large (but variable) numbers of species. At ANC between about 0 and 50 to 100 µeq/L, ANC is positively correlated with number of fish species (Bulger et al. 1999; Sullivan et al. 2006a).

The taxonomic richness of zooplankton varies with ANC in Adirondack lakes (Table 5.4; Sullivan et al. 2006a; Nierzwicki-Bauer et al. 2010). The taxonomic richness of crustaceans, rotifers, and total zooplankton all increased with increasing ANC. Adirondack lake water ANC explained nearly half of the variation in total zooplankton and crustacean taxonomic richness, but it explained only about one-third of the variation in rotifer richness. These results (Table 5.4) provided the basis for estimating changes in zooplankton richness in response to past or future changes in lake-water ANC. Model simulations by Sullivan et al. (2006a) suggested that some Adirondack lakes had lost up to four species of zooplankton in response to acidic deposition. Acidification has also decreased the complexity of the food web by decreasing the number of trophic links in Adirondack lakes (Havens and Carlson 1998).

Acidification causes decreased richness and diversity of phytoplankton communities in lakes and changes in the composition of the dominant taxa. These effects are most prevalent in the pH range of 5.0 to 6.0 (Baker et al. 1990a; Nierzwicki-Bauer et al. 2010). In the Killarney Park area of Ontario, lakes that were previously low in pH (5.0 to 5.5) and that had increased to above pH 6 in response to decreased acidic deposition shifted toward phytoplankton assemblages that are typical of circumneutral environments (Findlay 2003).

5.2.7.2. *Indices of Biotic Integrity*

Stream pH in the acid-sensitive upper Neversink River basin in the Catskill Mountains increased at several sites by about 0.01 units/yr from 1987 to 2003 (Burns et al. 2008a). To determine possible changes in stream biota, Burns et al. (2008a) sampled macroinvertebrates, fish, and periphytic diatoms. Statistical data comparisons and the AcidBAP derived from invertebrate data showed no significant differences between conditions in 1987 and those in 2003. These results

TABLE 5.4. Observed relationships between zooplankton species richness (R) and lake-water ANC

Taxonomic group	Equation	r^2	p
Total zooplankton	R = 15.65 + 0.089ANC	0.46	0.001
Crustaceans	R = 6.35 + 0.028ANC	0.47	0.001
Rotifers	R = 9.04 + 0.053ANC	0.30	0.001

Source: Sullivan et al. 2006a.

suggested that recovery of invertebrates in the most acidified parts of the upper Neversink River basin had not occurred to any substantial degree.

A multi-metric index such as the Index of Biotic Integrity (Moss et al. 1987) can be used for quantifying the biological condition of a whole community. It is developed by selecting the best metrics that quantify condition over a suite of different aspects of biotic integrity and then summing the individual metric scores. Interpretations of metric values at sampling sites are based on values observed at least-disturbed reference sites in similar settings. The predictive modeling approach often uses reference sites to assemble lists of taxa that appear to be indicative of least-disturbed reference condition (the expected, or "E," list). Taxa lists developed in the field from a specific study site constitute the observed, or "O," list. The proportion of the expected taxa found in the observed list (O/E ratio) is a measure of the proportion of the taxa expected to be at an undisturbed site that are actually present at the site. An O/E ratio of 1 suggests a high-quality site at which all expected taxa are present. An O/E ratio of < 0.5 means that less than half of the expected reference taxa are present at the study site. The E list can be developed for each study site by statistical modeling (such as cluster analysis and discriminate function analysis) of reference-site data to take into account natural differences in expected taxa distributions.

The index of biotic integrity has been used extensively in streams to characterize the condition of fish, macroinvertebrates, and periphytons. A number of different methods are used to calculate indices of biotic integrity, but they all follow a similar process. First, the metrics that best reflect condition are selected from the set of candidate metrics. Indices of biotic integrity typically are composed of 5 to 15 different metrics. Metric values are then scored to a consistent scale (e.g., 0–10 points) and summed to calculate the one overall for the index. Alternatively, the metric scores can be divided into classes. A wide variety of indices of biotic integrity have been developed for different stream types and regions around the world. For assessing stream benthos, the macroinvertebrate Index of Biotic Integrity developed by EPA for the National Wadeable Streams Assessment was designed for application nationwide (Stoddard et al. 2008). However, the Wadeable Streams Assessment Index of Biotic Integrity is formulated somewhat differently for each of nine different benthic habitat ecoregions in the United States. Candidate metrics were divided into six different categories and the best-performing metric in each category was selected for inclusion in the regional Index of Biotic Integrity. The six metrics selected for the Wadeable Streams Assessment region that includes the Adirondack and Catskill Mountains are listed in Table 5.5. The six metrics values were each scored on a 0–10 scale and summed into a final Index of Biotic Integrity score (Stoddard et al. 2008). Much less work has been done to develop indices of biotic integrity for lake systems.

The basic data analysis approach for studying the effects of stressors on biological condition involves plotting biological metric scores or index of biotic integrity scores against water chemistry. This analysis provides useful information on the extent to which invertebrate biological assemblages are associated with differences in water acid-base chemistry. It is important to note, however, that index of biotic integrity scores are also influenced by such factors as climate, vegetative cover, and disturbance. This variability has a great influence on our ability to estimate biological status or to detect trends over time. Stemberger et al. (2001) attempted to quantify the various contributions to the variance in zooplankton status as

TABLE 5.5. List of metrics in each category used in the EPA's National Wadeable Stream Assessment for the northern Appalachians aggregate ecological region

Metric category	Individual metrics
Composition	% EPT[a] taxa
Diversity	% Individuals in top 5 taxa
Feeding	Scraper richness
Habit[b]	% Clinger taxa
Richness	EPT taxa richness
Tolerance	% PTV[c] 0-5.9 taxa

Source: Stoddard et al. 2008.
[a] EPT taxa include Ephemeroptera, Plecoptera, and Tricoptera
[b] Habit reflects the life strategy of the various taxa with respect to maintaining position in the stream (i.e., burrowing, clinging)
[c] PTV = pollution tolerance value

part of the EPA's EMAP sampling program in the northeastern United States. Variance in zooplankton indicators was attributed primarily to four components of variance:

1. Lake variance: lake-to-lake variability in zooplankton indicators in the study population, depending on such factors as lake size, depth, fish presence/absence, pH, thermal characteristics, and productivity (Dodson et al. 2000)
2. Year variance: variation from year to year across all lakes, due, for example, to unusually warm or wet weather
3. Lake by year interaction variance: independent year-to-year variation at each lake due to site-specific forcing factors, such as variation in nutrient inflows or stratification
4. Index variance: local spatial and temporal variance due, for example, to temporal changes in the index period, measurement error, or differences among crews or laboratories in application of the protocols (Stemberger et al. 2001)

Stemberger et al. (2001) found lake variance to be the largest component of overall variability for zooplankton status in the northeastern United States, followed by index variance. Efforts to reduce the magnitude of the factors that contribute most to zooplankton variance can maximize our ability to detect differences among sites or trends over time.

5.3.   EFFECTS ON COASTAL AQUATIC BIOTA

Excessive N supply can degrade estuarine and marine habitat quality. Adverse impacts include algal blooms, toxicity, hypoxia, anoxia, fish kills, and decreased biodiversity (Paerl 2002). Atmospheric deposition of N can contribute to such effects. These largely chemical effects are reflected in the health of the aquatic ecosystem.

Eutrophication is one of the most significant environmental stressors in coastal waters. All estuaries in New York are affected to some degree. The amount of nutrient enrichment that an estuary can tolerate before substantial biological damage occurs depends to some extent on the size and shape of the estuary, the amount of tidal flushing, and the size of the watershed that drains into the estuary.

Atmospheric deposition delivers substantial quantities of N to coastal and estuarine areas in New York. Other significant sources of N input to these waters include discharges from wastewater treatment plants, industrial discharges, and runoff from agricultural and developed land. The combined influence of atmospheric and non-atmospheric sources of N has contributed to eutrophication in Long Island Sound, which is reflected in seasonal hypoxia in estuary waters.

A range of estuarine phytoplankton growth responses can occur as a consequence of N enrichment. In addition to differential responses that are attributable to estuary morphology, hydrology, transparency, and circulation, biotic adaptations can also be important determinants of phytoplankton dynamics (Paerl 1997). Various species of phytoplankton differ in their requirements for major nutrients and trace metals, nutrient uptake characteristics, and behavioral mechanisms that allow access to needed nutrient supplies. Such mechanisms include nutrient storage capabilities and buoyancy regulation.

Large inputs of N to near-coastal marine waters contribute to O_2 depletion. Low dissolved oxygen has occurred across large portions of Long Island Sound annually for more than a decade. Dissolved O_2 concentrations below about 3 mg/L contribute to adverse impacts on many aquatic life forms. Such low dissolved oxygen values typically cover at least 50% of the Long Island Sound estuary area for a portion of the warm season (Bricker et al. 2007). Reduced O_2 most seriously affects the western half of the sound. The primary sources of N that contribute to the algal production that causes O_2 depletion in Long Island Sound include effluent from wastewater treatment plants; urban runoff from New York City, Long Island, and Connecticut; and atmospheric deposition.

A national assessment of estuary condition by the NOAA in 2007 rated Long Island Sound as high for eutrophic condition. Other New York estuaries were rated somewhat better: Great South Bay was rated as moderately high, Hudson River–Raritan Bay as moderate, and Gardiners Bay as low (see Bricker et al. 2007).

In coastal marine ecosystems, the algal nutrients N, P, and Si and perhaps Fe are commonly associated with phytoplankton growth. Interactions among the supplies of these nutrients can affect the species composition of phytoplankton (Riegman 1992; Paerl et al. 2001). The proportions of key nutrient concentrations largely determine primary productivity, trophic structure, and energy flow (Dortch and Whitledge 1992; Justic et al. 1995a; Justic et al. 1995b; Turner et al. 1998).

Atmospheric inputs of N to estuarine and marine waters can alter nutrient ratios and affect nutrient limitation. Nitrogen loading from the land to estuarine and marine waters during high runoff conditions can contribute to periods of P limitation and N and P co-limitation (Boynton et al. 1995); the water returns to N limitation during low-flow periods (Paerl 2002). High loadings of both N and P can contribute to Si limitation and associated changes in diatoms. A shift in the Si-to-N atomic ratio to less than 1 might alter the marine food web, with decreasing ratios of diatoms to zooplankton to higher-tropic-level organisms (Officer and Ryther 1980; Turner et al. 1998; Paerl et al. 2001).

There are many environmental effects of estuarine and marine eutrophication in addition to depletion of dissolved oxygen in bottom waters. These include increased algal blooms and reduction in sea grass habitats and fisheries (Valiela and Costa 1988; Valiela et al. 1990; Boynton et al. 1995; Paerl 1995; Howarth et al. 1996; Paerl 1997). Eutrophication of coastal ecosystems also alters the biodiversity and species composition of aquatic biota. However, few data document the long-term response of coastal ecosystems in New York to N loading.

5.3.1. Phytoplankton in Coastal Waters

Phytoplankton taxa differ in their preference for and response to different forms of N (Stolte et al. 1994; Riegman 1998). Variable phytoplankton response to N input, in turn, can influence the distribution and abundance of zooplankton and herbivorous animals at higher trophic levels. In estuarine and marine waters, in particular, major algal functional groups, including diatoms, dinoflagellates, cyanobacteria, and chlorophytes, may have differing abilities to utilize the different forms of added N (Paerl et al. 2002). These can include oxidized inorganic N, reduced inorganic N, and dissolved organic N. Thus, the form of N contributed by human activities to coastal marine ecosystems can influence eutrophication and its effects on phytoplankton. Because of recent controls on atmospheric emissions of oxidized N, the proportion of atmospheric N deposition that is in reduced form has increased throughout the eastern United States (U.S. EPA 2008). This trend is expected to continue in the near future under existing and anticipated future N emissions controls. Large diatom species generally dominate coastal waters with a relatively high NO_3^- supply (Stolte et al. 1994; Paerl et al. 2001). It appears that smaller diatom species can more effectively take up NH_4^+ and tend to predominate under relatively high loading of reduced N. Ongoing decreases in NO_3^- deposition and increases in NH_4^+ deposition might eventually lead to changes in estuarine and/or near-coastal marine phytoplankton species and size distributions. These could cause further effects on trophic structure and element cycling (Paerl et al. 2001). Changes in phytoplankton in response to changes in the input ratio of reduced to oxidized N might cause changes in the species composition of zooplankton, herbivorous fish, and higher trophic levels of aquatic biota (U.S. EPA 2008).

Contributions of N from atmospheric deposition to the estuary surface may affect estuaries in ways that are different from watershed sources that contribute N to river flow that subsequently enters the estuary. This is because the upper reaches of estuaries are often N-limited and act as efficient sinks for riverine N sources (Billen et al. 1985; Paerl 1997). Much of the N that enters the estuary via river flow under low to moderate discharge conditions is typically assimilated, sedimented, or lost to the atmosphere in denitrification in the upper estuary (Nixon 1986; Paerl 1995; Paerl 1997). In contrast, much of the N that is atmospherically deposited to the estuary surface enters coastal waters downstream of this zone of N removal. Thus, the combination of the amount of N entering the estuary from direct atmospheric deposition and river discharge and the timing and amount of hydrological inputs can have substantial effects on estuarine and coastal primary production (Mallin et al. 1993; Paerl 1997).

Poorly grazed, bloom-forming algae and cyanobacteria may lead to more O_2 consumption in bottom waters than is found in more readily grazed species that support production at higher trophic levels (Paerl et al. 2002). Under light-limited conditions that prevail in turbid estuaries such as the Hudson River estuary (Cloern 1987), algae may respond more to NH_4^+ input than to NO_3^- input because of differing energy requirements associated with the uptake of these two forms of inorganic N (Dortch 1990).

During the 1990s, measurements of water quality in the Hudson River estuary reported by Howarth et al. (2000b) suggested that the estuary was often eutrophic. However, high rates of gross primary production occurred only when freshwater discharge from the watershed was relatively low (< 200 m^3/sec). At discharge > 300 m^3/sec, estimated water residence times were typically less than one day (Howarth et al. 2000b), and phytoplankton were likely to be transported out of the estuary almost as quickly as they could grow, a process that limited the size of algal blooms (Cloern 1996). As the water residence times increase beyond a residence time of about one day under conditions of lower discharge, the likelihood of phytoplankton blooms causing eutrophication increases in this estuary.

Freshwater input may also affect primary production in the Hudson River estuary via effects on light penetration and thermal stratification (Howarth et al. 2000b). Stratification in a deep turbid estuary such as the Hudson River estuary limits the extent to which water mixes and transports phytoplankton to deeper levels where light is too low to support photosynthesis. Stratification maintains phytoplankton at upper water column locations where there is more light, thereby increasing primary production. This estuary generally exhibits higher stratification under conditions of low freshwater discharge (Howarth et al. 2000b).

5.3.2. Submerged Aquatic Vegetation

The adverse impacts of eutrophication on estuarine and coastal marine ecosystems include a decreased extent of submerged aquatic vegetation. Sea grasses are flowering plants that are submerged in coastal areas. The principal sea grass species in New York is eelgrass (*Zostera marina*). It is found in coastal waters up to about 8 m depth and grows in small isolated patches to extensive meadows of several hectares or larger. The distribution of this species is strongly related to water quality (especially nutrient supply), light availability, sedimentation, temperature, and salinity (Burkholder et al. 1992; Dennison 1993). Eelgrass can grow in deeper water if the water is clear, thereby supporting photosynthesis. High nutrient loading and the resulting algal blooms inhibit the growth and occurrence of eelgrass.

Sea grasses have a very large influence on the ecology of coastal ecosystems and provide substantial ecosystem services. They alter water flow, recycle nutrients, increase dissolved oxygen, reduce wave turbulence, stabilize sediments, and provide critical habitat and nursery areas for a host of ecologically and commercially important fish and shellfish species. Many marine species depend on sea grass for attachment, refugia, spawning, and nursery areas. In particular, this habitat is critical for bay scallop (*Argopecten irradians*) and hard clam (*Mercenaria mercenaria*), two important shellfish species in New York. Sea grass habitat in Long Island Sound has been reduced to about 1% of its historical coverage (NYS Seagrass Taskforce 2009). The remaining sea grass in Long Island Sound is almost exclusively confined to the area around Fishers Island in the northeastern portion of the sound.

There are multiple known threats to sea grasses in Long Island Sound and elsewhere in the northeastern United States. The largest threats in Long Island Sound include excess N and human development (NYS Seagrass Taskforce 2009). Other potential threats include climate change, increases in sea levels, storm surges, invasive species, boating, disease, dredging, and toxic chemicals.

Atmospheric N deposition contributes to the total N load that stresses sea grasses in estuaries in New York and throughout the eastern United States. For example, Driscoll et al. (2003b) reported a strong negative relationship between modeled N loading and measured eelgrass area in Waquoit Bay, Massachusetts, based on repeated measurements of eelgrass coverage from 1951 to 1992. An increased total N load from about 15 to 30 kg N/ha/yr, largely in response to urban development in the watershed, virtually eliminated eelgrass meadows in Waquoit Bay (Bowen and Valiela 2001). Similarly, Hauxwell et al. (2001) found that the growth, areal extent, and biomass of eelgrass habitat decreased sharply across estuaries in the northeastern United States at N loads greater than about 20 kg N/ha/yr.

5.3.3. Shellfish and Fish

Many estuarine taxa, including commercially important fish and shellfish, depend heavily on the presence of sea grasses. Thus, changes in sea grass in response to N enrichment can affect the food web throughout the estuary. The bay scallop is a commercially important species that depends heavily on eelgrass habitat. The annual harvest of bay scallop in Waquoit Bay decreased by three orders of magnitude during the period of eelgrass loss since the 1960s (Bowen and Valiela 2001). This observed decline in scallop harvest was correlated with increases in estuarine N load. Ryther and Dunstan (1971) also found a loss of shellfish (oysters and hard clams) in Moriches Bay, New York, in association with nutrient enrichment.

Not all species are negatively affected by eutrophication. Nutrient enrichment can cause increased growth and abundance of some consumers, including finfish and shellfish, in estuaries that receive large contributions of nutrients (Kirby and Miller 2005; Nixon and Buckley 2005; Carmichael et al. 2012). For example, inputs of N and/or P have been associated with an increased abundance and biomass of finfish in Cape Cod estuaries (Nixon and Buckley 2005; Tober et al. 2005). Alternatively, nutrient enrichment can contribute to the loss of some species (Cloern 2001; Gray et al. 2002).

The effects of nutrient enrichment on bivalves can be positive in response to increased food supply or negative in response to adverse effects on habitat (Carmichael et al. 2012). Nutrient enrichment can also affect different species in different ways. Carmichael et al. (2012) reviewed data from five studies on Cape Cod to describe the influence of eutrophication on bivalves. The net effect of nutrient enrichment depended on the balance between positive and negative effects. Oysters and other species with high feeding and assimilation rates and high tolerance of hypoxia tended to be more successful in eutrophied waters. The effects of phytoplankton shading on eelgrass reduced the numbers of bay scallops because the scallops depend heavily on eelgrass habitat (Hauxwell et al. 2001). Eutrophication-driven hypoxia is likely the most significant way that eutrophication negatively impacts bivalves (Carmichael et al. 2012) and other benthic species (Breitburg 2002). Hypoxia negates the potential positive effects associated with the increased availability of phytoplankton food.

Although it is clear that estuarine eutrophication can kill fish, there is no clear evidence demonstrating that fish kills or other severe ecological consequences of eutrophication have been caused solely or even mostly by atmospheric N inputs to estuarine or marine waters in New York or elsewhere in the United States. It is clear, however, that atmospheric N deposition has contributed to the N supply that has caused such effects.

CHAPTER 6

Historical Patterns
of Effects

Much can be learned from examination of historical patterns in acidification, eutrophication, and Hg methylation and subsequent recovery in response to recent reductions in levels of S, N, and Hg emissions. The three most important tools available to environmental scientists for evaluating past changes in conditions of resources include paleoecological methods, process-based model hindcasting, and the collection of long-term monitoring data. Each of these approaches is discussed in the sections that follow.

6.1. PALEOECOLOGICAL STUDIES

The preindustrial chemistry of Adirondack lake water has been reconstructed based on functions derived from relationships between ambient lake-water chemistry and algal (typically diatom or chrysophyte) remains in upper lake sediments. Predictive relationships developed from regional lake data sets have been applied to algal data collected from horizontal slices of individual lake sediment cores to infer past lake-water conditions (Charles et al. 1989; Husar et al. 1991).

Diatoms typically constitute an important component of the phytoplankton in lakes and are good indicators of environmental change associated with acidity, nutrient status, salinity, and climatic change (Stoermer and Smol 1999). There are thousands of freshwater diatom species, and many have narrow ecological tolerances. Inferences about the history of lake chemistry based on diatom or chrysophyte fossil remains preserved in lake sediments have been widely used in the Adirondacks for quantifying historical acid-base chemical change (Charles and Norton 1986; Charles et al. 1989; Sullivan et al. 1990). The Paleoecological Investigation of Recent Lakewater Acidification research programs (PIRLA-I and PIRLA-II) were conducted wholly or partly in the Adirondack Mountains. Results showed that (1) Adirondack lakes had not acidified as much since preindustrial times as had been widely believed prior to 1990; (2) many Adirondack lakes with pH less than about 6.0 had decreased in pH historically; (3) many of the lakes with relatively high pH (> 6) and ANC (> 50 μeq/L) had increased (rather than decreased) in pH and ANC since the nineteenth century; and (4) the average F-factor (ratio of $\Delta[SBC]$ to $\Delta[SO_4^{2-} + NO_3^-]$) for acid-sensitive Adirondack lakes was near 0.8 (Charles et al. 1990; Sullivan et al. 1990). These results changed scientific understanding about the amount of lake acidification that had occurred in response to acidic deposition and modified expectations about continued acidification of soil and chemical recovery of surface waters as acidic deposition levels decreased (Sullivan 2000).

Paleolimnological reconstructions of lake-water pH and ANC for a statistically selected group of Adirondack lakes suggested that about one-fourth to one-third of the Adirondack lakes that are larger than 4 ha had acidified since preindustrial times (Cumming et al. 1992). Low-ANC lakes of the southwestern Adirondacks have acidified the most since the nineteenth century, probably because they had low initial buffering capacity and received high rainfall and acidic deposition. Cumming et al. (1992) also estimated that 80% of the Adirondack lakes that had pH ≤ 5.2 during the 1980s had experienced relatively large decreases in pH and ANC since the previous century and that 30% to 45% of Adirondack lakes with pH between 5.2 and 6.0 had also acidified since preindustrial times.

Cumming et al. (1994) reported chrysophyte inferences of pH in the upper slices of 20 low-ANC Adirondack lake sediment cores in order to assess the timing of historical acidification. The study was focused on lakes that were presumed to be highly sensitive. Various lakes began to acidify at different times throughout the twentieth century. Lakes that were inferred to have acidified the most since about 1900 tended to be small and located at high elevations. They had lower inferred preindustrial pH than the study lakes as a whole and are located in the High Peaks Wilderness Area and in the southwestern portion of Adirondack Park. Husar et al. (1991) estimated that S deposition was about 4 kg S/ha/yr at the time these most sensitive lakes began to acidify. Most lakes that were inferred to have not acidified during the twentieth century had preindustrial pH levels near 6.0 and are located at lower elevations.

Similarly, the Hg that has accumulated in lake sediments provides an index of changes over time in Hg inputs to the lake and its watershed from atmospheric deposition and other sources. Lake sediment records throughout North America and Europe show that atmospheric Hg deposition has increased substantially since preindustrial times (Bookman et al. 2008). Drevnick et al. (2012) analyzed 91 sediment cores collected from relatively undisturbed inland lakes in the Great Lakes region. Inferred rates of Hg accumulation in those lakes increased about sevenfold from preindustrial times to a peak in about the 1980s. Since then, Hg inputs inferred from the sediment record have declined by about 20%, a decrease that is consistent with measured trends in Hg wet deposition (Drevnick et al. 2012).

Bookman et al. (2008) quantified Hg fluxes to four lakes located near Syracuse, New York. The pre-disturbance regional Hg flux (~ 3.0 $\mu g/m^2/yr$) was estimated from sediment data derived from a seepage lake (Glacial Lake) that had no appreciable terrestrial watershed. Inferred atmospheric Hg deposition to this lake peaked during the latter part of the twentieth century at levels that were 3 to 30 times the preindustrial levels.

Van Metre (2012) evaluated the importance of local Hg emissions sources through an analysis of the Hg content of lake sediment cores collected from 12 lakes throughout the United States that were either relatively close (< 50 km) to urban source areas, or remote from (> 150 km) such source areas. The latter group of sites included one in New York (Arbutus Lake in the Adirondack Mountains). Mercury fluxes and flux ratios (modern to background) in the near-urban lakes (68 $\mu g/m^2/yr$ and 9.8, respectively) exceeded those for the remote lakes by a considerable margin (14 $\mu g/m^2/yr$ and 3.5). The Hg fluxes were correlated with distance from the nearest urban population center. The researchers further concluded, based on comparison to data collected by wet deposition monitors, that dry deposition constituted an important source of Hg to lakes located near urban areas.

6.2.   WATERSHED MODEL HINDCAST STUDIES

Paleoecological techniques have also been used as a way of testing and improving watershed models of acid-base chemistry. For example, inclusion of an organic acid representation in the Model of Acidification of Groundwater in Catchments (MAGIC) had a substantial effect on model predictions of past surface-water pH, even in Adirondack lakes with only moderate dissolved organic carbon concentrations. MAGIC model hindcasts of preindustrial Adirondack lake-water pH showed poor agreement with diatom inferences of preindustrial pH when organic acids were not considered in the model simulation (Sullivan et al. 1996a). Hindcast simulations of preindustrial lake-water pH that included an organic acid representation developed by Driscoll et al. (1994) showed improved agreement with diatom pH inferences (Figure 6.1). The mean difference between MAGIC and diatom estimates of preindustrial pH was reduced from 0.6 pH units to 0.2 pH units when organic acids were included in the model, and the agreement for some individual lakes improved by up to a full pH unit (Sullivan et al. 1996a).

Sullivan et al. (2006a) simulated historical changes in the acid-base chemistry of 70 Adirondack lake watersheds. Study lakes included 44 that had been statistically selected to be representative of the approximately 1,320 low-ANC (≤ 200 µeq/L) lakes in the Adirondacks that were included in EPA's EMAP statistical frame. Hindcast simulations were constructed using both the MAGIC and PnET-BGC models.

MAGIC simulations were extrapolated to the regional EMAP Adirondack lake population. Maximum historical acidification in response to acidic deposition occurred by about 1980 or 1990 in most lakes. The median ANC of the population of lakes with ambient ANC ≤ 200 µeq/L was estimated to have declined from pre-

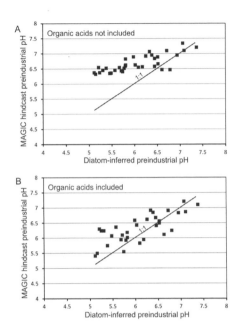

FIGURE 6.1. MAGIC model hindcast estimates of preindustrial pH versus diatom-inferred pH for 33 statistically selected Adirondack lakes: (A) without including organic acid representation in the MAGIC simulations; and (B) including a triprotic organic acid analog model in the MAGIC simulations. With kind permission from Springer Science+Business Media, Water, Air and Soil Pollution, Influence of organic acids on model projections of lake acidification, 91(1996), T. J. Sullivan et al., Figure 1. Copyright 1996 Kluwer Academic Publishers.

industrial times by 31 µeq/L (from 92 µeq/L to 61 µeq/L). By 1990, 10% of the population target lakes had decreased in ANC to below -16 µeq/L and 25% had ANC < 28 µeq/L. Simulated values in 2000 indicated that limited chemical recovery had occurred over the previous one to two decades. Most recovery estimates were about 3 to 5 µeq/L when compared with simulated values for the period 1980–1990.

The MAGIC model results indicated that none of the EMAP target lakes were chronically acidic (had ANC ≤ 0 µeq/L) under preindustrial conditions. However, by 1980 there were about 204 acidic Adirondack lakes. That number decreased by an estimated 14% between 1980 and 2000. MAGIC simulations suggested that there were no Adirondack lakes with ANC ≤ 20 µeq/L in 1850, but by 1990 there were 263 such lakes. An estimated 191 lakes had preindustrial ANC below 50 µeq/L; that number had increased threefold by 1990. As a result of limited chemical recovery during the 1990s, the estimated number of Adirondack lakes with ANC ≤ 50 μeq/L decreased to about 400 by the year 2000 (Sullivan et al. 2006a).

The PnET-BGC model simulation results for Adirondack lakes that Sullivan et al. (2006a) reported estimated that preindustrial lake-water SO_4^{2-} concentrations were much lower than current values. In 1850, simulated SO_4^{2-} concentrations in all of the study lakes were less than 25 μeq/L and the median value was about 15 μeq/L. By 1980, the median simulated SO_4^{2-} concentration had increased more than sixfold to about 100 μeq/L.

Results from PnET-BGC modeling in Zhai et al. (2008) suggested that the median Adirondack lake in the estimated 1,320 low-ANC (≤ 200 μeq/L) lakes in the EMAP population had preindustrial ANC near 80 μeq/L. An estimated 10% of the lake population had preindustrial ANC < 41 μeq/L. Estimates for the year 2000 suggested small increases in ANC during the 1990s for the lower-ANC lakes. None of the lakes in the Adirondack EMAP population that were modeled using PnET-BGC had preindustrial ANC below 20 μeq/L. PnET-BGC simulations indicated that by 1990, there were 289 lakes with ANC ≤ 20 μeq/L and 217 chronically acidic lakes. PnET-BGC estimated that 202 lakes in the EMAP population had preindustrial ANC below 50 μeq/L and that this number had increased 2.8 times by 1980.

6.3.   RECENT TRENDS IN MONITORING DATA

6.3.1.   Wet and Dry Deposition

The NADP/NTN has monitored wet deposition for about 30 years at several locations in New York (Figure 3.1) and elsewhere around the United States. This monitoring network continuously collects and analyzes samples of rain and snow.

In the 1970s and 1980s, watersheds in the Adirondack and Catskill Mountains received roughly twice as much S deposition as N deposition on a mass basis. Sulfur deposition started to decrease by a substantial amount in about the late 1980s and early 1990s; N deposition started to decrease about a decade later. Because of the more stringent emissions controls that the 1970 CAA and its amendments placed on S, the inputs of S are now lower than inputs of N at many locations in the eastern United States, including in the mountainous portions of New York (Plates 6 and 8; Figure 3.2).

6.3.2.   Soils

Relatively little information is available with which to quantify historical trends in soil acid-base chemistry in New York. Lawrence et al. (2013b) summarized the limited available data. In general, it appears that exchangeable base cations and percent base saturation of the upper mineral soil horizons in Adirondack watersheds have decreased substantially over the past century and have continued to decrease at some sites in more recent decades in response to lower (albeit still appreciable) levels of acidic deposition (Sullivan et al. 2006b; Warby et al. 2009). However, there is some evidence that some aspects of soil condition might be improving at some locations in the northeastern United States. Lawrence et al. (2012) re-sampled soils in six red spruce forest stands to determine if there were signs of improvement in soil acid-base chemistry between the initial soil sampling in 1992–1993 and the subsequent sampling in 2003–2004. The pH of the O_a horizon was lower at the time of the more recent sampling at the New York site but higher at four of five sites sampled in New England (and it was statistically higher [$p \le 0.01$] at three of those sites). Exchangeable Al in the O_a horizon decreased over time ($p \le 0.05$) by 20 to 40% at all sites except New York, which showed no change. Results for B-horizon samples were different; only the site in northeastern Maine (Kossuth) showed evidence of decreased B-horizon soil acidity, and this was limited to an observed decrease in exchangeable Al. None of the other sites, including the one New York site included in the study, showed any improvement in B-horizon soil chemistry.

6.3.3.   Surface Waters
6.3.3.1.   *Chemistry*

Surface-water chemistry monitoring in the EPA's Long Term Monitoring (LTM) project and the TIME project have documented significant changes in lake and stream chemistry in New York and elsewhere in the eastern United States in response to S and N emissions reductions. During the 1990s, surface-water SO_4^{2-} concentrations generally decreased. ANC increased modestly. Dissolved organic C increased, perhaps toward more natural pre-disturbance concentrations (Figures 6.2 and 6.3). Inorganic and potentially toxic Al_i concentrations appear to have decreased slightly in some of the more sensitive aquatic systems (U.S. EPA 2008).

The acid-base chemistry monitoring of lake water in New York occurs primarily in two programs: the LTM program (Ford et al. 1993; Stoddard et al. 1998) and the TIME program (Stoddard et al. 2003). Both programs are operated as cooperative efforts involving state and federal agencies and academic institutions.

The LTM program monitors a subset of the more acid-sensitive Adirondack lakes and Catskill streams and lakes and streams in other regions. Many sampling sites have relatively continuous water-chemistry monitoring dating from the early 1980s. Each water body is sampled 3 to 15 times per year. Data are used to characterize the response of the most sensitive waters to changing levels of acidic deposition. A small number of higher-ANC (> 100 µeq/L) sites are also sampled. Because of the long-term records at many sites, LTM trends provide better historical context than trends at TIME sites, where data are available only from the 1990s. Monitoring results from the LTM program have been widely published (Kahl et al. 1991; Driscoll and Van Dreason 1993; Kahl et al. 1993; Murdoch and Stoddard 1993; Stoddard and Kellogg 1993; Driscoll et al. 1995; Stoddard et al. 1998). Overall results, including for the Adirondacks and other regions, were summarized by Stoddard et al. (2003), Burns et al. (2008a), and Roy et al. (2013).

The available long-term water-quality monitoring record for one representative acid-sensitive stream site in the Catskill Mountains (Biscuit Brook) is depicted in

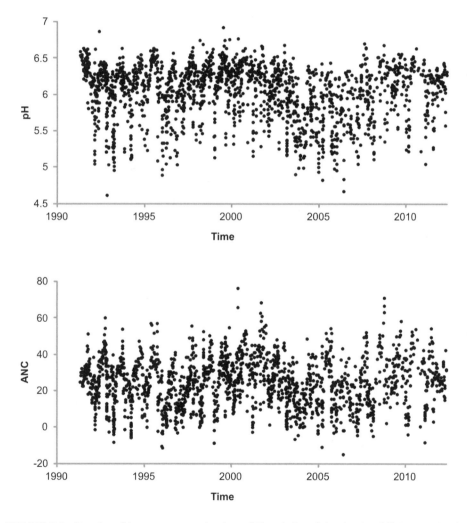

FIGURE 6.2. Results of long-term monitoring of Biscuit Brook in the Catskill Mountains for selected major surface-water chemical variables over the period of available data. One to several outliers were deleted from each plot. Values for ANC, SO_4^{2-}, NO_3^-, SBC, and Al_i, are expressed in µeq/L; values for dissolved organic carbon (DOC) are expressed in µmol/L. Data provided by D. Burns, U.S. Geological Survey.

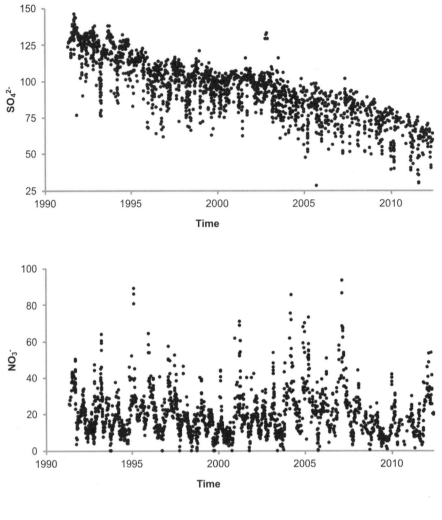

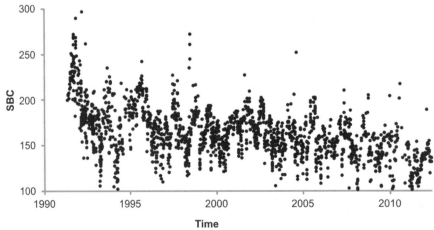

FIGURE 6.2—continued

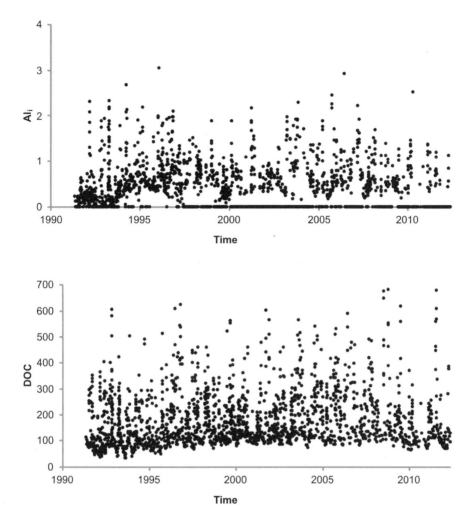

FIGURE 6.2—continued

Figure 6.2. Monitoring data at Biscuit Brook have exhibited substantial seasonal variability for all major parameters measured. The most pronounced pattern of change over time has been for SO_4^{2-}, which decreased by about 50% over the two-decade period of record. A rather pronounced decrease was also observed for SBC, especially during the early portion of the monitoring. No clear improvement in response to emissions controls was observed at Biscuit Brook for pH, ANC, NO_3^-, or Al_i, although there is some indication in the latter part of the period of record that ANC and pH may be increasing and that NO_3^- and Al_i concentrations may be decreasing. These data illustrate the fact that intra- and inter-annual variability is often high in comparison with the observed trends over time. This makes trend detection difficult.

Available long-term water-quality data for one representative lake (Big Moose Lake) in the Adirondack Mountains are shown in Figure 6.3. Monitoring data for

Big Moose Lake also revealed a large decrease in SO_4^{2-} concentration, with observations dating from 1982. Nitrate concentrations have declined since about 1990 (Figure 6.3). Decreases were also observed over time for SBC and Al_i in this lake. Concentrations of dissolved organic carbon and ANC increased, as did pH. The pattern of ANC recovery has been very pronounced, beginning in about 1990. Since then, the ANC of this lake has increased by about 20 µeq/L, and measurements below 0 µeq/L are now rarely observed. Big Moose Lake provides a clear and unambiguous example of recent chemical recovery from acidification. Many other Adirondack lakes show similar patterns.

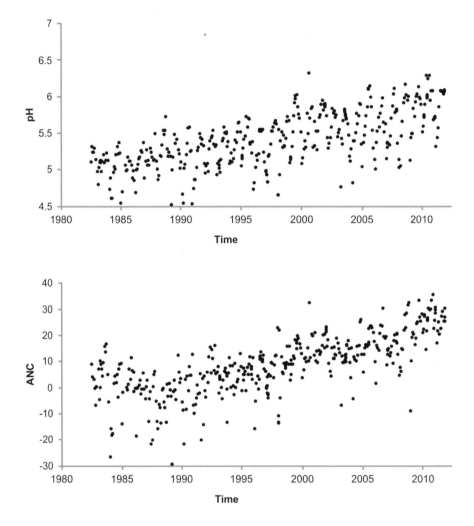

FIGURE 6.3. Results of long-term monitoring of Big Moose Lake in the Adirondack Mountains over the period of available data. One to several outliers were deleted from each plot. Values for ANC, SO_4^{2-}, NO_3^-, SBC, and Al_i, are expressed in µeq/L; values for dissolved organic carbon (DOC) are expressed in µmol/L. Data provided by the Adirondack Lakes Survey Corporation.

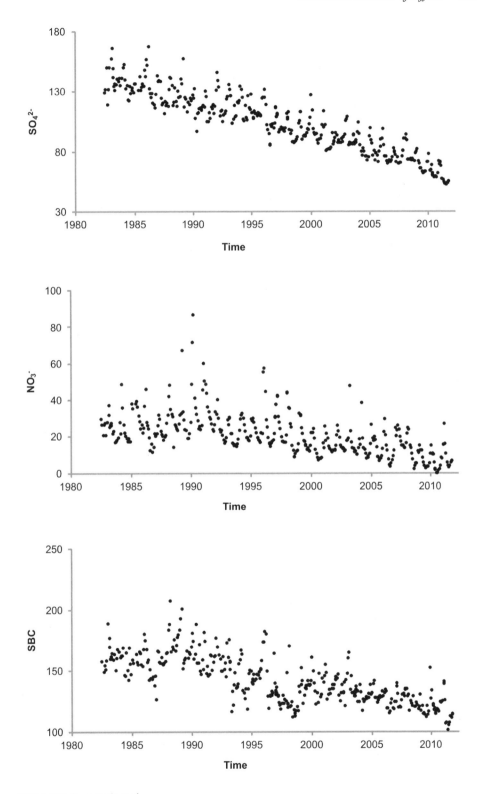

FIGURE 6.3—continued

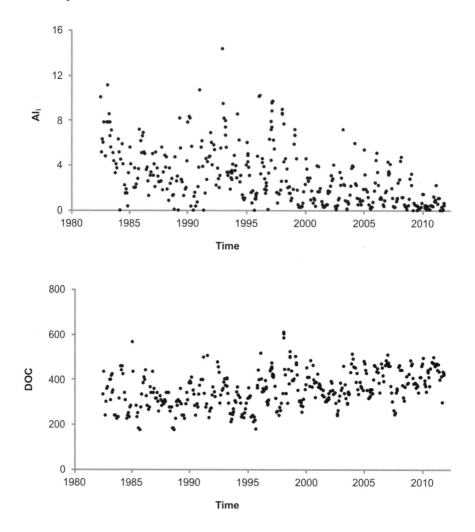

FIGURE 6.3—continued

Civerolo et al. (2011) compared lake-chemistry monitoring data measured by TIME and the Adirondack Long Term Monitoring Program over the period 1992–2008 for the six Adirondack lakes that are common to both monitoring programs. Although the TIME project provides the benefits of a statistical (probability) sampling frame, the Adirondack Long Term Monitoring Program provides higher temporal resolution (monthly, as opposed to annual) and it includes Al speciation for estimating the components of Al_i. TIME measured only total Al, some of which is organically complexed and nontoxic. Thus, the Adirondack Long Term Monitoring Program offers more biologically relevant data (Civerolo et al. 2011). Willys Lake, the lowest in ANC of the six Adirondack study lakes that are common to both the LTM program and the TIME program, has consistently shown $Al_i > 2$ µM over the full 17 years of monitoring.

Based on a rate of change in Gran ANC of +0.8 µeq/L/yr indicated by trend slopes for TIME and LTM data, Stoddard et al. (2003) estimated that approximately 8.1% (n = 149) of the population of Adirondack lakes remained chronically acidic in 2000. Their analyses suggested that more than one-third of the lakes in the Adirondacks that were chronically acidic in the early 1990s were not chronically acidic a decade later. Driscoll et al. (2007a) and Burns et al. (2011) reached similar conclusions.

Variability over time in surface-water and soil-solution chemistry and seasonal patterns in nutrient uptake by terrestrial and aquatic biota influence acidification processes and pathways. Because of complications associated with climate and other sources of variation, many years of data are required to document trends in surface-water chemistry. Developing an understanding of the causes of any documented trend is even more challenging.

A number of time series analyses have been conducted for Adirondack lakes since monitoring programs were initiated in 1982 (Roy et al. 2013). Each provides a snapshot of changes in lake chemistry for the various monitored lakes over a particular period of time. The first published time series analysis of the original 17 Adirondack Long Term Monitoring Program lakes found that most monitored lakes were showing decreases in SO_4^{2-} concentrations that were generally consistent with observed decreases in regional S emissions and deposition. Lake ANC did not show evidence of increase due to acidification recovery; rather, some lakes showed continued decreases in ANC that may have been caused, at least in part, by increased NO_3^- leaching (Driscoll and Van Dreason 1993). A subsequent trends analysis by Driscoll et al. (1995), this one for the period 1982–1994, documented continued declines in lake SO_4^{2-} concentrations at rates that were somewhat less than the observed decline in S deposition (Roy et al. 2013). Chemical recovery of pH and ANC had not occurred in a consistent fashion over that time period (Driscoll et al. 1995). Similar results were found for the period 1982–1997 (Driscoll et al. 1998).

The expanded Adirondack Long Term Monitoring Program data set for 48 lakes was first analyzed for trends in lake chemistry for the 1992 to 2000 time period (Driscoll et al. 2003a; Roy et al. 2013). All lakes showed decreases in SO_4^{2-} concentration. Some lakes also showed decreases in NO_3^- concentration. The ANC and pH increased in about half of the study lakes. Some lakes showed increased dissolved organic carbon and some showed decreased Al_i concentration. However, one-third of the monitored lakes still had Al_i > 2 µm in 2000, a level considered to be toxic to many species of aquatic biota. Decreases in Al_i occurred mainly in thin till drainage lakes. For those lakes that showed significant trends in ANC between 1992 and 2000, the extrapolated time frame of ANC recovery suggested variable periods of time to attain ANC = 50 µeq/L, ranging from about two decades or less for most lakes with ANC between about 20 and 47 µeq/L in 2000 to much longer periods of ANC recovery (> 20 years) in lakes with ANC < 20 µeq/L in the year 2000 (Roy et al. 2013). Decreases in dissolved organic carbon occurred in about one-third of the lakes (Driscoll et al. 2007a).

Roy et al. (2013) conducted seasonal Kendall tests in the Adirondack Long Term Monitoring Program using the methods of Raynal et al. (2004). Trend results are summarized in Table 6.1. Decreases in lake-water concentrations of SO_4^{2-} and NO_3^- were largely balanced by decreases in SBC and increases in ANC and dissolved organic carbon. Measured trends in lake base-cation surplus between 1994 and 2011 suggested increased base-cation surplus values over time in 88% of the monitored lakes. Only some of the lakes with ANC > 50 µeq/L and lakes that had been limed failed to show base-cation surplus recovery from 1994 to 2011 (Roy et al. 2013). The percent of lakes with base-cation surplus below 0 µeq/L decreased from about 60% in 1994 to 35% in 2011 (Lawrence et al. 2013a). Many lakes showed increases in base-cation surplus in the range of 20 to 40 µeq/L.

Roy (2013) reported water-chemistry trends for several Adirondack streams. The chemistry of three monitored streams (Buck Creek, Bald Mountain Brook, and Fly Pond Inlet) was evaluated for the period 1991–2001. Each stream showed significant increases in pH and ANC, but the chemical recoveries were variable. On average, the ANC of Buck Creek increased by about 10 µeq/L over the period

TABLE 6.1. Summary of trends for selected parameters in Adirondack lakes, 1992–2011

Parameter	Lakes increasing	Lakes decreasing	Range		Mean	Units
			Low	High		
SO_4^{2-}		48	-0.76	-4.09	-2.48	µeq/L/yr
NO_3^-		29	-0.03	-1.23	-0.42	µeq/L/yr
SBC		42	0.33	-3.06	-1.54	µeq/L/yr
ANC	38		2.01	0.39	0.93	µeq/L/yr
Lab pH	29		0.12	-0.76	0.02	units/yr
Al_i		41	0.0	-0.65	-0.11	µmol/L/yr
Al_m		32	-0.02	-0.64	-0.17	µmol/L/yr
DOC[a]	30		14.57	-9.37	4.07	µmol/L/yr

Source: Roy et al. 2013.
[a] Dissolved organic carbon.

of monitoring. The data for Buck Creek suggested that episodic decreases in ANC and pH continued to periodically reach values below 0 μeq/L and 5.0, respectively (Driscoll et al. 2001a). The ANC of Bald Mountain Brook increased over that time period by 31 μeq/L. The well-buffered Fly Pond Inlet showed an increase in ANC in 1997, with no subsequent change through 2001.

Sulfate. The concentrations of SO_4^{2-} in surface waters in the Adirondack and Catskill Mountains increased throughout much of the twentieth century (Husar et al. 1991) but have decreased substantially over the last two to four decades (Figure 6.3). Data collected in the LTM and TIME projects during the 1990s showed that most regions in the eastern United States, including the Adirondack and Catskill Mountains, experienced large declines in SO_4^{2-} concentrations in surface waters. Rates of change varied from about 1.5 to 3 μeq/L/yr (Figures 6.2 and 6.3).

This recent decrease in SO_4^{2-} concentrations has been caused by decreased emissions and deposition of S. Decreased concentrations of SO_4^{2-} in lakes and streams of about 50% have been commonly observed. Lake-water SO_4^{2-} concentrations have decreased steadily in Adirondack lakes at least since the early 1980s (Driscoll et al. 1995; Stoddard et al. 2003).

Stoddard et al. (2003) reported data on Adirondack lakes monitored in the LTM project and showed that the average rate of decrease in SO_4^{2-} concentrations from 1990 to 2000 was 2.26 μeq/L/yr. A decrease in SO_4^{2-} concentration that averaged 2.16 μeq/L/yr was also observed in 47 of 48 monitored Adirondack lakes from 1992 to 2004, and a similar decrease of 2.09 μeq/L/yr was observed in a subset of these lakes over the longer period from 1982 to 2004 (Driscoll et al. 2007a). This pattern has continued in more recent years.

Decreases in SO_4^{2-} concentration were generally somewhat more pronounced in precipitation than in lake and stream water in the Adirondack and northern Appalachian Mountains (Stoddard et al. 2003). This finding suggests that recovery of surface waters, at least with respect to SO_4^{2-} leaching, has lagged behind the recovery in wet deposition. Mitchell et al. (2011) drew similar conclusions based on S watershed budget calculations for Biscuit Brook in the Catskills and Arbutus Lake in the Adirondacks and 13 other surface waters in the larger region. The lakes and streams Stoddard et al. (2003) observed to have experienced the largest decreases in SO_4^{2-} concentration had similar decline rates to the decline rates in precipitation. Thus, the most responsive lakes and streams appear to have changed in SO_4^{2-} concentration approximately proportionately to precipitation.

Nitrate. During the 1980s, NO_3^- concentration increased in many lakes and streams in the Adirondack and Catskill Mountains (Driscoll and Van Dreason 1993; Murdoch and Stoddard 1993). Scientists expressed concern that perhaps forests were becoming N saturated in response to relatively high atmospheric N deposition. This was hypothesized to cause increased NO_3^- leaching from forest soils to drainage water throughout the region. Such a response could negate the benefits of decreasing SO_4^{2-} concentrations in lake and stream waters that were occurring subsequent to the S emissions reductions mandated by the CAAA. However, the trend of increasing surface-water NO_3^- concentration at many locations was reversed in about 1990 (e.g., Figures 6.2 and 6.3), and the reversal could not be attributed to a change in N deposition (U.S. EPA 2008). Lakes in the Adirondacks and streams in the northern Appalachian Plateau, which includes the Catskill Mountains, exhibited small but significant downward trends in NO_3^- concentration during the 1990s. These trends do not appear to reflect changes in N emissions or deposition. Strong *upward* trends in NO_3^- concentration in these same regions had earlier been observed during the 1980s (Murdoch and Stoddard 1992; Stoddard 1994). These increases and subsequent decreases in the concentrations of NO_3^- in lakes and streams in New York and elsewhere in the northeastern United States may have been due to multiple factors, especially climatic variation. However, trends measured on the scale of one or two decades may be insufficient for identifying long-term patterns of N cycling. Decadal (or longer) patterns caused by climatic conditions can obscure the multi-decade trends that reflect responses to emissions reductions. Momen et al. (2006) hypothesized that decreasing NO_3^- trends in some Adirondack lakes are correlated with increases in chlorophyll *a* and suggested that acidification recovery may be accompanied by increases in productivity, which drives NO_3^- concentrations down.

Studies by Mitchell et al. (1996b) and Murdoch et al. (1998) suggested that climate can substantially influence long-term patterns in NO_3^- concentrations in surface waters in the northeastern United States. Synchronous patterns in NO_3^- concentrations were observed from 1983 to 1993 in four small watersheds in New York, New Hampshire, and Maine (Mitchell et al. 2006). Unusually high NO_3^- concentrations during the snowmelt period of 1990 followed very cold temperatures the previous December. The cold weather was hypothesized to have disrupted soil root and microbial N cycling processes (Mitchell et al. 1996b), thereby contributing to NO_3^- leaching. Murdoch et al. (1998) also found that mean annual air temperature was correlated with average annual stream NO_3^- concentrations in a Catskill Mountain watershed that exhibited signs of N saturation.

The leaching of NO_3^- from soil to drainage water is the result of a complex set of biological, chemical, and hydrological processes. Key components include N uptake by plants and microbes, microbially mediated transformations among the various forms of inorganic and organic N, and patterns in local rainfall and snowpack dynamics. Most of the major processes are influenced by climate, including temperature, soil moisture, and snowpack development and melting. Insect defoliation can influence N export because the insects convert large stores of organic N in the foliage to reactive N on the forest floor. The effect can persist for multiple years (Eshleman et al. 2001). Thus, the concentrations of NO_3^- in surface waters change in response to a variety of factors in addition to N deposition. Monitoring programs of many decades may be needed to elucidate the major factors governing NO_3^- leaching in New York (Driscoll et al. 1995).

Base Cations. Stoddard (1991) reported trends in base cation concentrations in 12 Catskill streams over an approximately 70-year period. In five of the streams, concentrations of ($Ca^{2+} + Mg^{2+}$) increased from 1915 to 1922 and 1945 but decreased from 1945 to 1946 and 1990. In the other seven streams, concentrations increased during both periods but more slowly in the more recent time period in five of those streams. These measured ($Ca^{2+} + Mg^{2+}$) concentrations are consistent with the expectations of high cation leaching during early stages of acidification and relatively high atmospheric deposition of base cations in the middle of the twentieth century.

Likens et al. (1996) and Likens and Buso (2012) reported trends in base cation and ($SO_4^{2-} + NO_3^-$) concentrations in stream water in the Hubbard Brook Experimental Forest. An increasing ratio of base cation to ($SO_4^{2-} + NO_3^-$) concentration was observed from 1964 to 1969 and a decreasing ratio was observed from 1970 to 1994. The slope of the ratio during the increasing phase was steeper than the slope for the decreasing phase. This suggested lower base cation leaching per equivalent of mineral acid anion in more recent years. This pattern was attributed to depletion of exchangeable base cations from the soil.

Decreases in base cation concentrations in lakes and streams in New York have been well documented over the past several decades (Figures 6.2 and 6.3). Changes in base cation concentrations have been closely tied to trends in SO_4^{2-} concentrations. Regional declines in base cation concentrations in lakes and streams were found in the Adirondack Long Term Monitoring Program and were reported by Stoddard et al. (2003) and Driscoll et al. (2007a). Base cation concentrations decreased in 16 Adirondack lakes from 1982 to 2004, and similar rates of decrease were documented in 48 lakes (including the 16 original LTM lakes) from 1992 to 2004 (Driscoll et al. 2007a). Lawrence et al. (1999) showed decreases in base cation concentrations in Catskill streams from 1984 to 1997 that exceeded decreases in ($SO_4^{2-} + NO_3^-$).

Decreased lake-water SO_4^{2-} concentrations in Adirondack lakes during the 1980s were charge-balanced by a nearly equivalent decrease in the concentrations of base cations in many of the low-ANC lakes (Driscoll et al. 1995). Calculated F-factors for the nine monitored lakes that showed significant declines in the SBC, and ($SO_4^{2-} + NO_3^-$) concentrations ranged from 0.55 to greater than 1.0, with a mean of 0.93 (Driscoll et al. 1995). These high F-factor values for chemical recovery from acidification were similar to results of historical acidification rates Sullivan et al. (1990) obtained, based on diatom reconstructions of historical change over the previous century for 33 statistically selected Adirondack lakes.

Acid Neutralizing Capacity. Long-term monitoring data collected in glaciated regions in the eastern United States, including the Adirondacks and northern Appalachian Plateau, showed relatively consistent declines in base cation (Ca^{2+} +

Mg^{2+}) concentrations in stream and lake waters during the 1990s (Stoddard et al. 2003). The average decreases were in the range of −1.5 to −3.4 µeq/L/yr. All of the regional trends were highly significant. Surface-water SO_4^{2-} concentrations on glaciated terrain in the eastern United States decreased at a rate of about −2.5 µeq/L/yr (mean of regional median slopes), and NO_3^- decreased at a rate of −0.5 µeq/L/yr during the 1990s. These rates of change set an upper limit to expectations of ANC recovery of +3 µeq/L/yr (i.e., the sum of SO_4^{2-} and NO_3^- trend magnitudes). The Gran ANC increase Stoddard et al. (2003) reported was about one-third of this maximum, +1 µeq/L/yr. The difference between the observed Gran ANC trend and the maximum trend estimated from rates of acid anion (SO_4^{2-} + NO_3^-) change could be explained by the average regional median decline in (Ca^{2+} + Mg^{2+}) concentration, which was about −2.0 µeq/L/yr (Stoddard et al. 2003).

Moderate increases in surface-water ANC during the 1990s reduced the estimated number of acidic lakes and stream segments throughout New York and elsewhere in the northeastern United States. Stoddard et al. (2003) found increasing ANC in surface waters during the 1990s in all of the glaciated regions of the eastern United States, including the Adirondacks and northern Appalachian Plateau. Increases in ANC were relatively modest compared with decreases in SO_4^{2-} concentration, although the regional ANC increases in the Adirondacks and the northern Appalachian Plateau were statistically significant. Stoddard et al. (2003) interpreted the observed median increase of about +1 µeq/L/yr in these regions as an indication of significant movement toward ecological recovery from acidification. Increases in ANC were largest for lakes that were most acidic at the time of the study (Table 6.2).

Waller et al. (2012) analyzed monitoring data over the period 1991–2007 from TIME lakes in the Adirondack Mountains. The percent of acidic Adirondack lakes was judged to have decreased from 15.5% to 8.3% since implementation of the Acid Rain Program (under Title IV of the 1990 CAAA) and the Nitrogen Budget Tracking Program in 2003. They considered two measures of ANC, Gran analysis and calculated ANC. The calculated ANC measure increased over the study period at more than twice the rate of the Gran ANC measure. Waller et al. attributed this difference to compensatory increases in the concentrations of organic acids, as evidenced by increases in dissolved organic carbon in lakes.

The Neversink River watershed is the most acid-sensitive stream system in the Catskill Mountains. Its acid sensitivity stems from geologic and soil characteristics and topography. Many of the tributary streams in the upper parts of the Neversink watershed are chronically acidic. There have been some improvements in water-chemistry conditions in this watershed since about 1990 (Burns et al. 2006). The acidity of the most acidified tributary streams located furthest upstream has decreased slightly in recent years. Some streams also showed increased diversity of aquatic invertebrates and perhaps an expanded range for brook trout in association with the improved water quality (Burns et al. 2008a).

TABLE 6.2. Slopes of trends in Gran ANC in acidic, low-ANC, and moderate-ANC lakes and streams, 1990–2000

ANC class	Number of sites	Change in Gran ANC (μeq/L/yr)
Acidic (ANC < 0 μeq/L)	26	+1.29**
Low ANC (0 < ANC < 25 μeq/L)	51	+0.84**
Moderate ANC (25 < ANC < 200 μeq/L)	43	+0.32[ns]

Source: Stoddard et al. 2003.
Note: Analysis includes all sites in New England, the Adirondacks, the Appalachian Plateau, and the Upper Midwest; Ridge and Blue Ridge sites excluded.
[ns] Trend not significant ($p > 0.05$).
** $p < 0.01$

Aluminum. Periodic monitoring of Al_i concentrations in lake water was begun in 16 Adirondack long-term monitoring lakes in 1982. This effort expanded to 48 lakes in 1990. From 1982 to 2004, 5 of the original 16 Adirondack monitoring lakes showed decreasing trends in Al_i concentrations at rates that ranged from 0.02 μmol/L/yr to 0.18 μmol/L/yr. From 1992 to 2004, 24 of the 48 monitored lakes showed decreasing trends in Al_i concentrations (Driscoll et al. 2007a). Declining trends in Al_i concentration have mainly been demonstrated in the more acidic of the monitored lakes, which contained the highest concentrations of Al_i. Such Al_i concentration trends have been more pronounced in the Adirondack region than in other regions of the eastern United States (Table 6.3).

Dissolved Organic Carbon. Lake-water dissolved organic carbon and organic acid anion concentrations in the Adirondacks have apparently increased in association with decreased S deposition during recent decades (Table 6.3; Driscoll et al. 2007a; Monteith et al. 2007). Many TIME lakes showed increased dissolved organic carbon trends from 1991 to 2007 (Waller et al. 2012), but only four were significant. There was a wide range of responses across the study lakes. In general, dissolved organic carbon in lakes decreased during the last four years of the period Waller et al. analyzed. This result appears to be partly responsible for the rather limited lake-water ANC and pH recovery that has occurred at most study lakes.

All regions of the eastern United States that had sufficient dissolved organic carbon data for analysis in Stoddard et al.'s (2003) study of the responses of surface waters to implementation of the CAAA exhibited increases in dissolved organic carbon concentrations during the 1990s. All regional trends were significant with the exception of the northern Appalachian Plateau, the region with the low-

TABLE 6.3. Regional trend results for long-term monitoring lakes and streams, 1990–2000

Region	SO_4^{2-}	NO_3^-	Base cations	Gran ANC	H	DOC	Al
New England lakes	-1.77**	+0.01[ns]	-1.48**	+0.11[ns]	-0.01[ns]	+0.03*	+0.09[ns]
Adiron-dack lakes	-2.26**	-0.47**	-2.29**	+1.03**	-0.19**	+0.06**	-1.12**
Appa-lachian streams	-2.27*	-1.37**	-3.40**	+0.79*	-0.08*	+0.03[ns]	+0.56[ns]

Source: Stoddard et al. 2003.
Notes: Values are median slopes for the group of sites in each region. Units for sulfate (SO_4^{2-}), nitrate (NO_3^-), base cations ($Ca^{2+} + Mg^{2+}$), Gran ANC, and hydrogen (H) are μeq/L/yr. Units for dissolved organic carbon (DOC) are mg/L/yr. Units for aluminum (Al) are μg/L/yr.
[ns] Regional trend not significant ($p > 0.05$).
* $p < 0.05$
** $p < 0.01$

est median dissolved organic carbon concentration. The median increase in dissolved organic carbon of 0.05 mg/L/yr Stoddard et al. (2003) reported represented an overall increase of about 10% across study regions, similar to trends reported elsewhere in the northern hemisphere (Evans and Monteith 2001; Skjelkvåle et al. 2001). This suggests a common cause. Both climate warming and decreasing acidic deposition are possible causal agents (U.S. EPA 2008).

Dissolved Oxygen. Data are available regarding trends in estuarine water quality in New York associated with changes in nutrient loading. For example, subsequent to abatement of municipal and industrial discharge in the Hudson River–Raritan Bay estuary watershed, water quality improved substantially (O'Shea and Brosnan 2000). Such improvement included relatively large increases in dissolved oxygen.

O'Shea and Brosnan (2000) examined changes in indicators of eutrophication of western Long Island Sound over time. The minimum dissolved oxygen in bottom water had decreased since the mid-1980s and vertical stratification had increased, even though raw sewage discharge had been eliminated and decreases in biochemical oxygen demand point-source loads from municipalities had been observed. Parker and O'Reilly (1991) hypothesized that decreased loads of total suspended solids may have increased transparency and caused a vertical extension of the euphotic zone (the sunlit portion of the water column where net photosynthesis is positive). Such an increase in transparency could increase O_2 demand despite reductions in N loading.

Meteorological factors have also been linked with the quality of estuarine water (Clark et al. 1995; Valle-Levinson et al. 1995; Togersen et al. 1997). Nearly two-thirds of the 180 algal blooms documented in the Hudson River–Raritan Bay estuary from the 1950s to the 1980s could be attributed, at least in part, to hydrodynamic or climatological factors (Olha 1990). High water temperatures in the summer and diminished winds contribute to a weak summer stratification in the western Long Island Sound that persists until it is disrupted by autumn storms (Parker and O'Reilly 1991; Welsh and Eller 1991). Analysis of monitoring data collected from 1963 to 1993 by HydroQual, Inc. (1995) found that periods of dissolved oxygen depletion in Long Island Sound bottom water were most strongly correlated ($r^2 = 0.63$) with density stratification, mainly due to development of vertical temperature gradients. Vertical density stratification in Long Island Sound increased markedly in about 1986, with earlier onset of stratification and a longer period of stratification each year. This increased stratification restricts O_2 replenishment of bottom waters with O_2 derived from waters at the estuary surface.

The precise cause of the observed decrease in dissolved oxygen in western Long Island Sound bottom water during the latter decades of the twentieth century is not known. However, O'Shea and Brosnan (2000) concluded that it was associated more with changes in vertical temperature stratification than with changes in estuary nutrient loadings. They also speculated that warmer estuary surface temperatures in the 1980s and 1990s had a larger effect on vertical temperature stratification and dissolved oxygen depletion of bottom waters in the deeper western Long Island Sound than in the shallower Raritan Bay.

6.3.3.2. Biology

Published studies of long-term monitoring data for biological communities in acid-sensitive surface waters (e.g., Henriksen and Grande 2002) are rare. Effects of acidic deposition on the distribution of sensitive species are commonly evaluated based on changes in species richness with increasing pH and ANC for multiple water bodies across the landscape (Stoddard et al. 2003). Increasing richness of phytoplankton, zooplankton, and fish with increasing ANC and pH has been shown for lakes in the Adirondacks (Sullivan et al. 2006a; Nierzwicki-Bauer et al. 2010), and for both streams and lakes elsewhere (U.S. EPA 2008).

Concentrations of Hg in fish and birds have declined during recent decades in the Great Lakes region. For example, from 1967 to 2009, Hg concentrations in walleye, largemouth bass, and the eggs of herring gulls (*Larus argentatus*) de-

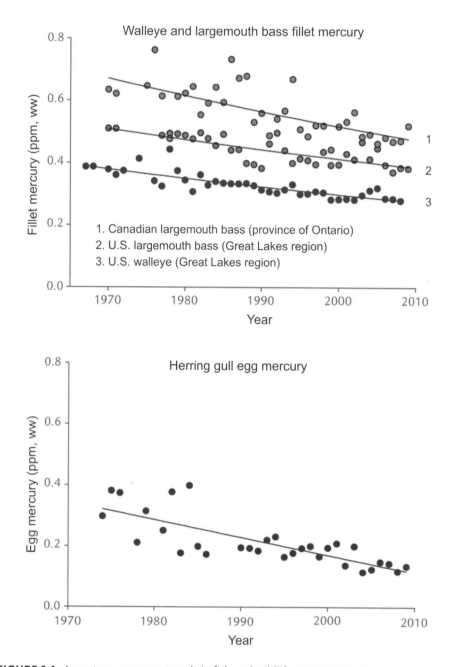

FIGURE 6.4. Long-term mercury trends in fish and wildlife, 1967–2009. Mercury concentrations in fish fillets (walleye and largemouth bass, averaged by year across multiple sites in the Great Lakes and inland water bodies in the U.S. Great Lakes states and the province of Ontario) and temporal trends in mercury concentrations in herring gull eggs (averaged by year across multiple sites in the Great Lakes region). These data are characteristic of the regional trend of decreasing mercury concentrations in fish and wildlife in recent decades. Much of this decrease has been attributed to reductions in regional mercury emissions, although there may be other contributing factors as well. Sources: Monson et al. 2011; Weseloh et al. 2011; Evers et al. 2011b.

creased substantially (Figure 6.4; Evers et al. 2011a; Monson et al. 2011; Weseloh et al. 2011). This decrease can be attributed largely to decreases in Hg deposition (Evers et al. 2011a). More mixed results were found for lakes in the Adirondacks (Dittman and Driscoll 2009).

In spite of the general decline in Hg concentrations in fish and wildlife of the Great Lakes region since the 1970s, there has also been evidence of a more recent increase in Hg concentrations for some species at some locations. This has been shown for walleye from Ontario, walleye and northern pike from Minnesota (Monson 2009; Monson et al. 2011), walleye from Lake Erie (Bhavsar et al. 2010; Zananski et al. 2011), loon blood from northern Wisconsin (Meyer et al. 2011), and bald eagles from Voyageurs National Park (Pittman et al. 2011). The causes of these recent increases in Hg concentrations in fish and wildlife are not known. It has been suggested that the observed response may be related to lower water levels and increased exposed shoreline associated with changing climatic conditions (Meyer et al. 2011), effects of the introduction of exotic species (Monson et al. 2011), or decreased nutrient loading (Zananski et al. 2011).

CHAPTER 7

Extrapolation of Site-Specific
Data to the Broader Region

7.1. METHODS OF REGIONALIZATION

In reporting numbers and percentages of lakes or streams above or below various water-chemistry or critical-load thresholds or showing biological impacts of air pollutants, the population of surface waters to which these numbers and percentages pertain must be considered. Lakes and streams can be defined in various ways, and the definition that is selected can have a considerable influence on the results reported for a regional assessment. Of particular importance in describing and defining a lake or stream population is the lower size limit for inclusion. For example, the EPA's ELS included lakes larger than about 4 ha in the Adirondack subregion of the northeastern region of the United States. The Adirondack subregion of the ELS included some landscape that is beyond the borders of the Adirondack ecoregion. The EPA's EMAP study surveyed lakes larger than 1 ha. The ALS included some lakes that were considerably smaller than 1 ha. The distribution of lake size in the Adirondacks and elsewhere tends to be skewed towards smaller-sized lakes. These smaller lakes frequently are more acid-sensitive or are more impacted by acidification than larger lakes. Thus, the choice of the minimum lake size to be included in the population of interest has a substantial influence on calculations of the number and percentage of acid-sensitive or acid-impacted lakes.

The overall population frame can be further restricted to fit the goals of a particular study. For example, Sullivan et al. (2006b) restricted the EMAP lake frame for the Adirondack ecoregion to include in their modeling study only those lakes that had ANC ≤ 200 μeq/L, as measured in the EMAP survey. This was done to focus the modeling effort on the lakes and their watersheds of greatest sensitivity to acidic deposition. There are an estimated 1,320 Adirondack lakes in the EMAP frame that have ANC ≤ 200 μeq/L.

The ALSC surveyed 1,469 Adirondack lakes during the period 1984–1990. Of the sampled lakes, 1,136 were larger than 1 ha in area and were located in the Adirondack ecoregion. The lakes the ALS sampled were not statistically selected from a mapped frame, but the ALS did survey the vast majority of the lakes in the Adirondack ecoregion that are larger than 1 ha. Thus, the ALS approached a census of all lakes larger than this size cutoff rather than a representative sampling of a subset of all lakes.

Several Adirondack population frames and lake or stream sets can be considered, depending on the objectives of the study. Results of model calculations and extrapolations based on a sampling of streams or lakes must always be considered in the context of the population of surface waters represented.

7.2.   REGIONALIZATION OF SURVEY DATA

Regional assessments have been conducted of lake acid-base chemistry in the Adirondack Mountains. For example, the regional EMAP probability sample selected by EPA included 115 Adirondack lakes and their watersheds. They were chosen from the 1,829 (standard error = 244) target Adirondack lakes included in the EMAP frame. These included lakes depicted on 1:100,000-scale U.S. Geological Survey maps that are larger than 1 ha, deeper than 1 m, and contain more than 1,000 m^2 of open water. Of those target lakes, an estimated 509 had summer index ANC > 200 μeq/L and could be considered insensitive to acidic deposition. The remaining 1,320 (standard error = 102) relatively low-ANC (≤ 200 μeq/L) lakes constitute the principal frame for statistically assessing Adirondack lake acid-base chemistry. Details of the EMAP design were given by Larsen et al. (1994). Whittier et al. (2002) presented an assessment of the relative effects of environmental stressors across lakes in the northeastern United States using the EMAP probability design and survey data. Based on field sampling, 42% of the lakes in the EMAP statistical frame for the Adirondack region had summer index ANC ≤ 50 μeq/L and another 30% had ANC between 50 and 200 μeq/L. These two strata of low-ANC lakes are thought to be most responsive to changes in air pollution.

Sullivan et al. (2006b, 2007a) used a random process to select candidate watersheds for soil sampling and modeling from among the 44 EMAP watersheds containing lakes with ANC ≤ 50 μeq/L and the 39 EMAP watersheds containing lakes with ANC between 50 and 200 μeq/L. County was used as a spatial clustering variable to achieve spatial balance in a manner identical to that used in the original EMAP probability design (Larsen et al. 1994). For lakes with ANC between 50 and 200 µeq/L, Sullivan et al. (2006b, 2007a) used a variable probability factor based on lake ANC class (50 to 100, 100 to 150, and 150 to 200 µeq/L). This resulted in obtaining more samples in the lower ANC range. No variable probability factors were used for the ANC ≤ 50 µeq/L lakes. Results of measurements and model projections for the selected EMAP watersheds were extrapolated to the populations of watersheds containing lakes with ANC ≤ 50 or ≤ 200 µeq/L, using the original EMAP sample weights adjusted for this random subsampling procedure. Soil survey data are more difficult to regionalize than surface-water data, probably in part because of the marked heterogeneity of soil chemical conditions. For example, C. Johnson et al. (2000) collected and analyzed soil samples from 72 locations in a single 214-ha watershed in the Catskill Mountains in order to evaluate landscape factors that regulate soil chemistry. For each sampling location, they determined the slope, aspect, elevation, topographic index, and flow accumulation in the watershed contributing area. These landscape attributes were not effective in predicting the chemical properties of organic or mineral soil. Collectively, they explained only about 4% to 25% of the variance in soil pH, effective cation exchange capacity, exchangeable bases, exchangeable acidity, total C, total N, and C-to-N ratio.

In the 1980s, there were nearly 600 Adirondack lakes with pH below 5.5, based on Sullivan et al.'s (2006a) subsampling from the EMAP statistical frame. Most of these low-pH lakes had probably been acidified to some extent by acidic deposition. About 40% of Adirondack lakes had chronic ANC below about 50 µeq/L. Most of those can be expected to become temporarily acidic or nearly so during rainstorms and spring snowmelt and have likely experienced some degree of biological harm from acidification. About 11% were chronically acidic and supported few or no fish or other acid-sensitive life forms.

A statistically based survey was conducted by the U.S. Geological Survey of 200 western Adirondack streams during spring, summer, and fall seasons (Lawrence et al. 2008a). The ANC was below 0 µeq/L in about 15% of the streams during summer, 25% during fall, and 29% during spring. In addition, 25% (summer) to 41% (spring) and 44% (fall) of the streams had ANC and pH low enough to cause Al$_i$ concentrations to increase above 2 µmol/L, which appears to be a threshold for killing young brook trout in the western Adirondacks (Baldigo et al. 2007).

The U.S. Geological Survey conducted a follow-up survey of over 200 streams in the Adirondacks during the period 2010–2011. It included streams in portions of the Adirondack Park that were not in the Oswegatchie and Black River basins that had been surveyed previously. Coupled with the earlier pilot study conducted in 2005 - 2006, the two U.S. Geological Survey stream surveys covered the entire Adirondack region. Of the streams surveyed during spring (April 2011), 18% had ANC ≤ 0 µeq/L. During summer (August 2010), about 8% of the streams had

ANC ≤ 0 µeq/L. The fractions of sampled streams having Al_i ≥ 2 µmol/L were 5% in August and 11% in April (Lawrence 2013).

Regional extrapolation of model simulation or assessment results can focus on (1) numbers and percentages of lakes and their watersheds projected to exceed certain thresholds or to belong to certain classes; or (2) maps showing the locations of lakes or watersheds in various classes. EMAP was the most statistically rigorous spatial frame available for quantitative extrapolation that covered the geographical extent of the Adirondack ecoregion and included lakes as small as 1 ha. Thus, for estimating numbers and percentages of Adirondack lakes in different response classes, EMAP is the preferred population frame. However, results from EMAP do not reveal where the various sensitive and impacted lakes are located. Spatial representation is best achieved using the ALS as the basis of extrapolation (Sullivan et al. 2012).

Driscoll et al. (2001a) used EMAP data collected during the period 1991–1994 to evaluate the number of acidic lakes in the Adirondacks. Because the EMAP survey was conducted during low-flow summer conditions, the results do not reflect the lowest ANC values that occur during the year. Nevertheless, the study documented that an estimated 10% of the population of Adirondack lakes was chronically acidic and that 31% had ANC between 0 and 50 µeq/L.

The spatial distribution of lake acid-base chemistry in the Adirondacks was revealed in the ANC values the ALS measured. The majority of the acidic and low-ANC lakes were located in the southwestern portions of the Adirondack Park and in the high peaks area of the north-central Adirondack Mountains.

Data on terrestrial resource sensitivities or impacts can also be regionalized. For example, Hallett et al. (2010) developed forest health and sensitivity indicators and baseline maps of potential sensitivity to disturbance for lands within the New York City water supply in the Catskill Mountains. Spatially explicit maps were constructed to reflect lands most likely to be susceptible to sugar maple decline, streams with low base-cation surplus, and overall ecosystem health. Data for this assessment were taken from Landsat 5 Thematic Mapper imagery. Stress-sensitive indices were evaluated to identify those that best predicted ecosystem condition on over 46 calibration plots with a range of tree species composition, health status, and topographic position. The best-fit model included mainly chlorophyll *a* and canopy water content as sensitive indices.

7.3.   REGIONALIZATION OF LONG-TERM MONITORING DATA

Monitoring results from the TIME and LTM programs have been used to extrapolate monitoring data from individual streams and lakes to the broader regional populations of surface waters (see Stoddard et al. 2003). The TIME project is a probability sampling; each site was chosen statistically to be representative of a portion of the target population of lakes or streams. In the Adirondacks, this target population includes lakes having Gran ANC ≤ 100 µeq/L. In the northern Appalachian Plateau, the target population is upland streams with ANC ≤ 100 µeq/L. These are the streams likely to be most responsive to changes in acidic deposition. Each selected lake or stream is sampled annually in TIME. Results are extrapolated to the target population of lakes or streams in each region (Larsen and Urquhart 1993; Larsen et al. 1994; Stoddard et al. 1996; Urquhart et al. 1998). The project began sampling 43 Adirondack lakes in 1991. Data collected for these waters can be extrapolated to the target population of about 1,000 Adirondack lakes with ANC ≤ 100 µeq/L out of the total population of 1,829 lakes with surface area > 1 ha that were included in the EMAP sampling frame.

CHAPTER 8

Projected Future Responses of Sensitive Resources to Reductions in Acidic Atmospheric Deposition

People differ in their view of what ecosystem recovery means. Societal goals for ecosystem recovery from acidification, nutrient enrichment, or Hg toxicity can be expressed in various ways. One goal might be to restore ecological conditions to those that prevailed before the industrial revolution. Such a goal might be impossible to achieve, however, because of the extent of past damage. Also, a variety of human activities besides air pollution emissions may have caused changes to the condition of natural resources. For example, logging, agriculture, urban development, fire, road building, and fish stocking have all changed the chemistry and biology of surface water and degraded water bodies to varying degrees. Climate change and the introduction of plant diseases and nonnative species may also have contributed to altered natural resource conditions.

Recovery might be viewed as restoring the health of one or more particular species that people highly value, such as brook trout or sugar maple trees. However, not all water bodies supported brook trout and not all forests supported sugar maple before the industrial revolution. The habitat may not have been suitable or accessible. It might be unreasonable to try to restore a species that did not occur in that watershed prior to the stress of air pollution. Even if the chemical conditions needed for supporting a species are restored at a particular site, it might still be impossible for that species or its food sources to recolonize without human intervention. There is no agreed-upon definition of what constitutes ecosystem recovery. One goal might be to restore water and soil chemistry to conditions capable of supporting the kinds of species that would likely have characterized that habitat in the absence of air pollution. This type of goal focuses on restoring habitat to conditions that are capable of supporting certain species rather than focusing on the actual restoration of specific species.

Some may prefer that recovery occur in a "natural" way, without species reintroduction or liming to reduce acidity and/or restore depleted nutrient supplies. Active steps may allow or hasten recovery but represent yet one more human intervention. It must be recognized that no restoration intervention is likely to fully restore conditions as they were prior to the industrial revolution.

Much of the S deposited on watersheds in the Adirondack and Catskill Mountains during the twentieth century moved more or less directly through soils and into the stream and lake water as SO_4^{2-}. This mobile SO_4^{2-} contributed to acidification of soil and water. However, a substantial fraction of the S deposited each year was stored in the soil (Mitchell et al. 2011). Now that S deposition inputs to the watersheds have decreased due to emissions reductions, some of that stored S is being released to drainage water (Mitchell et al. 2011). This process is partially delaying recovery from acidification.

Pools of base cations stored in the soil of acid-sensitive watersheds have gradually become depleted in response to acidic deposition. As a consequence, streams and lakes will probably acidify more in the future than they have so far, relative to the amount of acidic deposition received. The acidifying effects of acidic deposition are not completely reversible. Recovery will likely be slower than the initial rate of acidification. As a consequence, some damage may persist for decades or longer, even if deposition levels are substantially curtailed immediately.

8.1. MODELING APPROACHES

Several watershed models of acid-base chemistry are available for projecting future responses of soil and water to changing amounts of atmospheric deposition. Those most commonly applied models for simulating the recovery of upland aquatic and terrestrial ecosystems in New York have included the MAGIC and PnET-BGC models. Various other watershed models have been used to estimate N fluxes to estuaries, including SPARROW and WATERSN. A general model of estuarine eutrophication response takes into consideration issues as diverse as nutrient fluxes, tidal flushing, the presence of toxic algal species, dissolved oxygen, water residence time, and food web interactions. Such characteristics as nutrient loading and estuarine susceptibility to eutrophication are incorporated in the National Estuarine Eutrophication Assessment screening methodology (Bricker et al. 1999). The ASSETS eutrophication screening model has been used to score and rank estuaries based on eutrophication in response to nutrient addition. Other watershed models have been developed and applied in New York but they will not be discussed here.

8.1.1. MAGIC

MAGIC simulates the long-term effects of acidic deposition on soil and surface water chemistry (Cosby et al. 1985). A central concept in MAGIC is the size of the pool of exchangeable base cations on the soil. As the fluxes to and from this pool change over time in response to changes in atmospheric deposition of major ions, chemical equilibria between soil and soil-solution shift, resulting in changes in the simulated soil-solution and surface-water chemistry. The validity of the MAGIC model has been confirmed by comparison of its simulations with estimates of lake acidification inferred from paleolimnological reconstructions of historical changes in lake pH (Wright et al. 1986; Jenkins et al. 1990; Sullivan and Cosby 1995) and with the results of catchment-scale acidification and de-acidification experiments (e.g., Wright and Cosby 1987; Cosby et al. 1995; Cosby et al. 1996; Moldan et al. 1998). MAGIC has been used to reconstruct the history of acidification and to project future responses on a regional basis and in a large number of individual catchments in both North America and Europe (e.g., Wright et al. 1986; Lepistö et al. 1988; Whitehead et al. 1988; Hornberger et al. 1989; Cosby et al. 1990; Jenkins et al. 1990; Norton et al. 1992; Wright et al. 1994; Ferrier et al. 1995; Wright et al. 1998; Cosby et al. 2001; Sullivan et al. 2004, 2008).

Monthly and annual volume-weighted concentrations of major ions are simulated in soil solution and surface water. The model includes a component in which the concentrations of major ions are assumed to be governed by simultaneous reactions involving SO_4^{2-} adsorption, cation exchange, dissolution-precipitation-speciation of Al, and dissolution-speciation of inorganic C. It also contains a mass balance section in which the flux of major ions to and from the soil is controlled by atmospheric inputs, chemical weathering, net uptake and loss in biomass, and losses to runoff. Simulated change in surface water acidity depends on fluxes through the catchment and the characteristics of the affected soils.

Cation exchange is modeled using Gaines-Thomas equilibrium equations with selectivity coefficients for each base cation and Al. Sulfate adsorption is represented by a Langmuir isotherm. Aluminum dissolution and precipitation are controlled by equilibrium with gibbsite. Aluminum speciation is calculated from hydrolysis reactions and complexation with SO_4^{2-} and fluoride. Effects of CO_2 on pH and on the speciation of inorganic C are computed from equilibrium equations. Organic acids are represented as triprotic analogues. Element weathering and the uptake rate of N as a fraction of the N input are assumed to be constant and are calculated from the model calibration. A set of mass balance equations for base cations and strong acid anions are included. The model is described in greater detail in Cosby et al. (1985, 1989)

The aggregated nature of the MAGIC model requires that it be calibrated to observed data before it can be used to estimate past or future watershed responses. Given a description of the historical deposition pattern at a site, the model equations are solved numerically to give long-term reconstructions of soil and surface-water chemistry. Calibration is achieved by first setting the values of fixed parameters in the model that can be directly measured or observed. Then the model is run using observed and/or assumed atmospheric and hydrologic inputs. Variables for stream-water and soil chemistry are compared to observed values. If the observed and simulated values differ, the values of the "optimized" parameters are adjusted to improve the fit. After a number of iterations, the simulated-minus-observed values of the criterion variables usually converge to zero within some specified tolerance and the model is considered to be calibrated.

Because estimates of model-fixed parameters and inputs are uncertain, a "fuzzy optimization" procedure is implemented for calibrating the model. This approach consists of multiple calibrations using random values of the fixed parameters drawn from the observed possible range of values. In addition, random values of atmospheric deposition are assigned from a range that includes consideration of deposition uncertainty. Each calibration begins with a random selection of values of fixed parameters and deposition and a random selection of the starting values of the optimized parameters. Next, the optimized parameters are adjusted using the Rosenbrock (1960) algorithm to achieve a minimum-error fit to each of the target variables. This procedure is undertaken 10 times to yield 10 calibrations. The final calibrated model is represented by the range of parameter and variable values selected in each of the 10 calibrations.

Major sources of uncertainty in MAGIC acid-base chemistry model simulation and interpretation include the quality of data input, temporal variability in water chemistry, variability in biological response to water chemistry, validity and accuracy of the model, uncertainty about model calibration, errors associated with missing input data for the model, and errors associated with extrapolation of modeling results from individual watersheds to the region of interest. The relative magnitude of the effects of these sources of uncertainty has been evaluated for regional long-term MAGIC simulations in a series of uncertainty analyses using Monte Carlo methods (see Cosby et al. 1989; Hornberger et al. 1989; Cosby et al. 1990; Sullivan et al. 2002, 2003, 2004). The results of those analyses have suggested that the relative impacts on model projections may vary with data quality and/or quantity from application to application.

8.1.2. PnET-BGC

PnET-BGC is an integrated forest-soil-water model that assesses effects of air pollution and land disturbances on forest soil and aquatic ecosystems (Gbondo-Tugbawa et al. 2001). The model was developed by linking the PnET-CN

(PnET-carbon and nitrogen) model (Aber et al. 1997) and the BGC (BioGeo-Chemistry) model (Gbondo-Tugbawa et al. 2001). The main processes represented by the model include photosynthesis, growth, and productivity of trees; litter production and decay; mineralization of organic matter; immobilization of N; nitrification; abiotic soil processes; solution speciation; and surface-water processes (Aber et al. 1997; Gbondo-Tugbawa et al. 2001). The hydrologic algorithms used in PnET-BGC were developed by Aber and Federer (1992) and Chen and Driscoll (2005c). The model can use one or multiple soil layers (Chen and Driscoll 2005c). For lake simulations, it is assumed that the water column is completely mixed. The model predicts monthly concentrations and fluxes of major solutes in lake water, monthly concentrations and pools of exchangeable cations and adsorbed SO_4^{2-} on soil, and monthly fluxes of major solutes from soil and forest vegetation. A detailed description of the model is provided in Gbondo-Tugbawa et al. (2001).

Soil and surface-water projections from PnET-BGC tend to be most sensitive to parameter values of soil mass, partial pressure of CO_2 on soil, the cation exchange capacity of the soil, and cation exchange selectivity coefficients for Ca^{2+} and Al^{3+}. The model was applied at the Hubbard Brook Experimental Forest (Gbondo-Tugbawa et al. 2001) and to surface waters in the Adirondack and Catskill Mountains in order to predict the response of acid-sensitive forest ecosystems to future controls on atmospheric emissions (Gbondo-Tugbawa and Driscoll 2003; Chen and Driscoll 2005a, b).

Although the N cycle is complex, progress has been made over the past several decades in developing models such as PnET-BGC to use for making generalizations about how ecological and biogeochemical processes respond to N deposition. Significant scientific advancements in recent years have included the refinement of theoretical foundations of nutrient limitation, the development and improvement of analytical techniques, and an improved understanding of the role of N in influencing the cycling of C (U.S. EPA 2008).

8.1.3.   SPARROW

The transport of N in large river systems provides an important source of N to estuaries and coastal marine waters. Based on application of the SPARROW model to a number of large river systems, Smith et al. (1997) concluded that much of the United States probably exports less than 5 kg N/ha/yr in river discharge but that N export in watersheds of the northeastern United States is higher. For the watersheds that export more than 10 kg N/ha/yr, Smith et al. concluded that fertilizer was the largest source of N (48%), followed by atmospheric deposition (18%) and livestock waste (15%). In this analysis, fertilizer used for human food production was considered to be the ultimate source of the N that wastewater treatment plant point sources contribute to waterways.

SPARROW relates measured nutrient transport rates in streams to spatially referenced descriptors of nutrient sources. Nutrient transport is compared to characteristics of the watershed land surface and the stream channel. The method provides a mathematical estimate of the in-stream contaminant load at the downstream end of a defined stream reach as the sum of monitored and unmonitored load contributions from all upstream sources. The equations represent point sources, fertilizer applied to croplands and lawns, livestock waste, non-agricultural land, and atmospheric N deposition. Eight land-surface characteristics are included in the model; those typically found to be significant predictors of stream nutrient loads include soil permeability, stream density, and temperature (Smith et al. 1997).

Alexander et al. (2001) updated the SPARROW modeling approach by refining empirical in-stream nutrient-loss functions to better describe N attenuation in large rivers and measurements of detrending wet N deposition and by improving uncertainty estimates. Application of the model to 40 major U.S. coastal watersheds suggested that atmospheric N deposition contributions to riverine export varied over nearly two orders of magnitude. The model estimated that atmospheric N deposition contributed about 4% to 35% of the in-stream export of N into the study estuaries.

8.1.4.   WATERSN

The WATERSN model (Castro et al. 2000; Castro and Driscoll 2002; Whitall et al. 2004) is a watershed-scale mass balance model of N input and output. The model assumes that contribution atmospheric input makes to the total N output from upland forests is proportional to the atmospheric deposition component of the total N input. Forest export is estimated using a nonlinear regression between wet inorganic N deposition and stream export of inorganic N (Whitall et al. 2007). The WATERSN model has been used to simulate the relative contribution of N from various sources to coastal waters. It is based on calculated and assumed values of N inputs from crop fertilization, N fixation, atmospheric deposition, human wastewater discharge, and livestock waste storage and field application. Simulated outputs of N from the watershed include crop harvest, livestock grazing, NH_3 volatilization, and denitrification.

The model has been used to estimate the relative importance of sources of N to Long Island Sound and Raritan Bay and nine other East Coast estuaries (Whitall et al. 2007). Observed differences in N source types can be used to guide management strategies for protecting and restoring impacted water bodies.

8.1.5.   ASSETS

The ASSETS eutrophication screening model (Bricker et al. 2003; Nobre et al. 2005; Ferreira et al. 2007) was developed by the National Estuarine Eutrophication Assessment to assess estuarine eutrophication using three major indices:

1. Overall human influence, which combines human pressure and estuary susceptibility to nutrient enrichment
2. Overall eutrophic condition, based on five key variables: chlorophyll *a*, macroalgae, low dissolved oxygen, loss of submerged aquatic vegetation, and occurrence of nuisance and/or toxic algal blooms (Bricker et al. 2003; Ferreira et al. 2007)
3. Determination of future outlook, based on the capacity of the estuary to dilute and/or flush nutrients and predictions of future nutrient loading (Ferreira et al. 2007)

The overall human influence, overall eutrophic condition, and determination of future outlook ratings are combined into one eutrophication score for an estuary under study: high, good, moderate, poor, or bad (Bricker et al. 2003).

Thus, ASSETS combines elements of anthropogenic pollution pressure, eutrophication state, and system response due to inherent susceptibility to eutrophication with predicted changes in nutrient loading. Ferreira et al. (2007) applied the ASSETS model to Long Island Sound, yielding scores of bad condition in 1991 and moderate condition in 2002. The modeled improvement was attributable largely to reduced nutrient loading from the watershed and reduced hypoxia.

8.2. PROJECTIONS BASED ON EXISTING AND FUTURE EMISSIONS CONTROLS

Process-based models have been used to project future conditions of sensitive resources in New York in response to existing and modified air pollution emissions controls. For example, Sullivan et al. (2006b, 2007a) reported model projections of future acid-base chemistry under three scenarios of future atmospheric emissions controls for lakes in the Adirondack Mountains using the MAGIC and PnET-BGC models. The focus was on estimating the extent to which lakes might continue to increase in ANC in the future in response to varying levels of S and N emissions controls and deposition (Figure 8.1A). Model simulations for 44 statistically selected Adirondack lakes were extrapolated to the regional lake population. Cumulative distribution frequencies of ANC response projected by MAGIC are shown in Figure 8.1B for the past (1850), for the peak acidification period (1990), and for the future (2100). Results for the future are given for each of the emissions control scenarios. Both MAGIC and PnET-BGC simulations suggested that the ongoing trend of increasing lake-water ANC for the most acid-sensitive Adirondack lakes would not continue under emissions and deposition levels anticipated to occur in the future as of 2003 (Base Case Scenario). PnET-BGC projected that the numbers of Adirondack lakes with ANC below 20 and below 50 μeq/L

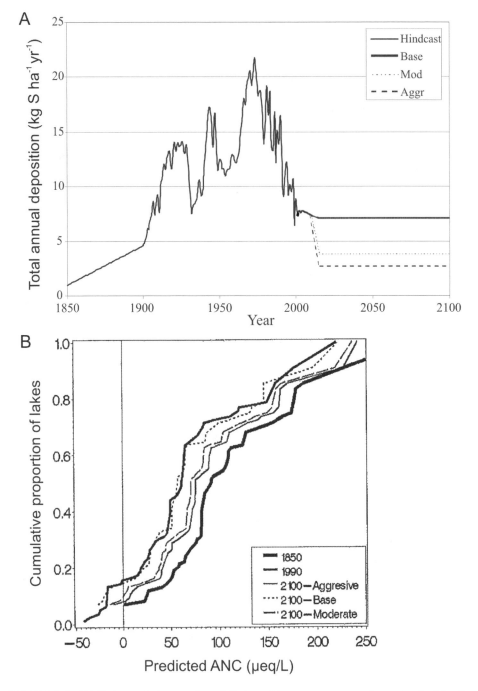

FIGURE 8.1. (A) Estimated time series of S deposition at one example watershed in the southwest Adirondack Mountains used by Sullivan et al. (2006a) as input to the MAGIC model for projecting past and future changes in lake-water chemistry attributable to acidic deposition. Future deposition estimates were based on three emissions control scenarios (Base Case, Moderate Additional Controls, Aggressive Additional Controls). (B) MAGIC model estimates of ANC distribution in Adirondack lakes with ANC < 200 µeq/L at three points in time. Source: Sullivan et al. 2006a.

would increase from 2000 to 2100 under that scenario and that the number of chronically acidic Adirondack lakes would stabilize at the level reached in 2000. This projected end to the ongoing chemical recovery of acid-sensitive Adirondack lakes was attributed to a projected decline in the pool of exchangeable base cations on soils. Model results suggested that re-acidification might be prevented with further emissions reductions.

MAGIC modeling was also conducted as part of the recent NAPAP assessment (Burns et al. 2011) to evaluate the ecological effects of additional S and N emissions controls on power generating facilities. Modeling of a base-case S emissions scenario indicated that implementation of existing rules and regulations through 2020 will likely reduce S deposition by 5 to 10 kg/ha/yr from 2010 levels throughout much of the northeastern United States. Further reductions in S deposition beyond the base case would provide additional ecological benefits but would not be sufficient to allow for full resource recovery (Burns et al. 2011). A base-case scenario for N deposition suggested that rules and regulations currently in effect will likely achieve reductions in N deposition across the United States in the range of 2.5 to 5 kg/ha/yr by 2020. The base-case S emissions control scenario projected that 8% of the Adirondack lakes would still have ANC ≤ 0 µeq/L in 2050. In addition, 40% of the Adirondack lakes would have ANC ≤ 50 µeq/L in 2050 under the most restrictive emissions control scenario (up to about an additional 60% decrease in S deposition). The number of lakes with ANC ≤ 50 µeq/L in 2050 would be closer under the restrictive emissions control scenario to modeled pre-acidification levels than under the other scenarios, approaching full recovery from acidic deposition.

There are few streams in New York for which future acid-base chemistry status has been modeled in response to atmospheric N deposition. However, Driscoll et al. (2003c) modeled some streams in the Catskill Mountains and at the Hubbard Brook Experimental Forest in New Hampshire. Calculations using the PnET-BGC model suggested that aggressive reductions in N emissions alone will not result in marked improvements in the acid-base status of forest streams. In response to a scenario of aggressive utility emissions control that specified a 75% reduction in utility N emissions beyond the CAAA, the ANC values of Watershed 6 at the Hubbard Brook Experimental Forest and Biscuit Brook in the Catskill Mountains were projected to increase only by 1 and 2 µeq/L, respectively, by the year 2050. Projected changes in water chemistry in response to differing levels of N deposition were small in comparison with model projections of variations resulting from climatic factors (Aber and Driscoll 1997; Driscoll et al. 2003c).

The MAGIC simulations suggested that the typical acid-sensitive Adirondack lake has lost about 50 µeq/L of ANC due to acidic deposition since preindustrial times (Sullivan et al. 2006a, 2006b). Modest chemical recovery of up to 20 µeq/L has occurred in some lakes since the 1980s in response to reduced acidic deposition. The model projections also suggested that continued chemical recovery of these lakes is unlikely unless emissions are reduced below levels that occurred in 2001 (Sullivan et al. 2006a). Unless air pollution emissions are further reduced, this recovery may be followed by a lengthy period of renewed acidification of lakes that were recently recovering from past acidification. Sullivan et al. (2006a) attributed this projected recovery reversal to soil processes. Even though recent levels of acidic deposition have been lower than during past decades, the simulations suggested that recent deposition may have been high enough to continue to remove base cations from soils. Both the MAGIC and PnET-BGC models estimated that the number of Adirondack lakes with ANC below 50 µeq/L would increase in the future with assumed modest continued reductions in emissions of S and N to levels below those that occurred in 2001 (Sullivan et al. 2006a).

Model simulations by Sullivan et al. (2006a) suggested that additional large future reductions in S and N deposition might result in biological improvements in many of the most severely acidified lakes. Projected biological improvements included the restoration of chemical conditions suitable for the return of one to two species of fish to the recovering lakes, improved conditions for brook trout, and an increase by one or two in the number of species of zooplankton. Simulations also suggested that even if S and N emissions are reduced to half or less of current values, the most acid-impacted Adirondack lakes are unlikely to increase to ANC above 50 µeq/L. Recovery to such an ANC level would likely be required for substantial biological recovery and to prevent episodic acidification to ANC below 0 µeq/L during snowmelt.

CHAPTER 9

Critical Load

The critical load is defined as the level, or load, of pollution below which significant harmful effects are not expected to occur to sensitive ecosystem elements (Nilsson and Grennfelt 1988). The critical load is generally specified on the basis of chemical indicator dose-response functions or tipping points expressed under long-term steady-state conditions (Henriksen and Posch 2001). Thus, the critical load specifies the pollutant load which, when applied to the ecosystem in question for a period of years to centuries, will eventually trigger a change in a chemical indicator of biological harm as the ecosystem comes into steady state with respect to that particular pollutant input level (Sullivan and Jenkins 2014).

9.1. APPROACHES

The critical load is calculated using either a steady-state or dynamic model, of which there are several to choose from (Henriksen and Posch 2001; U.S. EPA 2008). It is commonly calculated to protect against S, N, or S + N deposition damage caused by acidification or N nutrient enrichment. A critical load can also be calculated for Hg and/or S deposition to protect against Hg biomagnification. To date, however, the critical load approach has not been applied to Hg in the United States. In most critical-load analyses, the calculated deposition input that the ecosystem can tolerate without experiencing biological harm is quantitatively linked to a chemical indicator and a specified critical threshold value(s) for that indicator. Determination of the appropriate threshold value(s) is an important and somewhat subjective component of the critical-load process (Lovett 2013). Under the steady-state assumption, the period of resource protection is not specified and could be multiple decades or centuries into the future. A dynamic, as opposed to steady-state, pollutant load can also be defined, which is specific to a particular point in time. For example, the deposition load may be calculated that is sufficient to protect certain biological resources in a water body if pollutant inputs occur continuously at that level until the time specified (for example, the year 2050 or 2100). This dynamic load includes consideration of the temporal component of ecosystem damage or recovery. It is commonly called a target load.

The target-load concept can be used to describe the effects of time and can include various management perspectives. For example, a land manager may set a pollutant target load that is lower than the critical load in order to account for model uncertainty and to err on the side of resource protection. A target load can also be set below the critical load if the aim is to affect resource recovery in a shorter time period than the attainment of steady-state conditions. Conversely, a target load can be set that is higher than the critical load if the objective is to achieve only partial recovery of damaged resources by a certain date. Upon reaching that state of partial recovery, a new target load might be specified to achieve further recovery.

There is no single "definitive" critical load or target load for a natural resource or a watershed. The calculated value depends on what resource is being protected, the stress that is protected against, the time frame for protecting the sensitive resource, and the level of protection desired. It also depends on what indicator is used to represent biological harm and what is specified as that indicator's critical or threshold tipping point (Sullivan and Jenkins 2014). Critical load estimates reflect the current state of knowledge and policy priorities regarding ecosystem damage. Changes in scientific understanding in the future may include the development of new dose-response relationships, better resource maps and inventories, larger survey data sets, continuing results of time-series monitoring, and improved numerical models. Changes in the policy elements may include new mandates for resource protection, a focus on new pollutants and ecosystem services, and the inclusion of perceived new threats such as climate change or invasive species that may exacerbate the pollutant effects under consideration (U.S. EPA 2008).

Implementation of the critical-load process typically results in calculation of multiple critical loads for a given pollutant at a given location. Multiple critical-load values may also arise from an inability to agree on a single definition of "significant harm." Additional complications stem from the heterogeneity of natural ecosystems. Because of high spatial variability of soils and high temporal variability of surface waters, there will be a range of critical-load values for any sensitive indicator chosen.

The critical load specifies the point at which the ecosystem may begin losing ecosystem services and transition from a sustainable functioning ecosystem to one that is not functioning properly and is no longer considered sustainable (Sullivan 2012). The critical load is a resource management tool that enables decision making based on ecosystem services.

The critical load alone does not predict whether the ecosystem under investigation actually experiences or will in the future experience any biological harm. The ambient (and perhaps the expected future) pollutant load must also be considered. If the ambient deposition is higher than the critical load of deposition the ecosystem can tolerate, the ecosystem is in exceedance. However, transitioning from a condition of non-exceedance to a condition of exceedance does not mean that the ecosystem has been damaged. The transition from non-exceedance to exceedance (or vice versa) indicates a change in the probability that ecosystem services will be lost or recovered, depending on whether the starting point is an undamaged or a damaged state (Sullivan 2012). For undamaged systems, exceedance signifies that if deposition is continued at the current exceedance level, damage will occur at the time point specified for the analysis (e.g., 2100, eventual steady-state condition). Transitioning from exceedance to non-exceedance suggests that recovery will occur at the future time specified in the analysis. For steady-state critical-load modeling, this may be many decades or longer into the future.

Weathering is a key process in estimating the base-cation status of soils, critical load and target load, and recovery processes. It is a key element in both steady-state and dynamic critical-load calculations. As a consequence, substantial effort has been devoted to estimating in situ weathering using a variety of methods (Miller et al. 1993; Likens et al. 1996; Bailey et al. 2003). However, the complexity and variability of factors that affect weathering rates, including soil mineralogy, particle surfaces, soil organic matter, and moisture flux, add uncertainty to weathering estimates (White and Brantley 1995). Further complexity stems from the role of mycorrhizae that can penetrate silicate minerals to extract base cations while remaining isolated from the soil solution (van Breemen et al. 2000; Blum et al. 2002). Specification of the weathering rate is the most important and uncertain part of the critical-load, target-load, or exceedance calculation (Li and McNulty 2007; McDonnell et al. 2012, 2014).

9.2.   CRITICAL- AND TARGET-LOAD CALCULATIONS

Duarte et al. (2013) estimated critical-load exceedance of nutrient N and S + N deposition using a steady-state mass balance model at more than 4,000 monitoring sites located across the northeastern United States, including New York. They found negative correlations between critical-load exceedance and tree growth (19 species) and crown density (4 species). Positive correlations were found between exceedance and declining vigor (four species), crown dieback (six species) and crown transparency (seven species). The tree species that exhibited the most significant detrimental responses to critical-load exceedance were balsam fir, red spruce, quaking aspen, and paper birch (Duarte et al. 2013). Thomas et al. (2010) also found species-specific responses of different tree species across a gradient in atmospheric N deposition in the Upper Midwest and Northeast. Based on forest inventory data, they documented relationships between N deposition and tree growth, survival, and C storage during the 1980s and 1990s. Nitrogen deposition appeared to increase tree growth for 11 species and to reduce growth of 3 species (red pine [*Pinus resinosa*], red spruce, and white cedar [*Thuga occidentais*]). Deposition of N appeared to increase the growth of all species that had arbuscular mycorrhizal fungi associations (Thomas et al. 2010).

Sullivan et al. (2012) calculated target loads of atmospheric S and N deposition for a group of statistically selected lakes in the Adirondacks, using the MAGIC model. The target load was calculated for two future points in time (2050 and 2100) and for three critical ANC target levels (0, 20, 50 µeq/L). Results of the target loads and their exceedances were extrapolated to the regional population of Adirondack lakes. Model simulations suggested that the lakes were much more sensitive to changes in S deposition than to changes in N deposition. The study simulations demonstrated that nearly one-third of the lakes in the population had target loads below 50 meq S/m²/yr if future lake ANC was to be protected at a level of 50 µeq/L (Plate 11). About 600 Adirondack lakes received S deposition that was higher than their target load, some by more than a factor of 2 (Plate 12). Some lakes were estimated to be unable to reach an ANC level of 50 µeq/L by the target years considered in this analysis even if S deposition was decreased to zero and was held there throughout the duration of the simulation. This is not surprising because model hindcast simulations suggested that about 175 of the low-ANC lakes (13% of the 1,320 lakes with ANC ≤ 200 µeq/L) had preindustrial ANC below 50 µeq/L (Figure 9.1).

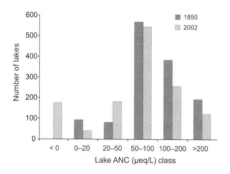

FIGURE 9.1. Histogram of extrapolated MAGIC simulations of lake-water ANC in 1850 and 2002 for the population of 1,320 Adirondack lakes included in the EMAP frame. Source: Sullivan et al. 2012.

Both the aquatic target load and the target-load exceedance exhibited clear spatial patterns across the Adirondack Park. Most of the acid-sensitive and acid-impacted lakes occur in the southwestern Adirondacks and in the high peaks area in the north-central Adirondacks (Plates 11 and 12; Sullivan et al. 2012).

Target loads of S deposition have also been calculated using MAGIC for protection of terrestrial resources in the Adirondack Mountains (Sullivan et al. 2011). Several chemical indicators were used in that analysis, including B-horizon soil base saturation and ratios of soil-solution base cations to Al (Ca to Al and Bc to Al). Modeling results were highly uncertain for estimating the target load to protect against poor acid-base chemistry of soil solution. This result was at least partly due to the unavailability of soil-solution field data with which to constrain the model calibrations. Simulation uncertainty was lower for target-load calculations intended to protect soil base saturation, but results were strongly influenced by selection of an appropriate base saturation threshold for indicating harm to terrestrial biota (Figure 9.2).

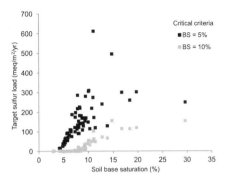

FIGURE 9.2. Target load of sulfur deposition for the year 2100 to protect the percent of soil base saturation from decreasing below two critical criteria thresholds compared to the average measured soil base saturation in the watershed in 2003. Each symbol represents the target load for a given modeled watershed. Source: With kind permission from Springer Science+Business Media, J. Environ. Stud. Sci., Target loads of atmospheric sulfur deposition protect terrestrial resources in the Adirondack Mountains, New York against biological impacts caused by soil acidification, 1(2011):305, T. J. Sullivan et al., Figure 2. Copyright AESS 2011.

About two-thirds or more of the modeled watersheds exhibited very low target loads (< 25 meq S/m²/yr) for achieving base saturation of 10% in the future (2050 or 2100). In contrast, a critical base saturation level of 5% was simulated to be readily achievable for most Adirondack watersheds. About 80% of the watersheds in the population of 1,320 acid-sensitive watersheds with low-ANC lakes (≤ 200 µeq/L) were estimated to be in exceedance of the target load for achieving base saturation of 10% (Sullivan et al. 2011). These results and analogous modeling results of Burns et al. (2011) confirm the importance of developing rigorous dose-response relationships when specifying appropriate tipping points for terrestrial effects.

9.3. UTILITY TO POLICY MAKERS

Recent reductions in atmospheric deposition of S, N, and Hg have yielded tangible benefits in New York. Surface-water ANC values in acid-impacted portions of New York have increased appreciably during the last two decades. Concentrations of highly toxic Al$_i$ in surface water and MeHg in fish have also decreased. Policy makers and resource managers now need to know how much further the emissions and deposition of S, N, and Hg will need to be reduced to support recovery of previously damaged resources. We need to know how much air pollution is too much. The critical load provides a management tool that can assist in answering such questions.

This tool has recently been applied by the U.S. EPA to develop an aquatic acidification index (Scheffe et al. 2014). This index incorporates critical-load concepts and is used to establish relationships between atmospheric S and N concentrations and the ANC of surface water. A NAAQS structure is used that relies on concentrations of air pollutants, as is specified by the CAA. The EPA is currently considering a revision of the NAAQS for oxides of S and N, and the EPA's ongoing NAAQS review will consider the aquatic acidification index approach as a way of incorporating critical-load concepts into air quality management.

9.4. LINKAGES TO BIOLOGICAL RESPONSE

Critical-load calculations rely on the availability of biological dose-response relationships. They depend upon identification of good (usually chemical) indicators that reflect biological harm, and such indicators are needed for all sensitive ecosystem elements. The most commonly used chemical indicators for calculations of critical loads of acidification are surface-water ANC, the ratio soil-solution base cations to Al_i, and soil base saturation. A comparable indicator for Hg bioaccumulation in New York has not been endorsed but may be the concentration of Hg in fish fillets. It could be YOY perch because they are widely distributed. However, YOY perch would not bioaccumulate Hg to the same extent as large piscivorous species that might be less widely distributed. Once an appropriate indicator has been selected, it will be important to determine the level of that indicator at which biological harm is manifested. Such data are available for only limited receptors in limited regions. Further work is needed to fully support a range of critical-load analyses in New York and other impacted areas.

CHAPTER 10

Climate Linkages

Ongoing climate change, caused in large part by greenhouse gas emissions, especially CO_2, has already had appreciable impacts on ecosystems in New York (Rosenzweig et al. 2011). However, in view of the much larger projections of likely future temperature increases over the next 100 years (cf. Fasullo and Trenberth 2012), future climate change has much greater potential for ecosystem disruption than what has been observed thus far. Climate change projections suggest multiple consequences, including warmer air and water temperatures, less snowpack, earlier snowmelt, variable effects on precipitation, and greater intensity and duration of extreme climatic events. All of these climate effects will interact with effects related to changes in S, N, and Hg emissions and deposition.

Effects of S and N deposition on natural ecosystems are strongly influenced by climatic conditions, especially temperature, precipitation, and groundwater flow. Both precipitation and temperature control nutrient uptake by vegetation, silicate mineral weathering, and watershed fluxes of base cations in streams (White and Blum 1995). Such linkages are briefly described below.

One of the most important implications of anticipated continuing changes in climate and atmospheric deposition relates to effects of N on forest growth and consequent C sequestration. The potential benefit of increased C sequestration, caused by atmospheric N input, must be weighed against the costs associated with nutrient enrichment, decreased biodiversity, and increased spread of invasive species that may result (Fenn et al. 2003c; Burns et al. 2011).

10.1. TEMPERATURE

Elemental cycling and the influence of air pollutants on natural resources are affected by change in temperature. A host of biogeochemical reactions and processes are sensitive to temperature, including weathering, mineralization, nitrification, and nutrient uptake from the soil into vegetation.

Air temperature also plays an important role in regulating the buildup and melting of the snowpack, which governs much of the episodic acidification that occurs during late winter and spring in the Adirondacks and Catskills. Snowmelt is also an important contributor to N flushing, Al mobilization, and base cation leaching.

Water temperature has a large influence on the quality of fisheries habitat. Cold-water fish species, such as brook trout, require cold water and the accompanying high levels of dissolved oxygen. In areas that have experienced stream acidification, such species are subject to the effects of water acidification at higher elevations and stress from warm water temperatures at lower elevations. This "habitat squeeze" may progressively reduce the size of suitable brook trout stream habitat if climate continues to warm and recovery from past water acidification fails to keep pace with the loss of suitable habitat caused by warming.

Weathering, which largely determines acid sensitivity, is well known to increase with temperature. The cycling of C and the soil C environment also help regulate the weathering flux (Williams et al. 2007). Natural organic and carbonic acids accelerate weathering by decreasing soil pH and by chelating Al (Raulund-Rasmussen et al. 1998).

Key steps in the N cycle, including mineralization, denitrification, and nitrification rates, are influenced by soil temperature. However, temperature effects appear to be most pronounced at the most extreme (i.e., wettest and warmest) sites with the highest and lowest C and N content (Knoepp and Vose 2007). Soil properties act together with climatic conditions to control N cycling in the soil. Maximum mineralization rates have been found to occur at soil temperatures between about 25 and 35 °C (Stark 1996) and moisture near field capacity (Stanford and Epstein 1974), conditions that are rarely encountered in the field (Knoepp and Vose 2007). The response of nitrification to temperature also depends on the environmental conditions at the site (Stark and Firestone 1996; Norton and Stark 2011).

Mitchell et al. (1996b) evaluated patterns of NO_3^- loss in four watersheds located across the northeastern United States, including Arbutus watershed in the Adirondacks and Biscuit Brook watershed in the Catskills. They observed a synchronous pattern in drainage-water NO_3^- concentration among sites from 1983 through 1993. In particular, very high NO_3^- levels occurred following an unusually cold period in December, 1989, followed by decreases in NO_3^- concentration during subsequent years at all sites studied. Such results suggest strong climate control on N cycling and loss in this region.

Pourmokhtarian et al. (2012) modeled potential hydrochemical responses of forests at Hubbard Brook Experimental Forest to climate change using the PnET-BGC model. Expected higher temperature during the twenty-first century stimulated net soil N mineralization and nitrification in the model simulation. The resulting simulated soil and drainage water acidification was partly mitigated by CO_2-induced increased plant growth.

10.2. WATER QUANTITY AND QUALITY

Multiple linkages connect eutrophication, acidification, and biomagnification processes with changing hydrologic conditions. For example, estuaries may vary in their responses to climate change and in the effects of climate change on trophic state. Howarth et al. (2000b) showed that primary production and eutrophication in the Hudson River Estuary can increase substantially in response to decreased freshwater discharge from the watershed. Low freshwater discharge increases water residence time, increases thermal stratification, and deepens the photic zone, and each of these changes can contribute to increased primary production and eutrophication.

In the future, freshwater discharge during summer in the northeastern United States may be lower in response to climate change (Wigley 1999). Estuaries may become more susceptible to eutrophication under conditions of reduced freshwater discharge, with the consequent increased freshwater residence times and increased light penetration (Howarth et al. 2000b). Responses are expected to be most pronounced for estuaries such as the Hudson River that have shorter residence times (Jay et al. 2000).

Acidification processes are also influenced by water quantity, and these interactions affect water quality. Lake and stream pH and ANC tend to be lower during periods of heavy precipitation and snowmelt runoff. Increased water flux dilutes base cation concentrations in drainage water and increases the flux of natural organic acids from soils and wetlands to surface waters. Hydrologic changes can create stress for cold water–adapted fish and invertebrates during low base-flow conditions. Consequent water warming can deplete dissolved oxygen, exacerbating the habitat squeeze. This occurs when stream water is too acidic in the headwaters and too warm in lowland zones, thereby constraining the amount of suitable habitat to a narrow (and mobile) band of stream length. Linkages between climate and acidification damage/recovery processes are poorly known. See the review of Burns et al. (2011) for further discussion of these issues.

CHAPTER 11

Linkages with Ecosystem Services

11.1. FOREST AND FRESHWATER AQUATIC RESOURCES

A general conceptual model was developed by Sullivan (2012) and Sullivan and McDonnell (2012) for incorporating critical loads and ecosystem service into a framework for assessing environmental effects of atmospheric S and N deposition and the preservation of sustainable ecosystems and their services (Figure 11.1). The model starts by identifying a forest, soil, lake, or stream as a potentially acid-sensitive resource for which we desire to develop a critical load to support resource damage assessment. Identification of services that are considered to be at risk or already damaged informs decisions about prioritizing which resources are to be protected or restored. The critical load provides a management tool to assist in the goal of protecting, maintaining, and restoring ecosystem services. Decisions about which resource(s) to protect and at what level of protection determine choices among available options in proceeding clockwise around the science/policy assessment loop shown in the schematic illustration in Figure 11.1 (Sullivan 2012).

First, a resource must be selected for protection or for recovery from previous damage. This might be the presence of brook trout for aquatic resources or the regeneration of sugar maple for terrestrial resources. The next step is to select an indicator of resource condition, which facilitates tracking of the biological health of the resource. Because of difficulties associated with monitoring biological conditions directly, a chemical indicator is most commonly used that reflects the

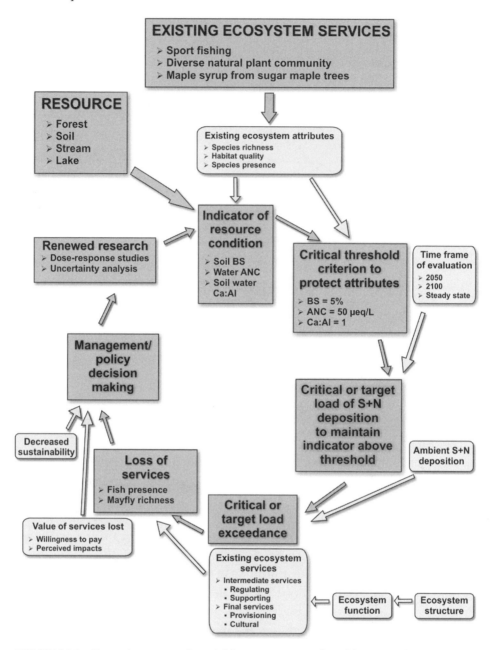

FIGURE 11.1. General conceptual model for assessment of acidification effects of atmospheric S and N deposition, incorporating the critical load and ecosystem service concepts. BS is percent base saturation of soil; ANC is acid-neutralizing capacity of surface water. Critical threshold criteria are provided only as possible examples. Source: Sullivan 2012.

likelihood of biological damage in response to the stressor under consideration. For example, surface-water ANC reflects the likelihood of acidification damage to brook trout; mineral soil base saturation reflects the likelihood of acidification damage to sugar maple. One or more critical threshold criteria are specified for each indicator, based on known tipping points or dose-response relationships. These threshold criteria connect the indicator with a biological response such as species presence or taxonomic richness. Appropriate tipping-point values might be ANC = 50 µeq/L and base saturation = 12% (Sullivan and McDonnell 2012). If the pollutant input level is found to be in exceedance of the critical load calculated to maintain the indicator below its tipping point, there is an increased likelihood that there will be a loss of one or more ecosystem services. Based on the estimated loss or gain of ecosystem services, management or policy decisions might be made to change emissions regulations, implement remediation actions, or make other decisions. If it is judged that the indicator threshold criteria are not sufficiently known or that the assessment uncertainty is too high, additional research or analysis may be warranted (Sullivan 2012).

The best-understood and most immediate incremental changes to ecosystem services in acidified regions of New York due to changes in emissions controls include benefits related to species composition and abundance of fish in lakes and streams, zooplankton in lakes, and benthic macroinvertebrates in streams. Additional important benefits relate to the health, growth, and regeneration of sensitive tree species, especially sugar maple and red spruce. Banzhaf et al. (2006) assigned values to natural resource improvements in the Adirondack Mountains that are attributed to air pollution control policies. A contingent-valuation survey of New York residents estimated the total economic value of expected ecological improvements in the range of $48 to $107 per household per year. This implies a statewide benefit of $336 million to $749 million per year.

The presence of brook trout and richness of fish species are important indicators of aquatic acidification response, in part because the public tends to place relatively high value on fisheries, especially native sport fisheries. Fish species found in the Adirondack and Catskill regions exhibit a range of sensitivity to surface-water ANC and associated elements of acid-base chemistry of water. Among the fish species typically found in upstate New York streams and lakes, brook trout is known to be relatively insensitive to acidification damage. Brook trout can tolerate ANC concentrations of 50 µeq/L or above with no expected effects on the likelihood of presence. Lakes and streams with ANC below about 0 µeq/L generally do not support brook trout or any other species of fish. There is often a positive relationship between pH or ANC and number of fish species for ANC values between about 0 and 50 to 100 µeq/L. Decreased pH and ANC and increased Al_i concentrations contribute to a sequential loss of fish species, beginning with the most acid-sensitive species.

A value of ANC of 50 or 100 µeq/L can be used to provide a chemical benchmark to indicate full acid-base chemistry support for both macroinvertebrate and fish populations. The value of 50 µeq/L is often chosen because physical lake or stream habitat features, rather than surface water acid-base chemistry, are expected to be the dominant factors that limit the number of available niches and associated biodiversity at ANC > 50 µeq/L and because model simulations suggest that many surface waters had preindustrial ANC between 40 and 100 µeq/L in the absence of acidic deposition (Sullivan and McDonnell 2012). Data from more than 1,400 Adirondack lakes support this ANC benchmark for fish in New York.

The combination of acidic deposition and other stressors is an important contributor to the decline of sugar maple trees in the Adirondack Mountains (Sullivan et al. 2013). Soil base saturation is one chemical indicator that provides insight into the level to which the terrestrial ecosystem has acidified and may be susceptible to associated biological effects. This chemical indicator may also be used to monitor the extent of acidification or recovery that occurs in forested ecosystems as deposition rates of S and N change. Low base-saturation values predominate in the soil B horizon in the areas where soil and surface water acidification from acidic deposition have been most pronounced.

Thus, for assessing critical load and its exceedance for the purpose of evaluating effects of acidification on ecosystem services in the Adirondack Mountains region, reasonable thresholds are surface-water ANC = 50 µeq/L for aquatic ecosystems and soil base saturation = 12% for terrestrial ecosystems. The ANC critical value of 50 µeq/L is generally protective of brook trout presence, fish species richness, and stream macroinvertebrate richness. The soil base saturation value of 12% is generally protective of sugar maple regeneration and vigor (Sullivan et al. 2013). Maintenance of stream ANC and soil base saturation above the tipping point values is expected to protect ecosystem services associated with fisheries and sugar maple resources. Nevertheless, some adverse impacts on both may occur at stream ANC between 50 and 100 µeq/L and at B-horizon soil base saturation between 12 and 20%.

A framework for valuation of ecosystem services is shown in Figure 11.2. Value can be expressed as ecological, sociocultural, or economic elements. Assignment of value is critical to an examination of trade-offs between ecosystem damage and preservation or restoration. The difficulty of translating services into a common currency remains an important limitation of the ecosystem services approach. The value to the public of a given ecosystem element must be weighed against the costs and/or loss of conventional benefits associated with natural resource protection, restoration, and preservation in informed environmental decision-making.

11.2. COASTAL RESOURCES

Impacts of air pollution and other pollution sources on ecosystem services are also important for coastal waters in New York. Ecological impacts on the quality of coastal waters in the eastern United States in response to nutrient inputs contribute to substantial economic losses associated with tourism and recreational

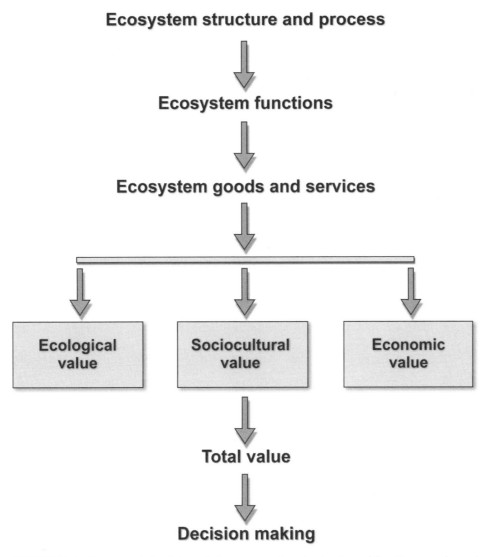

FIGURE 11.2. Framework for integrated assessment and valuation of function, goods, and services of an ecosystem. Adapted from de Groot et al. 2002.

and commercial fisheries (Lipton and Hicks 2003). Adverse effects of eutrophication on native biota can also increase opportunities for colonization by invasive species (Whitall et al. 2007).

There have been substantial losses of indirect and non-use values due to eutrophication (Turner et al. 1999). Morgan and Owens (2001) provided an example of estimated economic benefits associated with reducing the eutrophication of Chesapeake Bay. Many benefits relate to estuarine aesthetics. Quantification of cause-and-effect relationships can be difficult. In many cases, there is not a clear link between nutrient loading and symptoms of eutrophication. Estuaries with similar loading can differ widely in symptom expression (Tett et al. 2003).

CHAPTER 12

Active Intervention

Many Adirondack lakes have shown improvements in acid-base chemistry over the last few decades. A prime example is Brooktrout Lake, located in the southwestern portion of the Adirondack Park, which decreased in pH to values below 5 in the 1970s in response to acidic deposition. By the mid-1980s, fish had completely disappeared from the lake and water clarity had increased due to effects of acidification on zooplankton. In recent years, the pH has risen to near 6 and water clarity has decreased. Brook trout were reintroduced and are surviving (J. Sutherland, NYSDEC, and J. Duckett, ALSC, personal communication, November 2013).

Recent and ongoing reductions in acidic, nutrient, and toxic substance deposition have likely contributed to improved resource conditions and some degree of associated chemical and biological recovery in other surface waters. However, as described in detail in prior chapters of this book, some documented effects (especially of acidification) may be difficult (if not impossible) to reverse. Fish and other life forms that have been eliminated from various lakes and streams by water acidification may return on their own once chemical conditions are improved, but there is no guarantee that reduced deposition will fully restore suitable acid-base chemistry. Even if it does, there is no guarantee that any or all species will be able to return to a given water body. For example, there may be barriers to movement in the form of physical barriers or stream reaches that remain chemically unsuitable. Stocking of lakes and streams with previously extirpated fish species may be warranted. Reintroduction of other life forms may be more challenging, as standard practices may not exist for accomplishing such reintroductions. Reductions in N or Hg deposition would similarly be expected to restore nutrient balances and reduce the extent of Hg bioaccumulation and toxicity. It remains unclear, however, how long it might take, even under substantially reduced deposition levels, to restore ecosystems to generally unimpacted conditions. Soils and biota in impacted watersheds have built up stores of N and Hg in response to many decades of elevated deposition. Establishment of new equilibrium states at deposition levels that no longer cause biological harm may take considerable time. People can take steps to speed up the recovery process and reverse the effects of acidification. However, it is not known what steps might help speed up recovery processes for nutrient N enrichment and Hg neurotoxicity.

Liming of acidic soils has long been known to increase the bioavailability of nutrients to plants, including Ca, Mg, and P. Liming is a common tool used to speed up chemical recovery for acidification (Olem 1991). Some of the early work in this country on development of liming techniques was conducted in the Adirondack Mountains (Woods Lake). In-stream or in-lake liming can effectively increase water pH and ANC to desirable levels. But such liming requires an ongoing investment of effort and resources at frequent intervals. Whole watershed liming may be a more effective tool instead of liming a water body, as it may not need to be repeated and because it can promote the chemical recovery of both terrestrial and aquatic ecosystems.

Direct application of lime to acidified surface waters typically shows an immediate increase in pH and ANC and a decrease in Al_i in solution. Results of lime application to watershed soils differ. The effects of terrestrial lime application are more gradual and long lasting (Davis and Goldstein 1988).

An experimental watershed liming study was conducted at the Woods Lake watershed in the western Adirondacks over a three-year period (Driscoll et al. 1996). Calcite ($CaCO_3$) was applied by helicopter at an application rate of about 6.9 mg/ha to two experimental subwatersheds, one containing a wetland. Focused studies were conducted on the watershed responses, including soil and soil-water chemistry, stream and lake chemistry, forest and wetland vegetation, soil microbial processes, and aquatic biota. Selected results were compared to earlier results of direct-lake addition of $CaCO_3$. Driscoll et al. (1996) described the overall design and major outcomes of the project. Results of specific studies were provided in a series of papers, including Blette and Newton (1996), Simmons et al. (1996), and Burns (1996). Lake chemistry exhibited gradual increases in pH, ANC, and Ca^{2+} concentrations in the water column subsequent to watershed liming. These results were in marked contrast to the abrupt increases measured in response to direct application of $CaCO_3$ to the lake surface, which was subsequently followed by lake reacidification. The supply of ANC from the added Ca to surface waters was largely the result of $CaCO_3$ dissolution in wetlands; there was relatively little dissolution in upland soils. Effects on forest vegetation were relatively minor two years after treatment. Episodic acidification of stream water and near-shore lake water, an important biological stressor before the treatment, was essentially eliminated by the liming (Cirmo and Driscoll 1996; Newton et al. 1996). Transport of Al_i from the watershed to the lake was substantially reduced and concentrations of dissolved organic carbon increased. Improvements in water quality extended down to the lake outlet streams (Burns 1996). Fisheries in the lake and tributary streams improved. Prior to liming, brook trout spawning had been restricted to poor-quality lake substrate; spawning did not occur in the inlet tributaries because they were too acidic. The improved water quality extended the suitable spawning habitat for brook trout to the tributary streams (Schofield and Keleher 1996), suggesting that tributary spawning populations of brook trout can be reestablished where they have been eliminated by stream acidification.

Battles et al. (2013) added a Ca silicate material to a paired watershed at Hubbard Brook Experimental Forest and tracked the response of the treated forest to consequent reduction in soil acidity. The experimental Ca addition amounted to about 1,000 kg Ca/ha/yr. The largest increases in exchangeable soil Ca occurred at the highest elevations of the treated watershed. Over the decade after treatment, the increase in soil pH and exchangeable Ca moved down through the soil profile. Restoration of soil Ca caused a substantial increase in forest biomass accumulation. In the reference (untreated) watershed, live tree biomass remained more or less constant during the approximately 15 years before treatment. Forest biomass decreased subsequent to a large ice storm that occurred in 1998, affecting both the treatment and reference watersheds. From 2006 to 2011, there was a substantial increase in biomass in the treated watershed compared to the reference watershed. Above-ground net primary productivity and leaf area index were significantly higher on the treated watershed. These results suggest that improved Ca nutrition promoted above-ground net primary productivity by facilitating increased photosynthetic surface area. The most responsive tree species was sugar maple. Significant decreases in canopy dieback and crown transparency were observed. The strongest responses of sugar maple to the forest liming were observed at the higher elevations in the treated watershed, where soils had been most acidified prior to treatment. It is possible that species other than sugar maple also benefited from the liming (Battles et al. 2013).

The work of Battles et al. (2013) suggests that soil acidification at Hubbard Brook Experimental Forest caused in part by acidic deposition can be reversed with human intervention. This research suggests that biological as well as chemical recovery can be expected to occur. Recent research findings further suggest that such terrestrial ecosystem recovery will probably not be realized at many locations solely in response to reduced levels of acidic deposition. It is likely that on-the-ground restoration work will also be needed.

Liming may be counterproductive in lakes or streams that have been naturally acidified by organic acids (Bishop et al. 2001). Streams that are naturally acidic have rather unique assemblages of species that are well adapted to ambient chemical conditions (Petrin et al. 2007a, 2007b). Wetlands and their stream connections are important contributors to ecosystem health. Liming streams that are strongly influenced by wetlands is likely to disturb the natural structure and function of these important ecosystems (McKie et al. 2006; Pound et al. 2013). Pound et al. (2013) concluded that more research is needed to help develop an improved combination of approaches for reversing inorganic chemical acidification in such a way as to restore the diversity of species harmed by human-caused acidification while at the same time protecting the species that are adapted to natural organic acidity.

CHAPTER 13

Summary and Important Data Gaps and Recommendations

In areas of New York that are both acid sensitive and receive substantial acidic deposition, forest soils have been acidified and soil base saturation levels have decreased. Decreases in concentrations of exchangeable base cations in the O, A, and B soil horizons have been documented over the past several decades. Acidic deposition has been shown to be an important causal factor. Effects have been especially pronounced in the Adirondack and Catskill Mountains, and have included decreased Ca-to-Al ratios in soil solution.

Soil acidification, Al toxicity, and exposure of plant foliage to acidic deposition have collectively contributed to decline in red spruce, sugar maple, and perhaps other tree and understory plant species in New York. Effects have included reduced growth and increased stress and mortality in overstory trees, reduced regeneration, and likely changes in plant species distributions.

The natural movement of Al in the soil profile has been altered by acidic deposition, resulting in greater transport of Al_i from the soil to surface waters. In sufficiently high concentrations, Al_i in soil solution or surface water can be toxic to plant roots and to aquatic life.

Sulfate derived from atmospheric S deposition has been the principal anthropogenic agent of soil acidification in New York. Sulfate leaches base cations from the soil, reduces the soil base saturation, and decreases the ability of the soil to neutralize acidity deposited from the atmosphere. Sulfur retention through anion adsorption or incorporation into soil organic matter prevents or retards SO_4^{2-}

leaching temporarily. Nevertheless, most deposited SO_4^{2-} in New York eventually leaches to surface waters, where it can contribute to water acidification.

Atmospheric deposition of N has decreased the C-to-N ratio in soils, and has contributed to increased net nitrification. Nitrogen availability in excess of ecological demand has become more widespread. This N saturation is indicated by elevated NO_3^- concentrations in surface waters during the growing season. Such effects have been documented in both the Adirondack and Catskill Mountains.

Lakes and streams in some parts of New York are sensitive to episodic and chronic acidification in response to SO_x, NO_x, and NH_x deposition. Sensitive lakes and streams tend to occur at moderate to high elevations in areas that have base-poor bedrock, high relief, and shallow soils. Sensitive lakes and streams are numerous in some regions, including the Adirondack and Catskill Mountains, where more than one-fourth of the surface waters have been adversely affected by water acidification.

Aquatic biota have been impacted by acidification at virtually all levels of the food web. Effects have been most clearly documented for fish, aquatic insects, and algae. Some species and some life stages are more sensitive than others. The most sensitive species have been eliminated from many lakes and streams, and taxonomic richness has decreased. In extreme cases, all fish species have been eliminated from acidified waters.

Atmospheric inputs of N can alleviate nutrient deficiencies and increase growth of some plants at the expense of others. As a consequence, N deposition can alter competitive relationships among plant species and affect species composition and diversity. The extent to which such effects have occurred in New York is not known. It is well known, however, that atmospheric N deposition can increase the growth of plants in the short term. Enhanced plant growth generally occurs mainly above the ground level. This can cause changes in the shoot-to-root ratio, which can be detrimental to the plant because of decreased resistance to environmental stresses such as drought. Data are not sufficient with which to quantify any increase or decrease in forest growth that may have occurred in New York in response to N deposition, partly because trees have been exposed to multiple stresses simultaneously.

It appears that reduced tree growth and increased tree mortality attributable to soil acidification and/or N saturation have occurred in localized areas that have experienced depletion of soil Ca and increased Al_i in soil solution. Increased N availability from N deposition in New York has likely caused some plant species to flourish at the expense of other species. Some native species may have been replaced at some locations by other (often nonnative) opportunistic species. The extent and magnitude of these changes in New York are not known.

Estuaries and near-coastal marine waters tend to be sensitive to nutrient enrichment from N inputs, and many such waters in the United States, including some in New York, have experienced moderate to severe eutrophication. There is a scientific consensus that N-driven eutrophication of shallow estuaries has increased over the past several decades and that the degradation of coastal ecosystems is now a widespread occurrence. Atmospheric N deposition constitutes one source of N input to estuaries that have experienced eutrophication. Other non-atmospheric contributions have also been significant and in most cases have been dominant. Effects of severe eutrophication can include O_2 depletion, excessive growth of algae, fish kills, and damage to aquatic organisms at many trophic levels. Atmospheric N deposition contributes to these damages. The contribution of atmospheric N sources to estuarine and marine waters compared to that of non-atmospheric sources is poorly known. It is clear, however, that the atmospheric contribution is usually (but not always) a small percentage of the overall human-caused N input.

Fresh waters can be sensitive to nutrient enrichment effects from N deposition. In general, sensitive waters tend to be those that are highly oligotrophic. These are often found at high elevations in remote areas. Such lakes and streams are sometimes N limited, and atmospheric N deposition can cause or contribute to eutrophication in these waters. Primary productivity in N-limited freshwaters can be increased with the addition of even small amounts of N. Eutrophication increases algal biomass, alters algal species assemblages, and may affect food webs. The extent of such effects in the lakes and streams in New York is not known.

Great Lakes waters in New York have experienced eutrophication in the past as a consequence of nutrient inputs. This problem has been largely attributed to P addition, but N has also played a role. The majority of the nutrient inputs to the Great Lakes have been of non-atmospheric origin, but atmospheric N deposition has been a contributor. In more recent years, the introduction and massive proliferation of nonnative invasive mussels have further altered the trophic condition of these lakes and obfuscated our understanding of the role of atmospheric nutrient input.

Increased growth of silicate-utilizing diatoms as a result of NO_3^-- and PO_4^--induced eutrophication and the subsequent removal of fixed biogenic Si via sedimentation has brought about changes in the ratios of nutrient elements Si, N, and P in near-coastal marine waters. In turn, such changes can cause shifts from diatoms to nonsiliceous phytoplankton. Such changes are also expected to affect other levels of the food web.

Atmospheric deposition of N_r adds a critical element (N) to terrestrial, transitional, and aquatic environments. In many ecosystems, this element is growth limiting to algae and plants at the base of the food web. Thus, even relatively small additions of N in atmospheric deposition can alter the biotic makeup of sensitive ecosystems and affect key ecosystem processes. It is often difficult to quantify such responses because nutrient enrichment effects from N addition interact with the cycling of other elements and with the influence of climatic variability and both natural and human-caused disturbances.

Progress has been made in recent years in improving scientific understanding of nutrient enrichment effects from atmospheric N deposition. Knowledge gains have occurred in research areas such as ecological stoichiometry, characterization of ecosystem sensitivity, N vs. P limitation in aquatic ecosystems, quantification of de-nitrification, and use of stable isotopes to improve our understanding of N and other element cycling. Nutrient enrichment or eutrophication effects of N deposition are now known to be widespread and to occur in a variety of sensitive ecosystems.

Acidic deposition and associated effects have decreased in New York over the past few decades, largely in response to emissions controls mandated by the CAA and its amendments. Despite improvements in air quality and acidic deposition, however, damage to natural ecosystems has been only partially reversed. Hundreds of streams and lakes in New York are still acidic. Most are located in the Adirondack and Catskill Mountains. Many more streams and lakes are not quite acidic but have sustained some degree of biological damage. The health of some tree species, mainly red spruce and sugar maple, has been adversely affected by acidic deposition. Various species of fish and other aquatic animals can no longer exist in acidified waters.

Some recovery is under way. Sulfur deposition to acid-sensitive watersheds in New York has decreased by about half during the past three decades. The pH of rainfall has increased by about one-fourth of a pH unit. Nitrogen deposition has also decreased, but to a lesser extent. Hopes for chemical and biological recovery have varied among scientists. The expectation has shifted from the rapid recovery expected during the 1980s to the prevailing current scientific consensus of a long-term (decades or longer) gradual process of chemical recovery. This recovery will involve depleting the acids that have accumulated in the soils of sensitive watersheds through more than a century of acidic deposition and restoring some of the Ca and other base cations that have been lost from the soil.

Important conclusions from several decades of atmospheric deposition effects research in New York include the following:

- Atmospheric S deposition to acid-sensitive watersheds in New York has been declining for three decades and continues to decline. Nitrogen deposition was stable for many years but started to decline after the turn of the twenty-first century. Existing and anticipated federal and state regulations are expected to contribute to further reductions in S and N deposition during the next decade.
- Many natural ecosystems are able to buffer some level of air pollution. Sensitivity varies across the landscape. Only some of the soil, plant, and water resources in New York are sensitive to existing air pollution levels, and many remain largely unaffected.
- Even with the recent reductions in acidic deposition, sensitive regions of New York experience air pollution levels higher than many other parts of the United States.

- Despite some recent improvements in water chemistry, levels of acidity in many lakes and streams in the Adirondack and Catskill Mountains regions are still harmful to many species of fish and other aquatic life forms.
- Exposure of vegetation to continued relatively high levels of acidic and nutrient deposition and associated soil acidification will likely cause further damage to plants and a decline in the abundance of the more sensitive plant species.
- Estuaries and coastal marine waters in New York have experienced varying degrees of eutrophication from N addition, and atmospheric N deposition is partly responsible.
- Some of the damages that have occurred to soils and aquatic ecosystems are only partially reversible over the next century, even with additional reductions in pollutant emissions.
- Bioaccumulation of Hg in fish and piscivorous wildlife has been pronounced in many lakes and streams throughout New York. This has been attributed in part to relatively high levels of atmosphere Hg deposition, but more importantly to watershed characteristics that promote Hg methylation, transport, and bioaccumulation.

Monitoring data have supported the consensus that surface-water chemical recovery will be slow and limited. Model projections have suggested that the documented ongoing chemical recovery is not necessarily sustainable without further emissions reductions. The ability of watershed soils to neutralize acidic deposition changes over time. As atmospheric acidity is neutralized and S and N continue to be stored in the soil, the ability of a given watershed to neutralize deposited acids is progressively reduced. This affects the capacity of watersheds to recover from acidification. To the extent that emissions controls are delayed, those emissions controls will probably be less effective in mitigating the damage from soil and water acidification.

Soils can partially recover base cation reserves over time in response to reduced levels of acidic deposition. However, the recovery potential of soil-exchangeable base cations is dependent on weathering rates and operates over periods of many decades or longer. Modest emissions controls can decrease the rate of soil acidification but are unlikely to result in substantially improved soil condition for a long period of time.

Enhanced leaching of NO_3^- to surface waters in the northeastern United States generally occurs above a threshold of atmospheric N deposition of about 10 kg N/ha/yr. Estimated levels of N deposition in many parts of New York exceed this value. Relationships between atmospheric N deposition and NO_3^- leaching are complicated by past and current land use and climatic variations. Forest insect infestation and disease can also affect N cycling and leaching. Projections of changes in N retention and release from terrestrial ecosystems in response to changing levels of atmospheric N deposition are therefore highly uncertain.

Atmospheric deposition of Hg has contributed to increased concentrations of Hg in New York surface waters, especially in the Adirondack Mountains. Some of this deposited Hg becomes methylated in aquatic ecosystems and bioaccumulated and biomagnified in the food web. Because Hg methylation is largely carried out by bacteria that require S to sustain their metabolic activities, atmospheric S deposition has further contributed to Hg methylation, biomagnification, and toxicity in New York surface waters. The Adirondack region has been identified as a hotspot for Hg methylation owing to an abundance of lakes and wetlands and the common occurrence of low surface-water ANC and pH. High organic content of water, which is mainly a consequence of wetland influence, and increased acidity provide ideal conditions for Hg bioaccumulation in fish and wildlife. Such conditions contribute to high body burdens of Hg in fish and toxic concentrations in piscivorous wildlife, including common loon, bald eagle, river otter, and others.

Air pollution impacts appear to be generally improving in New York, or at least getting worse more slowly than previously. Emissions controls have reduced impacts in New York and throughout much of the eastern United States. More time is needed to allow various air pollution emissions control policies to have their full beneficial effect. In addition, our society faces ongoing decisions concerning the desirability of further cuts in emissions of S, N, and Hg and the possible need for additional remediation. Some surface waters that have been recovering during the past few decades may partially reacidify in the near future unless emissions levels are further reduced. Costs and benefits must be weighed to determine the levels of air pollution and acidic deposition that balance competing environmental, economic, and societal goals. The future of New York's outstanding natural resources, especially those in the coastal areas and the Adirondack and Catskill Mountains, will be affected by the decisions that are made.

As detailed in previous sections of this book, a great deal of research has been conducted on air pollution and its effects in New York. Much of that work has occurred within the last three decades. Scientific understanding of the effects of atmospherically deposited acid precursors, nutrients, and toxic materials has advanced to a relatively high level. However, many uncertainties and unknowns remain.

Existing knowledge has been formulated into various mathematical models of atmospheric processes, atmospheric deposition, nutrient cycling, biomagnification, and acid-base chemistry. Model projections shed light on preindustrial conditions and likely future changes in resource condition as pollutant emissions and deposition levels decline in response to emissions controls. In many cases, the more sensitive biological receptors have been identified. Chemical and biological monitoring programs have yielded critical information documenting changes over time in resource condition. Nevertheless, much remains to be done.

Some of the research areas in greatest need of further work in New York are highlighted below. This is not intended to be an exhaustive list. A great many other questions have yet to be answered. In many cases, we don't even know what the important questions are or will be. These suggestions for future work are not presented in any particular known order of importance.

- Chemical indicators of ecological damage in response to atmospheric deposition have been identified along with their tipping points or levels at which biological effects become apparent in sensitive regions of New York. These include, for example, water ANC of 50 μeq/L to protect fish biodiversity and B-horizon soil base saturation of 12% to protect sugar maple regeneration. Experimental and observational research is needed to identify additional sensitive biological resources, chemical indicators of adverse ecological effects, and tipping points for those indicators.

- Target loads of S deposition to protect fish biodiversity have been calculated within a management time frame for Adirondack lakes. These target-load results have been extrapolated to the full population of Adirondack lakes larger than 1 ha. Analogous target-load values have not been calculated for acid-sensitive streams, which are common in both the Catskill and Adirondack Mountains regions. It is likely that streams are generally more sensitive to acidification than lakes. Furthermore, such target-load estimates for protecting stream resources in acid-sensitive portions of New York need to be compared with ambient S deposition loads to determine the locations and magnitude of target-load exceedances. Calculations of target load and exceedance for protection against acidification could greatly benefit from an improved ability to predict mineral weathering across the landscape.

- Some Catskill and Adirondack Mountain watersheds leach considerable NO_3^- under ambient N deposition loads, suggesting partial N saturation. The climatic, management, and chemical factors that govern this apparent N saturation are poorly known. In addition, very little empirical information is available regarding the timing of transitions between conditions of saturation and nonsaturation under varying levels of atmospheric N deposition. More research is needed to discern these relationships and drivers of ecosystem change.

- Adirondack lakes that have shown signs of recent recovery from acidification exhibit a range of anticipated responses, including increased ANC and pH and decreased concentrations of base cations and Al. However, many lakes also show increased levels of dissolved organic carbon, a response that limits the extent of ANC recovery. Although this dissolved organic carbon response was predicted 25 years ago, the magnitude of the response has been surprising to many in the scientific community who study acidic deposition. Additional work is needed to better understand this response so that it can be effectively incorporated into mathematical acid-base chemistry models.

- Great progress has been made over the past three decades in developing a clear understanding of key processes that govern surface-water acidification and its effects and recovery processes in response to more recent decreases in acidic deposition. In addition, some (albeit more limited) progress has been made in understanding N cycling and saturation; biological responses to improvements in water acid-base chemistry; nutrient enrichment effects; and Hg deposition, methylation, and biomagnification. Chemical and biological long-term monitoring data have been critical to these scientific advancements. Sustained monitoring will be needed to continue to improve scientific understanding and tackle emerging scientific issues surrounding climate change and the effects of nonnative species and other threats and perturbations that interact with processes that govern ecosystem responses to atmospheric deposition.

- It has been established that Hg methylation and biomagnification in fish and wildlife are partly controlled by atmospheric Hg deposition. It has also been established, however, that watershed cycling of C and S are important drivers of Hg accumulation, especially in piscivorous predators. At many locations in New York, C and/or S cycling are likely more important than Hg deposition in this regard. Better information is needed to describe the influence of ongoing deacidification on Hg cycling in New York watersheds. In particular, scientists and policy makers need to know how Hg levels in fish will respond to continued decreases in S input and increases in C mobilization in watersheds that have relatively high Hg concentrations in fish.

- The concept of critical and target loads has not yet been successfully applied to atmospheric Hg deposition in New York. Scientists and policy makers need to know the levels of Hg input that lead to unacceptably high concentrations of Hg in fish and wildlife and the interactions between such critical Hg levels and watershed characteristics that control Hg cycling. We also need to identify, if they exist, the tipping points of S deposition below which Hg methylation becomes limiting for Hg biomagnification.

- Identification of ecosystem services that are impacted by atmospheric deposition and the extent to which these ecosystem services are influenced by emissions controls is at a very early stage of development in New York and elsewhere. Similarly, valuations of ecosystem services that may be lost or recovered in response to changes in atmospheric pollutant emissions have not been well quantified. Further research is needed at the interface between environmental science and policy to determine the values to human society that are expected to be lost or gained in response to existing and future emissions and controls on those emissions.

- It is well known that the effects of atmospheric S, N, and Hg deposition are modified by climatic conditions, especially temperature, soil moisture, high flows, drought, and snowpack dynamics. It is also well known that the regional climate is changing and that conditions in the coming decades and centuries may be quite different from the climate of today. More work is needed to quantify these linkages, allowing improved incorporation of climate change into mathematical models of ecosystem response to air pollutants.

- As lake and stream chemical conditions continue to improve in response to emissions controls and decreases in S, N, and Hg deposition, aquatic conditions may become increasingly hospitable for fish and other aquatic life and perhaps less conducive to Hg methylation and biomagnification. More work is needed to determine the effects of this chemical recovery on biological communities and the extent to which human intervention will be needed to reintroduce species to habitats from which they were extirpated in previous decades.

Glossary

acid anion. Negatively charged ion that does not interact to form an association with hydrogen ion in the pH range of most natural waters.

acid-neutralizing capacity (ANC). The capacity of a solution to neutralize strong acids. The components of ANC include weak bases (carbonate species, dissociated organic acids, alumino-hydroxides, borates, and silicates) and strong bases (primarily hydroxide). ANC can be measured in the laboratory by the Gran titration procedure or defined as the difference in the equivalent concentrations of the base cations and the mineral acid anions. ANC is a key indicator of the ability of water to neutralize acid or acidifying inputs and depends largely on associated biogeophysical characteristics of the watershed.

acid-base chemistry. The reaction of acids (proton donors) with bases (proton acceptors). In the context of this book, the reactions of natural and anthropogenic acids and bases, the result of which is described in terms of pH and the acid-neutralizing capacity of the system.

acidic deposition. Transfer of acids and acidifying compounds from the atmosphere to terrestrial and aquatic environments via rain, snow, sleet, hail, cloud droplets, particles, and gas exchange.

acidic lake or stream. A lake or stream in which the acid-neutralizing capacity is less than or equal to 0 μeq/L.

acidification. The decrease of acid-neutralizing capacity in water or base saturation in soil caused by acidic inputs derived from natural or anthropogenic processes.

acidified. Pertaining to a natural water that has experienced a decrease in acid-neutralizing capacity or a soil that has experienced a decrease in base saturation.

algae. Photosynthetic, often microscopic and planktonic, organisms occurring in marine and freshwater ecosystems.

algal bloom. A reproductive explosion of algae in a lake, river, or ocean.

alpine. The biogeographic zone made up of slopes above the tree line characterized by the presence of rosette-forming herbaceous plants and low, shrubby, slow-growing woody plants.

anion. A negatively charged ion.

anammox. Oxidation of NH_4^+ under aerobic conditions, yielding N_2.

anoxia. Absence of dissolved O_2.

anthropogenic. Of, relating to, derived from, or caused by humans or related to human activities or actions.

atmosphere. The gaseous envelope surrounding the earth. The dry atmosphere consists almost entirely of N and O_2 plus trace gases that include carbon dioxide and ozone.

atmospheric deposition. The transfer of substances from the atmosphere to terrestrial and aquatic environments via rain, snow, sleet, hail, cloud droplets, particles, and gas exchange.

base cation. An alkali or alkaline earth metal cation (Ca^{2+}, Mg^{2+}, K^+, Na^+).

base saturation. The proportion of total soil cation exchange capacity that is occupied by exchangeable base cations, i.e., by Ca^{2+}, Mg^{2+}, K^+, and Na^+.

benthic macroinvertebrates. Animals without backbones that inhabit the bottom substrates of streams and lakes.

bioaccumulation. The phenomenon wherein toxic chemicals are progressively amassed in greater qualities as individuals farther up the food chain ingest matter containing those chemicals.

biodiversity. The total diversity of all organisms and ecosystems at various spatial scales (from genes to entire biomes).

biological effects. Changes in biological (organism, population, community-level) structure and/or function in response to some causal agent; also referred to as biological response.

biomagnification. A progressive increase in concentration of a chemical with increasing trophic level.

buffering capacity. The ability of a body of water and its watershed to neutralize introduced acid.

calibration. (1) Process of checking, adjusting, or standardizing operating characteristics of instruments or coefficients in a mathematical model with empirical data of known quality. (2) The process of evaluating and adjusting if necessary the scale readings of an instrument with a known standard for the physical quantity to be measured.

carbon sequestration. The process of increasing the carbon storage of a reservoir/pool other than the atmosphere.

catchment. An area that collects and drains rainwater (also called a watershed).

cation. A positively charged ion.

cation exchange capacity. The sum total of exchangeable cations per unit mass that a soil can adsorb under a specified set of conditions.

chronic acidification. The decrease of acid-neutralizing capacity in a lake or stream over a period of decades or longer, generally in response to gradual leaching of ionic constituents.

circumneutral. Close to neutrality with respect to pH (neutral pH = 7); in natural waters, generally pH 6–8.

climate. Climate in a narrow sense is usually defined as the average weather," or more rigorously, as the statistical description in terms of the mean and variability of relevant quantities over a period of time ranging from months to thousands or millions of years. These quantities are most often surface variables such as temperature, precipitation, and wind. Climate in a wider sense is the state of the climate system, including a statistical description.

The classical period of time is 30 years, as defined by the World Meteorological Organization.

critical load. A quantitative estimate of an exposure to one or more pollutants below which significant harmful effects on specified sensitive elements of the environment do not occur according to present knowledge.

decomposition. The microbially mediated reaction that converts solid or dissolved organic matter into simpler elements or molecules (also called decay or mineralization).

deposition velocity. The value by which atmospheric concentrations at some specific height above a particular kind of surface are multiplied in order to compute the dry deposition rate to that surface. Deposition velocities change with time, the kind of surface, and the chemical in question.

dissolved organic carbon. Organic (derived from the breakdown of plant or animal material) carbon that is dissolved or filterable (typically 0.45-μm pore size) in a water sample.

drainage lake. A lake that has a permanent surface water inlet and outlet.

denitrification. Bacterially mediated anaerobic reduction of oxidized nitrogen (e.g., nitrate or nitrite) to gaseous nitrogen (e.g., N_2O or N_2, but not the reduction to NH_3).

dry deposition. The transfer of gases and particles from the atmosphere to surfaces in the absence of precipitation (e.g., rain or snow) or occult deposition.

ecosystem. The interactive system formed by all living organisms and their abiotic (physical and chemical) environment within a given area. Ecosystems cover a hierarchy of spatial scales and can consist of the entire globe, biomes at the continental scale, or small, well-circumscribed systems such as a small pond.

ecosystem service. An ecological process or function that has monetary or nonmonetary value to individuals or society at large. These include supporting services such as productivity or biodiversity maintenance; provisioning services such as food, fiber, or fish; regulating services such as climate regulation or carbon sequestration; and cultural services such as tourism or spiritual and aesthetic appreciation.

episodic acidification. The short-term decrease of acid-neutralizing capacity from a lake or stream. This process has a time scale of hours to weeks and is usually associated with hydrological events.

euphotic zone. The sunlit portion of the water column of a lake or estuary where net photosynthesis is positive.

eutrophication. The process whereby a body of water becomes overenriched in nutrients, resulting in increased productivity (of algae or aquatic plants), often accompanied by decreased dissolved oxygen levels and a variety of ecosystem effects.

evapotranspiration. The process by which water is returned to the atmosphere through direct evaporation or transpiration by vegetation.

fen. A wetland phase in the development of the natural succession from open lake to reedbed and fen to woodland as peat develops and its surface rises.

greenhouse gas. Gaseous constituents of the atmosphere, both natural and anthropogenic, that absorb and emit radiation at specific wavelengths within the spectrum of infrared radiation emitted by the earth's surface, the atmosphere, and clouds. These gases include water vapor (H_2O), CO_2, N_2O, methane (CH_4), and O_3.

gross primary production. The total C fixed by plants through photosynthesis.

groundwater. Water below the land surface in a saturated zone in soil or rock.

heathland. A wide-open landscape dominated by low-growing woody vegetation such as heathers and heathland grasses. Heathlands generally occur on acidic, nutrient-poor, and often sandy and well-drained soils.

hindcast. An estimate of the probability of some past event or condition as a result of rational study and analysis of available data, often with the assistance of a computer model.

hydrologic(al) event. Pertaining to increased water flow or discharge resulting from rainfall or snowmelt.

hydrology. The study of water, including its occurrence, circulation, and distribution; its chemical and physical properties; its interactions with the environment; and its relationships to living organisms.

hyporheic zone. The region below and alongside a stream in which the water has intermediate characteristics between those of surface water and groundwater.

hypoxia. Deficiency of dissolved O_2 in an aquatic system.

index of biotic integrity. An index that provides an assessment of biological condition, based on a combination of metrics.

invasive species. A nonnative species that aggressively expands its range and population density into a region, often through outcompeting or otherwise dominating native species.

labile. Reactive or mobile.

leaching. The removal of soil elements or applied chemicals by movement of water through the soil.

macrophyte. A rooted aquatic plant.

MAGIC. Model of Acidification of Groundwater in Catchments, a watershed ion balance model.

mercury methylation. The process of converting mercury from an inorganic form to an organic form that bioaccumulates in food webs; methylation is accomplished mainly by sulfate-reducing bacteria, largely in wetland soils, sediments, and anoxic bottom waters.

metaphyton. A group of loosely aggregated algae and cyanobacteria that are neither strongly attached to substrata nor fully suspended in the water column.

mineralization. The microbially mediated reaction that converts solid or dissolved organic matter into inorganic matter (also called decay or decomposition).

mitigation. Amelioration of adverse impacts caused by a stressor such as acidic deposition at the source (e.g., emissions reductions) or the receptor (e.g., lake liming).

model. An abstraction or representation of a system, generally on a smaller scale.

net primary production. The gross primary production minus autotrophic respiration, i.e., the sum of metabolic processes that contribute to plant growth and maintenance.

neutral salt effect. The process whereby a neutral salt, such as NaCl, is added to the soil and some of the neutral salt cation (i.e., Na^+) is exchanged for H^+ on the soil ion exchanger, thereby acidifying the drainage water.

nitrification. A biological process by which ammonia is oxidized to nitrite and then to nitrate. This process is primarily accomplished by autotrophic nitrifying bacteria.

nitrogen saturation. The condition whereby N inputs from atmospheric deposition and other sources exceed the biological requirements of the ecosystem.

occult deposition. The transfer of gases and particles from the atmosphere to surfaces by fog or mist.

ombrotrophic bog. An acidic peat-accumulating wetland that is isolated from surrounding groundwater and surface water and thus receives most of its water from precipitation, resulting in a lack of nutrients derived from soil and rock such as Ca.

organic acid. Heterogeneous group of acids that consists of organic molecules and generally possesses carboxyl (-COOH) groups or phenolic (C-OH) groups. In this book, these acids are derived from the decomposition of naturally occurring organic materials such as plants.

parameter. A term used to identify a characteristic, feature, or measurable factor that helps define a particular system.

pH. The negative logarithm of hydrogen ion activity. The pH scale is generally presented from 1 (most acidic) to 14 (most alkaline); a difference of one pH unit indicates a tenfold change in hydrogen ion activity.

phytoplankton. Photosynthesizing microscopic organisms that inhabit the upper sunlit layer of water bodies, which are often the principal agents of carbon fixation in aquatic ecosystems.

piscivore. An organism that feeds on fish.

plankton. Small (often microscopic) plant-like or animal species that spend part or all of their lives in open water.

PnET-BGC. Photosynthesis and Evapotranspiration–Biogeochemistry Model; A model of water, C, and nitrogen balance coupled with a biogeochemistry model.

pool. In ecological systems, the supply of an element or compound, such as exchangeable or weatherable cations or adsorbed sulfate, in a defined component of the ecosystem.

population. For the purpose of this book, (1) the total number of lakes or streams within a given geographical region or the total number of lakes or streams with a given set of defined chemical, physical, or biological characteristics; or (2) an assemblage of organisms of the same species inhabiting a given ecosystem.

precision. A measure of the capacity of a method to provide reproducible measurements of a particular analyte (often represented by variance).

primary productivity. All forms of production accomplished by plants and algae.

scenario. For this book, one possible atmospheric deposition sequence after implementation of a control or mitigation strategy and the effects associated with this deposition sequence.

sensitivity. For this book, the degree to which a system is affected, either adversely or beneficially, by NO_x, SO_x, and/or Hg pollution (e.g. acidification, N-nutrient enrichment, toxicity, etc.). The effect may be direct (e.g., a change in growth in response to a change in the mean, range, or variability of N deposition) or indirect (e.g., a change in growth due to the direct effect of N-altering competitive dynamics between species and decreased biodiversity.

species richness. The number of species occurring in a given ecosystem, generally estimated by the number of species caught and identified using a standard sampling regime.

stomata. Small openings (pores) in the leaf surfaces of plants through which gas exchange occurs.

sulfate-reducing bacteria. Bacteria that reduce sulfur from an oxidized (i.e., SO_4^{2-}) to a reduced (i.e., H_2S) state; such bacteria are common in wetlands, sediments, and other reducing environments.

surface runoff. The water that travels over the land surface to the nearest surface stream; runoff from a drainage basin that has not passed beneath the surface since precipitation.

target load. A critical load that includes a time and/or management component.

taxonomic richness. The number of different taxa (species, genera, family, etc.) within a community.

throughfall. The precipitation that passes through the canopy of a forest and/ or shrub/herbaceous plant community before reaching the earth's surface.

total maximum daily load. The maximum loading of a pollutant that a water body can receive from point and nonpoint (including atmospheric deposition) sources without exceeding the allowable standards.

trophic level. The position an organism occupies in a food chain.

turnover. The interval of time in which the density stratification of a lake is disrupted by seasonal temperature variation, generally resulting in complete mixing of the water mass.

valuation. The economic or non-economic process of determining either the value of maintaining a given ecosystem type, state, or condition or the value of a change in an ecosystem, its components, or the services it provides.

variable. A quantity that is designated by a numeric value during a mathematical calculation.

vulnerability. For this book, the degree to which a system is susceptible to and unable to cope with adverse effects of NO_x, Hg, and/or SO_x air pollution.

watershed. The geographic area from which surface water drains into a particular lake or point along a stream.

welfare effects. As defined by the Clean Air Act, effects of air pollutant oxides of S or N on soils, water, crops, vegetation, human-made materials, animals, wildlife, weather, visibility and climate; damage to and deterioration of property; hazards to transportation; and effects on economic values and on personal comfort and well-being, whether caused by transformation, conversion, or combination with other air pollutants.

wet deposition. The transfer of gases, particles, and dissolved solutes from the atmosphere to earth's surface by rain, snow or other forms of precipitation.

wetland. A transitional, regularly waterlogged area of poorly drained soils (often between an aquatic and a terrestrial ecosystem) that is fed from rain, surface water, or groundwater. Wetlands are characterized by a prevalence of vegetation adapted to live in saturated soil conditions.

zooplankton. The animal forms of plankton, including crustaceans, rotifers, pelagic (open water) insect larvae, and aquatic mites.

References Cited

Aber, J. D., and C. T. Driscoll. 1997. Effects of land use, climate variation, and N deposition on N cycling and C storage in northern hardwood forests. Glob. Biogeochem. Cycles 11(4):639–648.

Aber, J. D., and C. A. Federer. 1992. A generalized, lumped-parameter model of photosynthesis, evapotranspiration and net primary production in temperate and boreal forest ecosystems. Oecologia 92:463–474.

Aber, J. D., K. J. Nadelhoffer, P. Steudler, and J. M. Mellilo. 1989. Nitrogen saturation in northern forest ecosystems. BioScience 39(6):378–386.

Aber, J. D., J. M. Mellilo, K. J. Nadelhoffer, J. Pastor, and R. D. Boone. 1991. Factors controlling nitrogen cycling and nitrogen saturation in northern temperate forest ecosystems. Ecol. Appl. 1(3):303–315.

Aber, J. D., S. V. Ollinger, and C. T. Driscoll. 1997. Modeling nitrogen saturation in forest ecosystems in response to land use and atmospheric deposition. Ecol. Model. 101:61–78.

Aber, J. D., W. McDowell, K. Nadelhoffer, A. Magill, G. Berntson, M. Kamakea, S. McNulty, W. Currie, L. Rustad, and I. Fernandez. 1998. Nitrogen saturation in temperate forest ecosystems; hypotheses revisited. BioScience 48(11):921–933.

Aber, J. D., C. L. Goodale, S. V. Ollinger, M.-L. Smith, A. H. Magill, M. E. Martin, R. A. Hallett, and J. L. Stoddard. 2003. Is nitrogen deposition altering the nitrogen status of northeastern forests? BioScience 53(4):375–389.

Achermann, B., and R. Bobbink, eds. 2003. Empirical critical loads for nitrogen. 327. Environmental Documentation no. 164. Swiss Agency for the Environment, Forests, and Landscape, Berne, Switzerland.

Aiken, G., M. Haitzer, J. N. Ryan, and K. Nagy. 2003. Interactions between dissolved organic matter and mercury in the Florida Everglades. Journal du Physique IV 107:29–32.

Alberti, M., D. Booth, K. Hill, B. Coburn, C. Avolio, S. Coe, and D. Spirandelli. 2007. The impact of urban patterns on aquatic ecosystems: An empirical analysis in Puget lowland sub-basins. Landsc. Urban Plan. 80:345–361.

Aldous, A. R. 2002. Nitrogen retention by Sphagnum mosses: Responses to atmospheric nitrogen deposition and drought. Can. J. Bot. 80(7):721–731.

Alewell, C., and M. Gehre. 1999. Patterns of stable S isotopes in a forested catchment as indicators for biological S turnover. Biogeochemistry 47(3):319–333.

Alexander, R. B., R. A. Smith, G. E. Schwarz, S. D. Preston, J. W. Brakebill, R. Srinivasan, and P. A. Pacheco. 2001. Atmospheric nitrogen flux from the watersheds of major estuaries of the United States: An application of the SPARROW watershed model. *In* Nordic Council of Ministers R. Valigura, R. Alexander, M. Castro, T. Meyers, H. Paerl, P. Stacey, and R. E. Turner, eds., Nitrogen loading in coastal water bodies: An atmospheric perspective. American Geophysical Union, Washington, DC, 119–170.

Alexander, R. B., R. J. Johnes, E. W. Boyer, and R. A. Smith. 2002. A comparison of models for estimating the riverine export of nitrogen from large watersheds. Biogeochemistry 57/58:295–339.

Alvo, R., D. J. T. Hussell, and M. Berrill. 1988. The breeding success of common loons (*Gavia immer*) in relation to alkalinity and other lake characteristics in Ontario. Can. J. Zool. 66:746–752.

Anderson, D. M., Y. Kaoru, and A. W. White. 2000. Estimated annual economic impacts from harmful algal blooms (HABs) in the United States. WHOI-2000-11. Woods Hole Oceanographic Institute Technical Report.

Arnold, C. L., and C. J. Gibbons. 1996. Impervious surface coverage: Emergence of a key environmental indicator. J. Am. Planning Assoc. 62(2):243–258.

Arrigo, K. R. 2005. Marine microorganisms and global nutrient cycles. Nature 437:349–355.

Ashby, J. A., W. B. Bowden, and P. S. Murdoch. 1998. Controls on denitrification in riparian soils in headwater catchments of a hardwood forest in the Catskill Mountains, U.S.A. Soil Biol. Biogeochem. 30:853–864.

Auer, M. T., L. M. Tomlinson, S. N. Higgins, S. Y. Malkin, E. T. Howell, and H. A. Bootsma. 2010. Great Lakes *Cladophora* in the 21st century: Same algae, different ecosystem. J. Great Lakes Res. 36:248–255.

Axler, R. P., C. Rose, and C. Tikkanen. 1994. Phytoplankton nutrient deficiency as related to atmospheric nitrogen deposition in northern Minnesota acid-sensitive lakes. Can. J. Fish. Aquat. Sci. 51:1281–1296.

Baeseman, J. L., R. L. Smith, and J. Silverstein. 2006. Denitrification potential in stream sediments impacted by acid mine drainage: Effects of pH, various electron donors, and iron. Microb. Ecol. 51(2):232–241.

Bailey, S. W., S. B. Horsley, R. P. Long, and R. A. Hallett. 1999. Influence of geologic and pedologic factors on health of sugar maple on the Allegheny Plateau, U.S. *In* S. B. Horsley and R. P. Long, eds. Sugar maple ecology and health: Proceedings of an international symposium. USDA Forest Service, Radnor, PA, 63–65.

Bailey, S. W., D. C. Buso, and G. E. Likens. 2003. Implications of sodium mass balance for interpreting the calcium cycle of a forested ecosystem. Ecology 84(2):471–484.

Bailey, S. W., S. B. Horsley, and R. P. Long. 2005. Thirty years of change in forest soils of the Allegheny Plateau, Pennsylvania. Soil Sci. Soc. Am. J. 69(3):681–690.

Baker, J. P., and S. W. Christensen. 1991. Effects of acidification on biological communities in aquatic ecosystems. *In* D. F. Charles, ed. Acidic deposition and aquatic ecosystems: Regional case studies. Springer-Verlag, New York, 83–106.

Baker, J. P., and C. L. Schofield. 1982. Aluminum toxicity to fish in acidic waters. Water Air Soil Pollut. 18:289–309.

Baker, J. P., D. P. Bernard, S. W. Christensen, and M. J. Sale. 1990a. Biological effects of changes in surface water acid-base chemistry. State of Science/Technology Report 13. National Acid Precipitation Assessment Program, Washington, DC.

Baker, J. P., S. A. Gherini, S. W. Christensen, C. T. Driscoll, J. Gallagher, R. K. Munson, R. M. Newton, K. H. Reckhow, and C. L. Schofield. 1990b. Adirondack lakes survey: An interpretive analysis of fish communities and water chemistry, 1984–1987. Adirondack Lakes Survey Corporation, Ray Brook, NY.

Baker, L. A., P. R. Kauffman, A. T. Herlihy, and J. M. Eilers. 1990c. Current status of surface water acid-base chemistry. State of Science/Technology Report 9. National Acid Precipitation Assessment Program, Washington, DC.

Baker, J. P., J. Van Sickle, C. J. Gagen, D. R. DeWalle, W. E. Sharpe, R. F. Carline, B. P. Baldigo, P. S. Murdoch, D. W. Bath, W. A. Kretser, H. A. Simonin, and P. J. Wigington, Jr. 1996. Episodic acidification of small streams in the northeastern United States: Effects on fish populations. Ecol. Appl. 6(2):423–437.

Baker, L. A., and P. L. Brezonik. 1988. Dynamic model of in-lake alkalinity generation. Water Resour. Res. 24:65–74.

Baldigo, B. P., and G. B. Lawrence. 2000. Composition of fish communities in relation to stream acidification and habitat in the Neversink River, New York. Trans. Am. Fish. Soc. 129:60–76.

Baldigo, B. P., and G. B. Lawrence. 2001. Effects of stream acidification and habitat on fish populations of a North American river. Aquat. Sci. 63:196–222.

Baldigo, B. P., and P. S. Murdoch. 1997. Effect of stream acidification and inorganic aluminum on mortality of brook trout (*Salvelinus fontinalis*) in the Catskill Mountains, New York. Can. J. Fish. Aquat. Sci. 54(3):603–615.

Baldigo, B. P., P. S. Murdoch, and D. A. Burns. 2005. Stream acidification and mortality of brook trout (*Salvelinus fontinalis*) in response to timber harvest in Catskill Mountain watersheds, New York, USA. Can. J. Fish. Aquat. Sci. 42(5):1168–1183.

Baldigo, B. P., G. B. Lawrence, and H. A. Simonin. 2007. Persistent mortality of brook trout in episodically acidified streams of the southwestern Adirondack Mountains, New York. Trans. Am. Fish. Soc. 136:121–134.

Baldigo, B. P., G. B. Lawrence, R. W. Bode, H. A. Simonin, K. M. Roy, and A. J. Smith. 2009. Impacts of acidification on macroinvertebrate communities in streams of the western Adirondack Mountains, New York, USA. Ecol. Indicat. 9:226–239.

Banzhaf, S., D. Burtraw, D. Evans, and A. Krupnick. 2006. Valuation of natural resource improvements in the Adirondacks. Land Econ. 82(3):445–464.

Barbiero, R. P., M. L. Tuchman, and E. S. Millard. 2006. Post-dreissenid increases in transparency during summer stratification in the offshore waters of Lake Ontario: Is a reduction in whiting events the cause? J. Great Lakes Res. 32:131–141.

Baron, J. S. 2006. Hindcasting nitrogen deposition to determine ecological critical load. Ecol. Appl. 16(2):433–439.

Barr, J. F. 1996. Aspects of common loon (*Gavia immer*) feeding biology on its breeding ground. Hydrobiology 32:119–144.

Basu, N., A. M. Scheuhammer, S. Bursian, K. Rouvinen-Watt, J. Elliott, and H. M. Chan. 2007a. Mink as a sentinel in environmental health. Environ. Res. 103:130–144.

Basu, N., A. M. Scheuhammer, K. Rouvinen-Watt, N. Grochowina, R. D. Evans, M. O'Brien, and H. M. Chan. 2007b. Decreased N-methyl-daspartic acid (NMDA) receptor levels are associated with mercury exposure in wild and captive mink. Neurotoxicology 28:587–593.

Battles, J. J., T. J. Fahey, C. T. Driscoll, J. D. Blum, and C. E. Johnson. 2013. Restoring soil calcium reverses forest decline. Environ. Sci. Technol. Lett. 1(1):15–19.

Bechard, M. J., D. N. Perkins, G. S. Kaltenecker, and S. Alsup. 2007. Mercury contamination in Idaho bald eagles, (*Haliaeetus leucocephalus*). Bull. Environ. Contam. Toxicol. 83:698–702.

Beckvar, N., T. M. Dillon, and L. B. Read. 2005. Approaches for linking whole-body fish tissue residues of mercury or DDT to biological effects thresholds. Environ. Toxicol. Chem. 24:2094–2105.

Bedford, B. L., M. R. Walbridge, and A. Aldous. 1999. Patterns in nutrient availability and plant diversity of temperature North American wetlands. Ecol. Soc. Am. 80(7):2151–2169.

Bedison, J. E., and B. E. McNeil. 2009. Is the growth of temperate forest trees enhanced along an ambient nitrogen deposition gradient? Ecology 90(7):1736–1742.

Beeton, A. M., and W. T. Edmondson. 1972. The eutrophication problem. J. Fish. Res. Board Can. 29:673–682.

Bennett, E. B. 1986. The nitrifying of Lake Superior. Ambio 15(5):272–275.

Berendse, F., N. van Breemen, H. Rydin, A. Buttler, M. Heijmans, M. R. Hoosbeek, J. A. Lee, A. Mitchell, T. Saarinen, H. Vassander, and B. Wallen. 2001. Raised atmospheric CO_2 levels and increased N deposition cause shifts in plant species composition and production in Sphagnum bogs. Glob. Change Biol. 7(5):591–598.

Bergeron, C. M., J. F. Husak, J. M. Unrine, C. S. Romanek, and W. A. Hopkins. 2007. Influence of feeding ecology on blood mercury concentrations in four species of turtles. Environ. Toxicol. Chem. 26(8):1733–1741.

Bergström, A., and M. Jansson. 2006. Atmospheric nitrogen deposition has caused nitrogen enrichment and eutrophication of lakes in the northern hemisphere. Glob. Change Biol. 12:635–643.

Bernhardt, E. S., and G. E. Likens. 2002. Dissolved organic carbon enrichment alters nitrogen dynamics in a forest stream. Ecology 83:1689–1700.

Bhavsar, S. P., S. B. Gewurtz, D. J. McGoldrick, M. J. Keir, and S. M. Backus. 2010. Changes in mercury levels in Great Lakes fish between 1970s and 2007. Environ. Sci. Technol. 44:3273–3279.

Billen, G., M. Somville, E. DeBecker, and P. Servais. 1985. A nitrogen budget of the Scheldt hydrographical basin. Neth. J. Sea Res. 19:223–230.

Billett, M. F., F. Parker-Jervis, E. A. Fitzpatrick, and M. S. Cresser. 1990. Forest soil chemical changes between 1949/50 and 1987. J. Soil Sci. 41(1):133–145.

Billings, W. D. 1978. Plants and the ecosystem. Wadsworth Publishing Company, Inc., Belmont, CA.

Bishop, K., H. Laudon, J. Hruska, P. Kram, S. Köhler, and L. Löfgren. 2001. Does acidification policy follow research in northern Sweden? The case of natural acidity during the 1990's. Water Air Soil Pollut. 130:1415–1420.

Blackwell, B. D., and C. T. Driscoll. 2011. Deposition and fate of mercury in forests of the Adirondacks. Paper presented at the Annual Environmental Monitoring, Evaluation, and Protection meeting, November 15–16, Albany, NY. http://www.nyserda.ny.gov/Environmental-Research/EMEP/Conferences/2011-EMEP-Conference/2011-Conference-Presentations.aspx.

Blair, R. B. 1990. Water quality and the summer distribution of common loons in Wisconsin. Passenger Pigeon 52:119–126.

Blancher, P. J., and D. K. McNicol. 1988. Breeding biology of tree swallows in relation to wetland acidity. Can. J. Zool. 66:842–849.

Blancher, P. J., and D. K. McNicol. 1991. Tree swallow diet in relation to wetland acidity. Can. J. Zool. 69:2629–2637.

Blette, V. L., and R. M. Newton. 1996. Effects of watershed liming on the soil chemistry of Woods Lake, New York. Biogeochemistry 32:175–194.

Blum, J. D., C. E. Johnson, T. G. Siccama, C. Eagar, T. J. Fahey, G. E. Likens, A. Klaue, C. A. Nezat, and C. T. Driscoll. 2002. Mycorrhizal weathering of apatite as an important calcium source in base-poor forest ecosystems. Nature 417(6890):729–731.

Bobbink, R., M. Hornung, and J. G. M. Roelofs. 1998. The effects of air-borne nitrogen pollutants on species diversity in natural and semi-natural European vegetation. J. Ecol. 86:717–738.

Bobbink, R., M. Ashmore, S. Braun, W. Flückiger, and I. J. J. van den Wyngaert. 2003. Empirical nitrogen critical loads for natural and semi-natural ecosystems: 2002 update. *In* B. Achermann and R. Bobbink, eds. Empirical critical loads for nitrogen. Swiss Agency for Environment, Forest and Landscape SAEFL, Berne, 43–170. http://www.iap.ch/publikationen/nworkshop-background.pdf

Bobbink, R., K. Hicks, J. Galloway, T. Spranger, R. Alkemade, M. Ashmore, M. Bustamante, S. Cinderby, E. Davidson, F. Dentener, B. Emmett, J.-W. Erisman, M. Fenn, F. S. Gilliam, A. Nordin, L. Pardo, and W. De Vries. 2010. Global assessment of nitrogen deposition effects on terrestrial plant diversity: A synthesis. Ecol. Appl. 20(1):30–59.

Bookman, R., C. T. Driscoll, D. R. Engstrom, and S. W. Effler. 2008. Local to regional emission sources affecting mercury fluxes to New York lakes. Atmos. Environ. 42(24):6088–6097.

Bormann, F. H., and G. E. Likens. 1979. Pattern and process in a forested ecosystem. Springer-Verlag, New York.

Bormann, F. H., G. E. Likens, D. W. Fisher, and R. S. Pierce. 1968. Nutrient loss accelerated by clear cutting of a forest ecosystem. Science 159:882–884.

Bormann, F. H., G. E. Likens, and J. M. Melillo. 1977. Nitrogen budget for an aggrading northern hardwood forest ecosystem. Science 196:981–983.

Bowen, J. L., and I. Valiela. 2001. The ecological effects of urbanization of coastal watersheds: Historical increases in nitrogen loads and eutrophication of Waquoit Bay estuaries. Can. J. Fish. Aquat. Sci. 58:1489–1500.

Bowman, W. D. 1994. Accumulation and use of nitrogen and phosphorus following fertilization in two alpine tundra communities. Oikos 70:261–270.

Bowman, W. D., and M. C. Fisk. 2001. Primary production. *In* W. D. Bowman and T. R. Seastedt, eds. Structure and function of an alpine ecosystem: Niwot Ridge, Colorado. Oxford University Press, Oxford, UK, 177–197.

Bowman, W. D., T. A. Theodose, J. C. Schardt, and R. T. Conant. 1993. Constraints of nutrient availability on primary production in two alpine tundra communities. Ecology 74:2085–2097.

Bowman, W. D., J. R. Gartner, K. Holland, and M. Wiedermann. 2006. Nitrogen critical loads for alpine vegetation and terrestrial ecosystem response: Are we there yet? Ecol. Appl. 16(3):1183–1193.

Boyd, J., and S. Banzhaf. 2006. What are ecosystem services? Discussion Paper RFF DP 06-02. Resources for the Future, Washington, DC.

Boyd, J., and S. Banzhaf. 2007. What are ecosystem services? The need for standardized environmental accounting units. Ecol. Econ. 63:616–626.

Boyer, E. W., C. L. Goodale, N. A. Jaworski, and R. W. Howarth. 2002. Anthropogenic nitrogen sources and relationships to riverine nitrogen export in the northeastern U.S.A. Biogeochemistry 57/58:137–169.

Boyer, G. 2006. Toxic cyanobacteria in the Great Lakes: More than just the western basin of Lake Erie. Great Lakes Research Review 7:2–7.

Boynton, W. R., J. H. Garber, R. Summers, and W. M. Kemp. 1995. Inputs, transformations, and transport of nitrogen and phosphorus in Chesapeake Bay and selected tributaries. Estuaries 18:285–314.

Bradley, P. M., D. A. Burns, K. Riva-Murray, M. E. Brigham, D. T. Button, L. C. Chasar, M. Marvin-DiPasquale, M. A. Lowery, and C. A. Journey. 2011. Spatial and seasonal variability of dissolved methylmercury in two stream basins in the eastern United States. Environ. Sci. Technol. 45:2048–2055.

Branfireun, B. A., N. T. Roulet, C. A. Kelly, and W. M. Rudd. 1999. *In situ* sulphate stimulation of mercury methylation in a boreal peatland: Toward a link between acid rain and methylmercury contamination in remote environments. Glob. Biogeochem. Cycles 13(3):743–750.

Breitburg, D. 2002. Effects of hypoxia, and the balance between hypoxia and enrichment, on coastal fishes and fisheries. Estuaries 25:767–781.

Brezonik, P. L., J. G. Eaton, T. M. Frost, P. J. Garrison, T. K. Kratz, C. E. Mach, J. H. McCormick, J. A. Perry, W. A. Rose, C. J. Sampson, B. C. L. Shelley, W. A. Swenson, and K. E. Webster. 1993. Experimental acidification of Little Rock Lake, Wisconsin: Chemical and biological changes over the pH range 6.1 to 4.7. Can. J. Fish. Aquat. Sci. 50:1101–1121.

Bricker, O. P., and K. C. Rice. 1989. Acidic deposition to streams: A geology-based method predicts their sensitivity. Environ. Sci. Technol. 23:379–385.

Bricker, S. B., C. G. Clement, D. E. Pirhalla, S. P. Orlando, and D. G. G. Farrow. 1999. National estuarine eutrophication assessment: Effects of Nutrient enrichment in the nation's estuaries. Special Projects Office and the National Centers for Coastal Ocean Science, National Ocean Service, National Oceanic and Atmospheric Administration, Silver Spring, MD.

Bricker, S. B., J. G. Ferreira, and T. Simas. 2003. An integrated methodology for assessment of estuarine trophic status. Ecol. Model. 169:39–60.

Bricker, S., B. Longstaff, W. Dennison, A. Jones, K. Boicourt, C. Wicks, and J. Woerner. 2007. Effects of nutrient enrichment in the nation's estuaries: A decade of change. NOAA Coastal Ocean Program Decision Analysis Series No. 26. National Centers for Coastal Ocean Science, Silver Spring, MD. http://ccma.nos.noaa.gov/publications/eutroupdate/.

Britton, A. J., and J. M. Fisher. 2007. Interactive effects of nitrogen deposition, fire and grazing on diversity and composition of low-alpine prostrate *Calluna vulgaris* heathland. J. Appl. Ecol. 44:123–135.

Brumbaugh, W. G., D. P. Krabbenhoft, D. R. Helsel, J. G. Weiner, and K. R. Echols. 2001. A national pilot study of mercury contamination of aquatic ecosystems along multiple gradients: Bioaccumulation in fish. Biological Science Report USGS/BRD/BSR-P2001-0009. U.S. Dept. of the Interior, U.S. Geological Survey.

Buckler, D. R., P. M. Mehrle, L. Cleveland, and F. J. Dwyer. 1987. Influence of pH on the toxicity of aluminum and other inorganic contaminants to East Coast striped bass. Water Air Soil Pollut. 35:97–106.

Bulger, A. J., B. J. Cosby, C. A. Dolloff, K. N. Eshleman, J. R. Webb, and J. N. Galloway. 1999. SNP:FISH, Shenandoah National Park: Fish in Sensitive Habitats. Project Final Report to National Park Service. University of Virginia, Charlottesville, VA.

Bulger, A. J., B. J. Cosby, and J. R. Webb. 2000. Current, reconstructed past, and projected future status of brook trout (*Salvelinus fontinalis*) streams in Virginia. Can. J. Fish. Aquat. Sci. 57(7):1515–1523.

Bunyak, J. 1993. Permit application guidance for new air pollution sources. Natural Resources Report NPS/NRAQD/NRR-93/09. U.S. Department of the Interior, National Park Service, Denver, CO.

Burger, J., and M. Gochfeld. 1997. Risk, mercury levels and birds: Related adverse laboratory effects to field monitoring. Environ. Res. 75:160–172.

Burges, S. J., M. S. Wigmosta, and J. M. Meena. 1998. Hydrological effects of land-use change in a zero-order catchment. J. Hydro. Engin. 3:86–97.

Burgess, N., D. C. Evers, and J. D. Kaplan. 1998. Mercury and reproductive success of common loons breeding in the Maritimes. *In* Mercury in Atlantic Canada: A progress report. Environment Canada, Atlantic Region, Sackville, NB, Canada, 104–109.

Burgess, N. M., and M. W. Meyer. 2008. Methylmercury exposure associated with reduced productivity in common loons. Ecotoxicology 17(2):83–91.

Burgin, A. J., and S. K. Hamilton. 2013. Have we overemphasized the role of denitrification in aquatic ecosystems? A review of nitrate removal pathways. Front. Ecol. Environ. 5(2):89–96.

Burkholder, J. M., and H. B. J. Glasgow. 1997. *Pfiesteria piscicida* and other *Pfiesteria*-like dinofagellates: Behavior, impacts, and environmental controls. Limnol. Oceanogr. 42:1052–1075.

Burkholder, J. M., K. M. Mason, and H. B. Glasgow Jr. 1992. Water-column nitrate enrichment promotes decline of eelgrass *Zostera marina*: Evidence from seasonal mesocosm experiments. Mar. Ecol. Prog. Ser. 81:163–178.

Burns, D. A. 1996. The effects of liming an Adirondack lake watershed on downstream water chemistry. Biogeochemistry 32:339–362.

Burns, D. A. 1998. Retention of NO_3^- in an upland stream environment: A mass balance approach. Biogeochemistry 40:73–96.

Burns, D. A., and P. S. Murdoch. 2005. Effects of a clearcut on the net rates of nitrification and N mineralization in a northern hardwood forest, Catskill Mountains, New York, USA. Biogeochemistry 72:123–146.

Burns, D. A., K. Riva-Murray, R. W. Bode, and S. I. Passy. 2006. Changes in stream chemistry and aquatic biota in response to the decreased acidity of atmospheric deposition in the Neversink River Basin, Catskill Mountains, New York, 1987 to 2003. Report 06-16. New York State Energy Research and Development Authority, Albany, NY.

Burns, D. A., J. Klaus, and M. R. McHale. 2007. Recent climate trends and implications for water resources in the Catskill Mountain region, New York, USA. J. Hydrol. 336:155–170.

Burns, D. A., T. Blett, R. Haeuber, and L. Pardo. 2008a. Critical loads as a policy tool for protecting ecosystems from the effects of air pollutants. Front. Ecol. Environ. 6(3):156–159.

Burns, D. A., K. Riva-Murray, R. W. Bode, and S. I. Passy. 2008b. Changes in stream chemistry and biology in response to reduced levels of acid deposition during 1987–2003 in the Neversink River Basin, Catskill Mountains. Ecol. Indicat. 8:191–203.

Burns, D. A., J. A. Lynch, B. J. Cosby, M. E. Fenn, J. S. Baron, and U.S. EPA Clean Air Markets Division. 2011. National Acid Precipitation Assessment Program report to Congress 2011: An integrated assessment. National Science and Technology Council, Washington, DC.

Burns, D. A., K. Riva-Murray, P. M. Bradley, G. R. Aiken, and M. E. Brigham. 2012. Landscape controls on total and methyl Hg in the upper Hudson River basin, New York, USA. J. Geophys. Res. 117(G1): G01034.

Bushey, J. T., A. G. Nallana, M. R. Montesdeoca, and C. T. Driscoll. 2008. Mercury dynamics of a northern hardwood canopy. Atmos. Environ. 42:6905–6914.

Cai, W.-J, X. Hu, W.-J. Huang, M. C. Murrell, J. C. Lehrter, S. E. Lohrenz, W.-C. Chou, W. Zhai, J. T. Hollibaugh, Y. Wang, P. Zhao, X. Guo, K. Gundersen, M. Dai, and G.-C. Gong. 2011. Acidification of subsurface coastal waters enhanced by eutrophication. Nature Geosci. 4(11):766–770.

Cairns, J., Jr., and J. R. Pratt. 1993. A history of biological monitoring using benthic macroinvertebrates. *In* D. M. Rosenberg and V. H. Resh, eds. Freshwater biomonitoring and benthic macroinvertebrates. Chapman and Hall, New York, 10–28.

Camargo, J. A., and A. Alonso. 2006. Ecological and toxicological effects of inorganic nitrogen pollution in aquatic ecosystems: A global assessment. Environ. Int. 32:831–849.

Campbell, J. L., J. W. Hornbeck, W. H. McDowell, D. C. Buso, J. B. Shanley, and G. E. Likens. 2000. Dissolved organic nitrogen budgets for upland, forested ecosystems in New England. Biogeochemistry 49:123–142.

Campbell, J. L., L. E. Rustad, E. W. Boyer, S. F. Christopher, C. T. Driscoll, I. J. Fernandez, P. M. Groffman, D. Houle, J. Kiekbusch, A. H. Magill, M. J. Mitchell, and S. V. Ollinger. 2009. Consequences of climate change for biogeochemical cycling in forests of northeastern North America. Can. J. For. Res. 39:264–284.

Canham, C. D., M. L. Pace, M. J. Papaik, A. G. B. Primack, K. M. Roy, R. J. Maranger, R. P. Curran, and D. M. Spada. 2004. A spatially-explicit watershed-scale analysis of dissolved organic carbon in Adirondack lakes. Ecol. Appl. 14:839–854.

Canham, C. D., M. L. Pace, K. C. Weathers, E. W. McNeil, B. L. Bedford, L. Murphy, and S. Quinn. 2012. Nitrogen deposition and lake nitrogen concentrations: A regional analysis of terrestrial controls and aquatic linkages. Ecosphere 3(7): Article 66.

Cape, J. N., I. D. Leith, D. Fowler, M. B. Murray, L. J. Sheppard, D. Eamus, and R. H. F. Wilson. 1991. Sulfate and ammonium in mist impair the frost hardening of red spruce seedlings. New Phytol. 118:119–126.

Capps, T., S. Mukhi, J. J. Rinchard, C. W. Theodorakis, V. S. Blazer, and R. Patino. 2004. Exposure to perchlorate induces the formation of macrophage aggregates in the trunk kidney of zebrafish and mosquitofish. J. Aquat. Anim. Hlth. 16:145–151.

Carmichael, R. H., A. C. Shriver, and I. Valiela. 2012. Bivalve response to estuarine eutrophication: The balance between enhanced food supply and habitat alterations. J. Shellfish Res. 31(1):1–11.

Carpenter, S. R., T. M. Frost, J. F. Kitchell, T. M. Kratz, D. W. Schindler, J. Schearer, W. G. Sprules, M. J. Vanni, and A. P. Zimmerman. 1991. Patterns of primary production and herbivory in 25 North American lake ecosystems. *In* J. Cole, G. Lovett, and S. Findley, eds. Comparative analyses of ecosystems: Patterns, mechanisms, and theories. Springer-Verlag, New York, 67–96.

Carpenter, S. R., N. F. Caraco, D. L. Correll, R. W. Howarth, A. N. Sharpley, and V. H. Smith. 1998. Nonpoint pollution of surface waters with phosphorus and nitrogen. Ecol. Appl. 8(3):559–568.

Castro, M. S., and C. T. Driscoll. 2002. Atmospheric nitrogen deposition to estuaries in the Mid-Atlantic and Northeastern United States. Environ. Sci. Technol. 36(15):3242–3249.

Castro, M. S., C. T. Driscoll, T. E. Jordan, W. G. Reay, W. R. Boynton, S. P. Seitzinger, R. V. Styles, and J. E. Cable. 2000. Contribution of atmospheric deposition to the total nitrogen loads to thirty-four estuaries on the Atlantic and Gulf Coasts of the United States. *In* R. A. Valigura, R. B. Alexander, M. S. Castro, T. P. Meyers, H. W. Paerl, P. E. Stacey, and R. E. Turner, eds. Nitrogen Loading in Coastal Water Bodies: An atmospheric perspective American Geophysical Union, Washington, DC, 77–106.

Castro, M. S., C. T. Driscoll, T. E. Jordan, W. G. Reay, and W. R. Boynton. 2003. Sources of nitrogen to estuaries in the United States. Estuaries 26(3):803–814.

Chapin, F. S., P. A. Matson, and H. A. Mooney. 2002. Principles of terrestrial ecosystem ecology. Springer-Verlag, New York.

Charbonneau, R., and G. M. Kondolf. 1993. Land use change in California: Nonpoint source water quality impacts. Environ. Manage. 17:453–460.

Charles, D. F., ed. 1991. Acidic deposition and aquatic ecosystems: Regional case studies. Springer-Verlag, New York.

Charles, D. F., and S. A. Norton. 1986. Paleolimnological evidence for trends in atmospheric deposition of acids and metals. *In* Committee on monitoring and assessment of trends in acid deposition. Acid deposition: Long-term trends. National Academy Press Washington, DC, 335–435.

Charles, D. F., R. W. Battarbee, I. Renberg, H. V. van Dam, and J. P. Smol. 1989. Paleoecological analysis of lake acidification trends in North America and Europe using diatoms and chrysophytes. *In* S. A. Norton, S. E. Lindberg, and A.

L. Page, eds. Soils, aquatic processes and lake acidification. Springer-Verlag, New York, 207–276.

Charles, D. F., M. W. Binford, E. T. Furlong, R. A. Hites, M. J. Mitchell, S. A. Norton, F. Oldfield, M. J. Paterson, J. P. Smol, A. J. Uutala, J. R. White, D. F. Whitehead, and R. J. Wise. 1990. Paleoecological investigation of recent lake acidification in the Adirondack Mountains, N.Y. J. Paleolimnol. 3:195–241.

Charlton, M. N., R. Le Sage, and J. E. Milne. 1999. Lake Erie in transition: The 1990's. *In* M. Munawar, T. Edsall, and I. F. Munawar, eds. State of Lake Erie: Past, present and future. Backhuys Publishers, Leiden, The Netherlands, 97–124.

Charlton, M. N., and J. E. Milne. 2004. Review of thirty years of change in Lake Erie water quality. Contribution no. 04-167. National Water Research Institute, Burlington, ON, Canada.

Chen, C. W., S. A. Gherini, N. E. Peters, P. S. Murdoch, R. M. Newton, and R. A. Goldstein. 1984. Hydrologic analyses of acidic and alkaline lakes. Water Resour. Res. 20(12):1875–1882.

Chen, C. Y., R. S. Stemberger, N. C. Kamman, B. M. Mayes, and C. L. Folt. 2005. Patterns of Hg bioaccumulation and transfer in aquatic food webs across multi-lake studies in the Northeast US. Ecotoxicology 14:135–147.

Chen, L., and C. T. Driscoll. 2004. Modeling the response of soil and surface waters in the Adirondack and Catskill regions of New York to changes in atmospheric deposition and historical land disturbance. Atmos. Environ.:1–35.

Chen, L., and C. T. Driscoll. 2005a. Regional application of an integrated biogeochemical model to northern New England and Maine. Ecol. Appl. 15(3):1783–1797.

Chen, L., and C. T. Driscoll. 2005b. Regional assessment of the response of acid-base status of lake-watersheds in the Adirondack region of New York to changes in atmospheric deposition using PnET-BGC. Environ. Sci. Technol. 39:787–794.

Chen, L., and C. T. Driscoll. 2005c. A two-layered model to simulate the seasonal variations in surface water chemistry draining a northern forest watershed. Water Resour. Res. 41:W09425.

Chorover, J., P. M. Vitousek, D. A. Everson, A. M. Esperanze, and D. Turner. 1994. Solution chemistry profiles of mixed-conifer forests before and after fire. Biogeochemistry 26:115–144.

Christensen, S. W., J. J. Beauchamp, J. A. Shaakir-Ali, J. Coe, J. P. Baker, E. P. Smith, and J. Gallagher. 1990. Patterns of fish distribution in relation to lake/watershed characteristics: regression analyses and diagnostics. *In* J. P. Baker, S. A. Gherini, S. W. Christensen, C. T. Driscoll, J. Gallagher, R. K. Munson, R. M. Newton, K. H. Reckhow, and C. L. Schofield, eds. Interpretative analysis of the Adirondack Lakes Survey. Adirondack Lakes Survey Corporation, Ray Brook, NY, A-1–A-62.

Cirmo, C. P., and C. T. Driscoll. 1996. The impacts of a watershed $CaCO_3$ treatment on stream and wetland biogeochemistry in the Adirondack Mountains. Biogeochemistry 32:265–297.

Civerolo, K. L., K. M. Roy, J. L. Stoddard, and G. Sistla. 2011. A comparison of the temporally integrated monitoring of ecosystems and Adirondack long-term monitoring programs in the Adirondack Mountain Region of New York. Water Air Soil Pollut. 222(1–4):285–296.

Clair, T. A., I. F. Dennis, R. Vet, and H. Laudon. 2008. Long-term trends in catchment dissolved organic carbon and nitrogen from three acidified catchments in Nova Scotia, Canada. Biogeochemistry 87:83–97.

Clair, T. A., I. F. Dennis, R. Vet, and G. Weyhenmeyer. 2011. Water chemistry and dissolved organic carbon trends in lakes from Canada's Atlantic Provinces: No recovery from acidification measured after 25 years of lake monitoring. Can. J. Fish. Aquat. Sci. 68(4):663–674.

Clark, C. M., and D. Tilman. 2008. Loss of plant species after chronic low-level nitrogen deposition to prairie grasslands. Nature 451:712–715.

Clark, J. F., H. J. Simpson, R. F. Bopp, and B. L. Deck. 1995. Dissolved oxygen in the lower Hudson Estuary: 1978–93. J. Environ. Eng. 121:760–763.

Clarke, J. F., E. S. Edgerton, and B. E. Martin. 1997. Dry deposition calculations for the Clean Air Status and Trends Network. Atmos. Environ. 31:3667–3678.

Clarkson, T. W. 1992. Mercury: Major issues in environmental health. Environ. Health Perspect. 100:31–38.

Cloern, J. 1987. Turbidity as a control on phytoplankton biomass and productivity in estuaries. Cont. Shelf Res. 7:1367–1382.

Cloern, J. E. 1996. Phytoplankton bloom dynamics in coastal ecosystems: A review with some general lessons from sustained investigation of San Francisco Bay, California. Rev. Geophys. 34:127–168.

Cloern, J. E. 2001. Our evolving conceptual model of the coastal eutrophication problem. Mar. Ecol. Prog. Ser. 210:223–253.

Cole, D., and M. Rapp. 1981. Elemental cycling in forest ecosystems. *In* D. Reichle, ed. Dynamic properties of forest ecosystems. Cambridge University Press, Cambridge, UK, 341–409.

Coleman-Wasik, J. K., C. P. J. Mitchell, D. R. Engstrom, E. B. Swain, B. A. Monson, S. J. Balogh, J. D. Jeremiason, B. A. Branfireun, S. L. Eggert, R. K. Kolka, and J. E. Almendinger. 2012. Methylmercury declines in a boreal peatland when experimental sulfate deposition decreases. Environ. Sci. Technol. 46(12):6663–6671.

Colquhoun, J. R., W. A. Kretser, and M. H. Pfeiffer. 1984. Acidity status update of lakes and streams in New York State. New York State Department of Environmental Conservation, Albany, NY.

Compeau, G. C., and R. Bartha. 1985. Sulfate-reducing bacteria: principal methylators of mercury in anoxic estuarine sediment. Appl. Environ. Microbiol. 50(2):498–502.

Conley, D. J., C. L. Schelske, and E. F. Stoermer. 1993. Modification of the biogeochemical cycle of silica with eutrophication. Mar. Ecol. Prog. Ser. 101:179–192.

Cosby, B. J., G. M. Hornberger, J. N. Galloway, and R. F. Wright. 1985. Modelling the effects of acid deposition: Assessment of a lumped parameter model of soil water and streamwater chemistry. Water Resour. Res. 21(1):51–63.

Cosby, B. J., G. M. Hornberger, P. F. Ryan, and D. M. Wolock. 1989. MAGIC/DDRP final report. U.S. Environmental Protection Agency, Corvallis, OR.

Cosby, B. J., A. Jenkins, R. C. Ferrier, J. D. Miller, and T. A. B. Walker. 1990. Modelling stream acidification in afforested catchments: Long-term reconstructions at two sites in central Scotland. J. Hydrol. 120:143–162.

Cosby, B. J., R. F. Wright, and E. Gjessing. 1995. An acidification model (MAGIC) with organic acids evaluated using whole-catchment manipulations in Norway. J. Hydrol. 170:101–122.

Cosby, B. J., S. A. Norton, and J. S. Kahl. 1996. Using a paired catchment manipulation experiment to evaluate a catchment-scale biogeochemical model. Sci. Total Environ. 183:49–66.

Cosby, B. J., R. C. Ferrier, A. Jenkins, and R. F. Wright. 2001. Modeling the effects of acid deposition: refinements, adjustments and inclusion of nitrogen dynamics in the MAGIC model. Hydrol. Earth Syst. Sci. 5(3):499–517.

Cosby, B. J., J. R. Webb, J. N. Galloway, and F. A. Deviney. 2006. Acidic Deposition Impacts on Natural Resources in Shenandoah National Park. NPS/NER/NRTR-2006/066. U.S. Department of the Interior, National Park Service, Northeast Region, Philadelphia, PA.

Cosper, E. M., and J. C. Cerami. 1996. Assessment of historical phytoplankton characteristics and bloom phenomena in the New York Harbor estuarine and New York bight ecosystems. U.S. Environmental Protection Agency, Region II, and the New York/New Jersey Harbor Estuary Program, Bohemia, NY.

Courchesne, F., B. Cote, J. W. Fyles, W. H. Hendershot, P. M. Biron, A. G. Roy, and M.-C. Turmel. 2005. Recent changes in soil chemistry in a forested ecosystem of southern Quebec, Canada. Soil Sci. Soc. Am. J. 69:1298–1313.

Cowardin, L. M., V. Carter, F. C. Golet, and E. T. LaRoe. 1979. Classification of wetlands and deepwater habitats of the United States. http://www.npwrc.usgs.gov/resource/wetlands/classwet/index.htm (version 4 December 1998).

Cowling, E. B., and L. S. Dochinger. 1980. Effects of acidic deposition on health and productivity of forests. *In* Effects of air pollutants on Mediterranean and temperate forest ecosystems. General Tech. Rep. PSW-43. USDA Forest Service, Berkeley, CA, 165–173.

Craig, B. W., and A. J. Friedland. 1991. Spatial patterns in forest composition and standing dead red spruce in montane forests of the Adirondacks and northern Appalachians. Environ. Monitor. Assess. 18:129–140.

Cronan, C. S., and D. F. Grigal. 1995. Use of calcium/aluminum ratios as indicators of stress in forest ecosystems. J. Environ. Qual. 24:209–226.

Cronan, C. S., and C. L. Schofield. 1979. Aluminum leaching response to acid precipitation: Effects on high elevation watersheds in the Northeast. Science 204:304–306.

Cronan, C. S., W. A. Reiners, R. C. J. Reynolds, and G. E. Lang. 1978. Forest floor leaching: contributions from mineral, organic, and carbonic acids in New Hampshire subalpine forests. Science 200(4339):309–311.

Croteau, M. N., S. N. Luoma, and A. R. Stewart. 2005. Trophic transfer of metals along freshwater food webs: Evidence of cadmium biomagnification in nature. Limnol. Oceanogr. 50(5):1511–1519.

Cumming, B. F., J. P. Smol, J. C. Kingston, D. F. Charles, H. J. B. Birks, K. E. Camburn, S. S. Dixit, A. J. Uutala, and A. R. Selle. 1992. How much acidification has occurred in Adirondack region lakes (New York, USA) since preindustrial times? Can. J. Fish. Aquat. Sci. 49(1):128–141.

Cumming, B. F., K. A. Davey, J. P. Smol, and H. J. B. Birks. 1994. When did acid-sensitive Adirondack lakes (New York, USA) begin to acidify and are they still acidifying? Can. J. Fish. Aquat. Sci. 51:1550–1568.

Dahlgren, R. A., and C. T. Driscoll. 1994. The effects of whole-tree clear cutting on soil processes at Hubbard Brook Experimental Forest, New Hampshire, USA. Plant Soil 58:239–262.

Dail, D. B., E. A. Davidson, and J. Chorover. 2001. Rapid abiotic transformation of nitrate in an acid forest soil. Biogeochemistry 54(2):131–146.

Dale, V. H., L. A. Joyce, S. McNulty, R. P. Neilson, M. P. Ayres, M. D. Flannigan, P. J. Hanson, L. C. Irland, A. E. Lugo, C. J. Peterson, D. Simberloff, F. J. Swanson, B. J. Stocks, and B. M. Wotton. 2001. Climate change and forest disturbances. BioScience 51:723–734.

Dalias, P., J. M. Anderson, P. Bottner, and M. M. Coûteaux. 2002. Temperature responses of net nitrogen mineralization and nitrification in conifer forest soils incubated under standard laboratory conditions. Soil Biol. Biogeochem. 34:691–701.

Dangles, O., B. Malmqvist, and H. Laudon. 2004. Naturally acid freshwater ecosystems are diverse and functional: Evidence from boreal streams. Oikos 104:149–155.

Danz, N. P., G. J. Niemi, R. R. Regal, T. Hollenhorst, L. B. Johnson, J. M. Hanowski, R. P. Axler, J. J. H. Ciborowski, T. Hrabik, V. J. Brady, J. Kelly, J. Morrice, J. C. Brazner, R. W. Howe, C. A. Johnston, and G. E. Host. 2007. Integrated measures of anthropogenic stress in the U.S. Great Lakes Basin. Environ. Manage. 39(5):631–647.

David, M. B., M. J. Mitchell, and T. J. Scott. 1987. Importance of biological processes in the sulfur budget of a northern hardwood ecosystem. Biol. Fertil. Soils 5:258–264.

Davidson, E. A., and R. W. Howarth. 2007. Environmental science: Nutrients in synergy. Nature 449(7165):1000–1001.

Davidson, E. A., and S. Seitzinger. 2006. The enigma of progress in denitrification research. Ecol. Appl. 16(6):2057–2063.

Davis, J. E., and R. A. Goldstein. 1988. Simulated response of an acidic Adirondack Lake Watershed to various liming mitigation strategies. Water Resour. Res. 24(4):525–532.

Davis, R. B., D. S. Anderson, D. F. Charles, J. N. Galloway, W. J. Adams, G. A. Chapman, and W. G. Landis. 1988. Two-hundred-year pH history of Woods, Sagamore, and Panther Lakes in the Adirondack Mountains, New York State. *In* Aquatic toxicology and hazard assessment: 10th Volume. American Society for Testing and Materials, Philadelphia, 89–111.

de Witt, H. A., J. Mulder, P. H. Nygaard, and D. Aamlid. 2001. Testing the aluminum toxicity hypothesis: A field manipulation experiment in mature spruce forest in Norway. Water Air Soil Pollut. 130(1–4 Iii):995–1000.

DeConinck, F. 1980. Major mechanisms in formation of spodic horizons. Geoderma 24:101–128.

Dederen, L. H. T., S. E. Wendelaar Bonga, and R. S. E. W. Leuven. 1987. Ecological and physiological adaptations of the acid-tolerant mudminnow *Umbra pygmaea* (De Kay). Ann. Soc. R. Zool. Belgium 117:277–284.

DeHayes, D. H., P. G. Schaberg, G. J. Hawley, and G. R. Strimbeck. 1999. Acid rain impacts on calcium nutrition and forest health. BioScience 49(1):789–800.

Demers, J. D., J. B. Yavitt, C. T. Driscoll, and M. R. Montesdeoca. 2013. Legacy mercury and stoichiometry with C, N, and S in soil, pore water, and stream water across the upland-wetland interface: The influence of hydrogeologic setting. J. Geophys. Res.: Biogeosciences 118(2):825–841.

Dennis, R. 1997. Using the regional acid deposition model to determine the nitrogen deposition airshed of the Chesapeake Bay watershed. *In* J. E. Baker, ed. Atmospheric deposition of contaminants to the Great Lakes and Coastal Waters. Society of Environmental Toxicology and Chemistry Press, Pensacola, FL, 393–413.

Dennison, W. C. 1993. Assessing water quality with submersed aquatic vegetation. BioScience 43:86–94.

DesGranges, J.-L., and M. Darveau. 1985. Effect of lake acidity and morphometry on the distribution of aquatic birds in southern Quebec. Holarct. Ecol. 8:181–190.

De Groot, R. S., M. A. Wilson, and R. M. J. Boumans. 2002. A typology for the classification, description and valuation of ecosystem functions, goods and services. Ecol. Econ. 41:393–408.

DeWalle, D. R., R. S. Dinicola, and W. E. Sharpe. 1987. Predicting baseflow alkalinity as an index to episodic stream acidification and fish presence. Water Resour. Bull. 23:29–35.

DeWalle, D. R., J. N. Kochenderfer, M. B. Adams, G. W. Miller, F. S. Gilliam, F. Wood, S. S. Odenwald-Clemens, and W. E. Sharpe. 2006. Vegetation and acidification. *In* M. B. Adams, D. R. DeWalle, and J. L. Hom, eds. The Fernow Watershed Acidification Study. Springer, Dordrecht, The Netherlands, 137–188.

Dickson, W. T. 1978. Some effects of the acidification of Swedish lakes. Int. Ver. Theor. Angew. Limnol. Verh. 20:851–856.

Dicosty, R. J., M. A. Callaham Jr, and J. A. Stanturf. 2006. Atmospheric deposition and re-emission of mercury estimated in a prescribed forest-fire experiment in Florida, USA. Water Air Soil Pollut. 176(1–4):77–91.

Dietze, M. C., and P. R. Moorcroft. 2011. Tree mortality in the eastern and central United States: patterns and drivers. Glob. Change Biol. 17(11):3312–3326.

Dillon, T., S. Beckvar, and J. Kern. 2010. Residue-based mercury dose-response in fish: An analysis using lethality-equivalent endpoints. Environ. Toxicol. Chem. 29:2559–2565.

Dise, N. B. 1984. A synoptic survey of headwater streams in Shenandoah National Park, Virginia, to evaluate sensitivity to acidification by acid deposition. MS thesis, Department of Environmental Sciences, University of Virginia, Charlottesville.

Dise, N. B., E. Matzner, and P. Gundersen. 1998. Synthesis of nitrogen pools and fluxes from European forest ecosystems. Water Air Soil Pollut. 105:143–154.

Dittman, J. S., and C. T. Driscoll. 2009. Factors influencing changes in mercury concentrations in lake water and yellow perch (*Perca flavescens*) in Adirondack lakes. Biogeochemistry 93:179–196.

Dodds, W. K. 2003. Misuse of inorganic N and soluble reactive P concentrations to indicate nutrient status of surface waters. J. North Am. Benthol. Soc. 22:171–181.

Dodds, W. K. 2006. Eutrophication and trophic state in rivers and streams. Limnol. Oceanogr. 51(1, part 2):671–680.

Dodds, W. K., W. W. Bouska, J. L. Eitzmann, T. J. Pilger, K. L. Pitts, A. J. Riley, J. T. Schloesser, and D. J. Thornbrugh. 2009. Eutrophication of U.S. freshwaters: Analysis of potential economic damages. Environ. Sci. Technol. 43(1):12–19.

Dodson, S. I., S. E. Arnott, and K. L. Cottingham. 2000. The relationship in lake communities between primary productivity and species richness. Ecology 81:2662–2679.

Dolan, D. M. 1993. Point source loadings of phosphorus to Lake Erie: 1986–1990. J. Great Lakes Res. 19:212–223.

Dortch, Q. 1990. The interaction between ammonium and nitrate uptake in phytoplankton. Mar. Ecol. Prog. Ser. 61:183–201.

Dortch, Q., and T. E. Whitledge. 1992. Does nitrogen or silicon limit phytoplankton production in the Mississippi River plume and nearby regions? Cont. Shelf Res. 12:1293–1309.

Dove, A. 2009. Long-term trends in major ions and nutrients in Lake Ontario. Aquat. Ecosyst. Health Manage 12(3):281–295.

Downing, J. A., and E. McCauley. 1992. The nitrogen-phosphorus relationship in lakes. Limnol. Oceanogr. 37:936–945.

Drevnick, P. E., D. E. Canfield, P. R. Gorski, A. L. C. Shinneman, D. R. Engstrom, D. C. G. Muir, G. R. Smith, P. J. Garrison, L. B. Cleckner, J. P. Hurley, R. B. Noble, R. R. Otter, and J. T. Oris. 2007. Deposition and cycling of sulfur controls mercury accumulation in Isle Royale fish. Environ. Sci. Technol. 41:7266–7272.

Drevnick, P. E., D. R. Engstrom, C. T. Driscoll, E. B. Swain, S. J. Balogh, N. C. Kamman, D. T. Long, D. G. C. Muir, M. J. Parsons, K. R. Rolfhus, and R. Rossmann. 2012. Spatial and temporal patterns of mercury accumulation in lacustrine sediments across the Laurentian Great Lakes region. Environ. Pollut. 161:252–260.

Driscoll, C. T., and J. J. Bisogni. 1984. Weak acid/base systems in dilute acidified lakes and streams of the Adirondack region of New York State. *In* J. L. Schnoor, ed. Modeling of total acid precipitation. Butterworth Publishers, Boston, 53–72.

Driscoll, C. T., and R. M. Newton. 1985. Chemical characteristics of Adirondack lakes. Environ. Sci. Technol. 19:1018–1024.

Driscoll, C. T., and R. Van Dreason. 1993. Seasonal and long-term temporal patterns in the chemistry of Adirondack lakes. Water Air Soil Pollut. 67:319–344.

Driscoll, C. T., J. P. Baker, J. J. Bisogni, and C. L. Schofield. 1980. Effect of aluminum speciation on fish in dilute acidified waters. Nature 284:161–164.

Driscoll, C. T., N. Van Breemen, and J. Mulder. 1985. Aluminum chemistry in a forested spodosol. Soil Sci. Soc. Am. J. 49(2):437–444.

Driscoll, C. T., N. M. Johnson, G. E. Likens, and M. C. Feller. 1988. Effects of acidic deposition on the chemistry of headwater streams: A comparison between Hubbard Brook, New Hampshire, and Jamieson Creek, British Columbia. Water Resour. Res. 24:195–200.

Driscoll, C. T., G. E. Likens, L. O. Hedin, J. S. Eaton, and F. H. Bormann. 1989. Changes in the chemistry of surface waters. Environ. Sci. Technol. 23:137–143.

Driscoll, C. T., R. M. Newton, C. P. Gubala, J. P. Baker, and S. W. Christensen. 1991. Adirondack Mountains. *In* D. F. Charles, ed. Acidic deposition and aquatic ecosystems: Regional case studies. Springer-Verlag, New York, 133–202.

Driscoll, C. T., M. D. Lehtinen, and T. J. Sullivan. 1994. Modeling the acid-base chemistry of organic solutes in Adirondack, New York, lakes. Water Resour. Res. 30:297–306.

Driscoll, C. T., K. M. Postek, W. Kretser, and D. J. Raynal. 1995. Long-term trends in the chemistry of precipitation and lake water in the Adirondack region of New York, USA. Water Air Soil Pollut. 85:583–588.

Driscoll, C. T., C. P. Cirmo, T. J. Fahey, V. L. Blette, P. A. Bukaveckas, D. A. Burns, C. P. Gubala, D. J. Leopold, R. M. Newton, D. J. Raynal, C. L. Schofield, J. B. Yavitt, and D. B. Porcella. 1996. The experimental watershed liming study: Comparison of lake and watershed neutralization strategies. Biogeochemistry 32(3):143–174.

Driscoll, C. T., K. M. Postek, D. Mateti, K. Sequeira, J. D. Aber, W. J. Kretser, M. J. Mitchell, and D. J. Raynal. 1998. The response of lake water in the Adirondack region of New York to changes in acidic deposition. Environ. Sci. Policy 1:185–198.

Driscoll, C. T., G. B. Lawrence, A. J. Bulger, T. J. Butler, C. S. Cronan, C. Eagar, K. F. Lambert, G. E. Likens, J. L. Stoddard, and K. C. Weathers. 2001a. Acidic deposition in the northeastern United States: Sources and inputs, ecosystem effects, and management strategies. BioScience 51(3):180–198.

Driscoll, C. T., G. B. Lawrence, A. J. Bulger, T. J. Butler, C. S. Cronan, C. Eagar, K. F. Lambert, G. E. Likens, J. L. Stoddard, and K. C. Weathers. 2001b. Acid rain revisited: Advances in scientific understanding since the passage of the 1970 and 1990 Clean Air Act amendments. Hubbard Brook Research Foundation, Hanover, NH.

Driscoll, C. T., K. M. Driscoll, K. M. Roy, and M. J. Mitchell. 2003a. Chemical response of lakes in the Adirondack region of New York to declines in acidic deposition. Environ. Sci. Technol. 37:2036–2042.

Driscoll, C. T., D. Whitall, J. Aber, E. Boyer, M. Castro, C. Cronan, C. Goodale, P. Groffman, C. Hopkinson, K. Lambert, G. Lawrence, and S. Ollinger. 2003b. Nitrogen pollution: sources and consequences in the U.S. northeast. Environment 45(7):9–22.

Driscoll, C. T., D. Whitall, J. Aber, E. Boyer, M. Castro, C. Cronan, C. L. Goodale, P. Groffman, C. Hopkinson, K. Lambert, G. Lawrence, and S. Ollinger. 2003c. Nitrogen pollution in the northeastern United States: Sources, effects, and management options. BioScience 53(4):357–374.

Driscoll, C. T., K. M. Driscoll, K. M. Roy, and J. Dukett. 2007a. Changes in the chemistry of lakes in the Adirondack region of New York following declines in acidic deposition. Water Air Soil Pollut. 22(6):1181–1188.

Driscoll, C. T., Y.-J. Han, C. Y. Chen, D. C. Evers, K. F. Lambert, T. M. Holsen, N. C. Kamman, and R. K. Munson. 2007b. Mercury contamination in forest and freshwater ecosystems in the northeastern United States. BioScience 57(1):17–28.

Drohan, J. R., and W. E. Sharpe. 1997. Long-term changes in forest soil acidity in Pennsylvania, U.S.A. Water Air Soil Pollut. 95:299–311.

Duarte, N., L. H. Pardo, and M. J. Robin-Abbott. 2013. Susceptibility of forests in the northeastern USA to nitrogen and sulfur deposition: Critical load exceedance and forest health. Water Air Soil Pollut. 224:1355.

Duchesne, L., and R. Ouimet. 2009. Present-day expansion of American beech in northeastern hardwood forests: Does soil base status matter? Can. J. For. Res. 39(12):2273–2282.

Dueck, T. A., L. J. M. Van der Eerden, B. Breemsterboer, and J. Elderson. 1991. Nitrogen uptake and allocation by *Calluna vulgaris* (L.) Hull and *Deschampsia flexuosa* (L.) Trin. exposed to $^{15}NH_3$. Acta Bot. Neerl. 40:257–267.

Dukett, J. E., N. Aleksic, N. Houck, P. Snyder, P. Casson, and M. Cantwell. 2011. Progress toward clean cloud water at Whiteface Mountain New York. Atmos. Environ. 45(37):6669–6673.

Dukett, J. E., N. Houck, P. Snyder, and S. Capone. 2013. An examination of fish presence using the relationship between base cations (BC) and strong organic anions (RCOO⁻s) in waters collected by the Adirondack Lakes Survey Corporation. Poster presented at New York State Energy Research and Development Authority conference Environmental Monitoring, Evaluation, and Protection in New York: Linking Science and Policy. November 6–7, Albany, NY.

Durance, I., and S. J. Ormerod. 2007. Climate change effects on upland stream macroinvertebrates over a 25-year period. Glob. Change Biol. 13:942–957.

Eagar, C., H. Van Miegroet, S. B. McLaughlin, and N. S. Nicholas. 1996. Evaluation of effects of acidic deposition to terrestrial ecosystems in Class I Areas of the southern Appalachians. A Technical Report to the Southern Appalachian Mountains Initiative.

Egerton-Warburton, L., and E. B. Allen. 2000. Shifts in arbuscular mycorrhizal communities along an anthropogenic nitrogen deposition gradient. Ecol. Appl. 10(2):484–496.

Eilers, J. M., G. E. Glass, K. E. Webster, and J. A. Rogalla. 1983. Hydrologic control of lake susceptibility to acidification. Can. J. Fish. Aquat. Sci. 40:1896–1904.

Eimers, M. C., S. A. Watmough, J. M. Buttle, and P. J. Dillon. 2007. Drought induced sulphate release from a wetland in south-central Ontario. Environ. Model. Assess. 127:399–407.

Ellison, A. M., and N. J. Gotelli. 2002. Nitrogen availability alters the expression of carnivory in the northern pitcher plant *Sarracenia purpurea*. Proc. Nat. Acad. Sci. 99:4409–4412.

Elser, J. J., and C. R. Goldman. 1991. Zooplankton effects on phytoplankton in lakes of contrasting trophic status. Limnol. Oceanogr. 36:64–90.

Elser, J. J., E. R. Marzolf, and C. R. Goldman. 1990. Phosphorus and nitrogen limitation of phytoplankton growth in the freshwaters of North America: A review and critique of experimental enrichments. Can. J. Fish. Aquat. Sci. 47:1468–1477.

Elser, J. J., M. E. S. Bracken, E. E. Cleland, D. S. Gruner, W. S. Harpole, H. Hillebrand II, J. T. Ngai, E. W. Seabloom, J. B. Shurin, and J. E. Smith. 2007. Global analysis of nitrogen and phosphorus limitation of primary producers in freshwater, marine, and terrestrial ecosystems. Ecol. Lett. 10(12):1135–1142.

Elser, J. J., M. Kyle, L. Steger, K. R. Nydick, and J. S. Baron. 2009. Nutrient availability and phytoplankton nutrient limitation across a gradient of atmospheric nitrogen deposition. Ecology 90(11):3062–3073.

Elvir, J. A., G. B. Wiersma, A. S. White, and I. J. Fernandez. 2003. Effects of chronic ammonium sulfate treatment on basal area increment in red spruce and sugar maple at the Bear Brook Watershed in Maine. Can. J. For. Res. 33:862–869.

Emmett, B. A. 1999. The impact of nitrogen on forest soils and feedbacks on tree growth. Water Air Soil Pollut. 116:65–74.

Emmett, B. A., D. Boxman, M. Bredemeier, P. Gunderson, O. J. Kjrnaas, F. Moldan, P. Schleppi, A. Tietema, and R. F. Wright. 1998. Predicting the effects of atmospheric nitrogen deposition in conifer stands: Evidence from the NITREX ecosystem-scale experiments. Ecosystems 1:352–360.

Engstrom, D. R., and E. B. Swain. 1997. Recent declines in atmospheric mercury deposition in the Upper Midwest. Environ. Sci. Technol. 31.

Environment Canada and U.S. Environmental Protection Agency. 2005. State of the Great Lakes 2005. EPA 905-R-06-001.

Ericksen, J. A., M. S. Gustin, D. E. Schorran, D. W. Johnson, S. E. Lindberg, and J. S. Coleman. 2003. Accumulation of atmospheric mercury in forest foliage. Atmos. Environ. 37(12):1613–1622.

Eriksson, E. 1981. Aluminum in groundwater possible solution equilibria. Nord. Hydrol. 12:43–50.

Eriksson, M. O. G. 1983. The role of fish in the selection of lakes by non-piscivorous ducks: Mallard, teal and goldeneye. Wildfowl 34:27–32.

Eriksson, M. O. G. 1986. Fish delivery, production of young, and nest density of Osprey (*Pandion haliaetus*) in southwest Sweden. Can. J. Zool. 64:1961–1965.

Eshleman, K. N. 1988. Predicting regional episodic acidification of surface waters using empirical techniques. Water Resour. Res. 24:1118–1126.

Eshleman, K. N., T. D. Davies, M. Tranter, and P. J. Wigington, Jr. 1995. A two-component mixing model for predicting regional episodic acidification of surface waters during spring snowmelt periods. Water Resour. Res. 31(4):1011–1021.

Eshleman, K. N., R. P. Morgan II, J. R. Webb, F. A. Deviney, and J. N. Galloway. 1998. Temporal patterns of nitrogen leakage from mid-Appalachian forested watersheds: Role of insect defoliation. Water Resour. Res. 34(8):2005–2116.

Eshleman, K. N., R. H. Gardner, S. W. Seagle, N. M. Castro, D. A. Fiscus, J. R. Webb, J. N. Galloway, F. A. Deviney, and A. T. Herlihy. 2000. Effects of disturbance on nitrogen export from forested lands of the Chesapeake Bay watershed. Environ. Monitor. Assess. 63:187–197.

Eshleman, K. N., D. A. Fiscus, N. M. Castro, J. R. Webb, and F. A. Deviney Jr. 2001. Computation and visualization of regional-scale forest disturbance and associated dissolved nitrogen export from Shenandoah National Park, Virginia. The Scientific World 1 (Suppl. 2):539–547.

Eshleman, K. N., D. A. Fiscus, N. M. Castro, J. R. Webb, and A. T. Herlihy. 2004. Regionalization of disturbance-induced nitrogen leakage from mid-Appalachian forests using a linear systems model. Hydrol. Process. 18:2713–2725.

Evans, C. D., and D. T. Monteith. 2001. Chemical trends at lakes and streams in the UK Acid Waters Monitoring Network, 1988–2000: Evidence for recent recovery at a national scale. Hydrol. Earth Syst. Sci. 5:351–366.

Evans, C. D., D. T. Monteith, and D. M. Cooper. 2005. Long-term increases in surface water dissolved organic carbon: Observations, possible causes and environmental impacts. Environ. Pollut. 137(1):55–71.

Evans, C. D., P. J. Chapman, J. M. Clark, D. T. Monteith, and M. S. Cresser. 2006. Alternative explanations for rising dissolved organic carbon export from organic soils. Glob. Change Biol. 12:2044–2053.

Evers, D. C. 2005. Mercury connections: The extent and effects of mercury pollution in northeastern North America. BioDiversity Research Institute, Gorham, ME.

Evers, D. C., K. M. Taylor, A. Major, R. J. Taylor, R. H. Poppenga, and A. M. Scheuhammer. 2003. Common loon eggs as bioindicators of methylmercury availability in North America. Ecotoxicology 12:69–81.

Evers, D. C., O. P. Lane, L. J. Savoy, and W. Goodale. 2004. Assessing the impacts of methylmercury on piscivores using a wildlife criterion value based on the common loon, 1998–2003. Report BRI 2004–2005. Submitted to the Maine Department of Environmental Protection, BioDiversity Res. Inst, Falmouth, ME.

Evers, D. C., N. M. Burgess, L. Champoux, B. Hoskins, A. Major, W. M. Goodale, R. J. Taylor, and R. Poppenga. 2005. Patterns and interpretation of mercury exposure in freshwater avian communities in northeastern North America. Ecotoxicology 14:193–222.

Evers, D. C., Y. Han, C. T. Driscoll, N. C. Kamman, M. W. Goodale, K. F. Lambert, T. M. Holsen, C. Y. Chen, T. A. Clair, and T. Butler. 2007. Biological mercury hotspots in the northeastern United States and southeastern Canada. BioScience 57(1):29–43.

Evers, D. C., L. J. Savoy, C. R. DeSorbo, D. E. Yates, W. Hanson, K. M. Taylor, L. S. Siegel, J. H. Cooley, Jr, M. S. Bank, A. Major, K. Munney, B. F. Mower, H. S. Vogel, N. Schoch, M. Pokras, M. W. Goodale, and J. Fair. 2008. Adverse effects from environmental mercury loads on breeding common loons. Ecotoxicology 17:69–81.

Evers, D., M. Duron, D. E. Yates, and N. Schoch. 2009. An exploratory study of methylmercury availability in terrestrial wildlife of New York and Pennsylvania, 2005–2006. Final report 10–03. New York State Energy Research and Development Authority, Albany, NY.

Evers, D. C., J. D. Paruk, J. W. Mcintyre, and J. F. Barr. 2010. Common loon (*Gavia immer*). *In* A. Poole, ed. The birds of North America. Cornell Laboratory of Ornithology, Ithaca, NY.

Evers, D. C., J. G. Wiener, C. T. Driscoll, D. A. Gay, N. Basu, B. A. Monson, K. F. Lambert, H. A. Morrison, J. T. Morgan, K. A. Williams, and A. G. Soehl. 2011a. Great Lakes mercury connections: The extent and effects of mercury pollution in the Great Lakes region. Report BR1 2011-18. Biodiversity Research Institute, Gorham, ME.

Evers, D. C., K. A. Williams, M. W. Meyer, A. M. Scheuhammer, N. Schoch, A. Gilbert, L. Siegel, R. J. Taylor, R. Poppenga, and C. R. Perkins. 2011b. Spatial gradients of methylmercury for breeding common loons in the Laurentian Great Lakes region. Ecotoxicology 20:1609–1625.

Falconer, I. R. 2005. Is there a human health hazard from microcystins in the drinking water supply? Acta Hydrochimica et Hydrobiologica 33(1):64–71.

Falkengren-Grerup, U., and H. Eriksson. 1990. Changes in soil, vegetation and forest yield between 1947 and 1988 in beech and oak sites of southern Sweden. For. Ecol. Manage. 38(1–2):37–53.

Fangmeier, A., A. Hadwiger-Fangmeier, L. Van der Eerden, and H.-J. Jäger. 1994. Effects of atmospheric ammonia on vegetation: A review. Environ. Pollut. 86:43–82.

Fasullo, J. T., and K. E. Trenberth. 2012. A less cloudy future: The role of subtropical subsidence in climate sensitivity. Science 338(6108):792–794.

Federer, C. A., J. W. Hornbeck, L. M. Tritton, C. W. Martin, R. S. Pierce, and C. T. Smith. 1989. Long-term depletion of calcium and other nutrients in eastern US forests. Environ. Manage. 13:593–601.

Fenn, M. E., and M. A. Poth. 1998. Indicators of nitrogen status in California. PSW-GTR-166. USDA Forest Service.

Fenn, M. E., M. A. Poth, J. D. Aber, J. S. Baron, B. T. Bormann, D. W. Johnson, A. D. Lemly, S. G. McNulty, D. F. Ryan, and R. Stottlemyer. 1998. Nitrogen excess in North American ecosystems: Predisposing factors, ecosystem responses, and management strategies. Ecol. Appl. 8:706–733.

Fenn, M. E., J. S. Baron, E. B. Allen, H. M. Rueth, K. R. Nydick, L. Geiser, W. D. Bowman, J. O. Sickman, T. Meixner, D. W. Johnson, and P. Neitlich. 2003a. Ecological effects of nitrogen deposition in the western United States. BioScience 53(4):404–420.

Fenn, M. E., R. Haeuber, G. S. Tonnesen, J. S. Baron, S. Grossman-Clark, D. Hope, D. A. Jaffe, S. Copeland, L. Geiser, H. M. Rueth, and J. O. Sickman. 2003b. Nitrogen emissions, deposition, and monitoring in the western United States. BioScience 53(4):391–403.

Fenn, M. E., M. A. Poth, A. Bytnerowicz, J. O. Sickman, and B. Takemoto. 2003c. Effects of ozone, nitrogen deposition, and other stressors on montane ecosystems in the Sierra Nevada. *In* A. Bytnerowicz, M. J. Arbaugh, and R.

Alonso, eds. Ozone air pollution in the Sierra Nevada: Distribution and effects on forests. Vol. 2 of Developments in environmental sciences. Elsevier, Amsterdam, Netherlands, 111–155.

Fenn, M. E., K. F. Lambert, T. Blett, D. A. Burns, L. H. Pardo, G. M. Lovett, R. A. Haeuber, D. C. Evers, C. T. Driscoll, and D. S. Jefferies. 2011. Setting limits: Using air pollution thresholds to protect and restore U.S. Ecosystems. Issues in Ecology, Report no. 14. Ecological Society of America.

Ferrari, J. B. 1999. Fine-scale patterns of leaf litterfall and nitrogen cycling in an old-growth forest. Can. J. For. Res. 29:291–302.

Ferreira, J. G., S. B. Bricker, and T. C. Simas. 2007. Application and sensitivity testing of a eutrophication assessment method on coastal systems in the United States and European Union. J. Environ. Manage. 82:433–445.

Ferrier, C. A., R. F. Wright, B. J. Cosby, and A. Jenkins. 1995. Application of the MAGIC model to the Norway spruce stand at Solling, Germany. Ecol. Model. 83:77–84.

Field, C. B., and H. A. Mooney. 1986. The photosynthesis-nitrogen relationship in wild plants. *In* T. J. Givnish, ed. On the economy of plant form and function. Cambridge University Press, Cambridge, UK, 25–55.

Findlay, D. L. 2003. Response of phytoplankton communities to acidification and recovery in Killarney Park and the Experimental Lakes Area, Ontario. Ambio 32(3):190–195.

Finlay, J. C., R. W. Sterner, and S. Kumar. 2007. Isotopic evidence for in-lake production of accumulating nitrate in Lake Superior. Ecol. Appl. 17:2323–2332.

Finzi, A. C., N. Van Breemen, and C. D. Canham. 1998. Canopy tree-soil interactions within temperate forests: Species effects on soil carbon and nitrogen. Ecol. Appl. 8:440–446.

Fisher, D. C., and M. Oppenheimer. 1991. Atmospheric nitrogen deposition and the Chesapeake Bay Estuary. Ambio 20:102–108.

Fisk, M. C., D. R. Zak, and T. R. Crow. 2002. Nitrogen dynamics of second- and old-growth northern hardwood forests: A test of the nutrient retention hypothesis. Ecology 83:73–87.

Fitzgerald, W. F., D. R. Engstrom, R. P. Mason, and E. A. Nater. 1998. The case for atmospheric mercury contamination in remote areas. Environ. Sci. Technol. 32:1–7.

Fitzhugh, R. D., G. E. Likens, C. T. Driscoll, M. J. Mitchell, P. M. Groffman, T. J. Fahey, and J. P. Hardy. 2003. Role of soil freezing events in interannual patterns of stream chemistry at the Hubbard Brook Experimental Forest, New Hampshire. Environ. Sci. Technol. 37:1575–1580.

Fleming, E. J., E. E. Mack, P. G. Green, and D. C. Nelson. 2006. Mercury methylation from unexpected sources: Molybdate-inhibited freshwater sediments and an iron-reducing bacterium. Appl. Environ. Microbiol. 72(1):457–464.

Ford, J., J. L. Stoddard, and C. F. Powers. 1993. Perspectives in environmental monitoring: An introduction to the U.S. EPA Long-Term Monitoring (LTM) project. Water Air Soil Pollut. 67:247–255.

Fournie, J. W., J. K. Summers, L. A. Courtney, V. D. Engle, and V. S. Blazer. 2001. Utility of splenic macrophage aggregates as an indicator of fish exposure to degraded environments. J. Aquat. Anim. Hlth. 13:105–116.

Fremstad, E., J. Paal, and T. Möls. 2005. Impacts of increased nitrogen supply on Norwegian lichen-rich alpine communities: A 10-year experiment. J. Ecol. 93:471–481.

Frenette, J. J., Y. Richard, and G. Moreau. 1986. Fish responses to acidity in Québec lakes: A review. Water Air Soil Pollut. 30(1–2):461–475.

Frescholtz, T. F., M. S. Gustin, D. E. Schorran, and G. C. J. Fernandez. 2003. Assessing the source of mercury in foliar tissue of quaking aspen. Environ. Toxicol. Chem. 22:2114–2119.

Fujinuma, R., J. Bockheim, and N. Balster. 2005. Base-cation cycling by individual tree species in old-growth forests of upper Michigan, USA. Biogeochemistry 74(3):357–376.

Gabriel, M. C., and D. G. Williamson. 2004. Principal biogeochemical factors affecting the speciation and transport of mercury through the terrestrial environment. Environ. Geochem. Health 26(3–4):421–434.

Gagen, C. J., W. E. Sharpe, and R. F. Carline. 1993. Mortality of brook trout, mottled sculpins, and slimy sculpins during acidic episodes. Trans. Am. Fish. Soc. 122:616–628.

Gallagher, J., and J. L. Baker. 1990. Current status of fish communities in Adirondack lakes. *In* J. P. Baker, S. A. Gherini, S. W. Christensen, C. T. Driscoll, J. Gallagher, R. K. Munson, R. M. Newton, K. H. Reckhow, and C. L. Schofield, eds. Adirondack Lakes Survey: An interpretative analysis of fish communities and water chemistry, 1984–87. Adirondack Lake Survey Corporation, Ray Brook, NY, 3-11–13-48.

Galloway, J. N. 1998. The global nitrogen cycle: Changes and consequences. Environ. Pollut. 102 (suppl. 1):15–24.

Galloway, M. E., and B. A. Branfireun. 2004. Mercury dynamics of a temperate forested wetland. Sci. Total Environ. 325:239–254.

Galloway, J. N., and E. B. Cowling. 2002. Reactive nitrogen and the world: 200 years of change. Ambio 31:64–71.

Galloway, J., H. Levy, and P. Kasibhatia. 1994. Year 2020: Consequences of population growth and development on deposition of oxidized nitrogen. Ambio 23:120–123.

Galloway, J. N., J. D. Aber, J. W. Erisman, S. P. Seitzinger, R. W. Howarth, E. B. Cowling, and B. J. Cosby. 2003. The nitrogen cascade. BioScience 53(4):341–356.

Garcia, E., and R. Carignan. 2000. Mercury concentrations in northern pike (*Esox lucius*) from boreal lakes with logged, burned, or undisturbed catchments. Can. J. Fish. Aquat. Sci. 57(Suppl. 2):129–135.

Garner, J. J. B. 1994. Nitrogen oxides, plant metabolism, and forest ecosystem response. *In* R. G. Alscher and A. R. Wellburn, eds. Plant responses to the gaseous environment: molecular, metabolic and physiological aspects. 3rd International Symposium on Air Pollutants and Plant Metabolism. Chapman & Hall, Blacksburg, VA, 301–314.

Gbondo-Tugbawa, S. S., and C. T. Driscoll. 2003. Factors controlling long-term changes in soil pools of exchangeable basic cations and stream acid neutralizing capacity in a northern hardwood forest ecosystem. Biogeochemistry 63:161–185.

Gbondo-Tugbawa, S. S., C. T. Driscoll, J. D. Aber, and G. E. Likens. 2001. Evaluation of an integrated biogeochemical model (PnET-BGC) at a northern hardwood forest ecosystem. Water Resour. Res. 37(4):1057–1070.

Geiser, L. H., S. E. Jovan, D. A. Glavich, and M. K. Porter. 2010. Lichen-based critical loads for atmospheric nitrogen deposition in western Oregon and Washington forests, USA. Environ. Pollut. 158:2412–2421.

Gensemer, R. W., and R. C. Playle. 1999. The bioavailability and toxicity of aluminum in aquatic environments. Crit. Rev. Environ. Sci. Tech. 29(4):315–450.

Gibbs, J. P., A. R. Breisch, P. K. Ducey, Johnson, G., J. L. Behler, and R. C. Bothner. 2007. The amphibians and reptiles of New York State: Identification, natural history, and conservation. Oxford University Press, New York.

Gilliam, F. S. 2006. Response of the herbaceous layer of forest ecosystems to excess nitrogen deposition. J. Ecol. 94(6):1176–1191.

Gilmour, C. C., and E. A. Henry. 1991. Mercury methylation in aquatic systems affected by acid deposition. Environ. Pollut. 71:131–169.

Gilmour, C. C., E. A. Henry, and R. Mitchell. 1992. Sulfate stimulation of mercury methylation in freshwater sediments. Environ. Sci. Technol. 26(11):2281–2287.

Gimona, A., and D. van der Horst. 2007. Mapping hotspots of multiple landscape functions: a case study on farmland afforestation in Scotland. Landsc. Ecol. 22:1255–1264.

Gitay, H., S. G. Brown, W. Easterling, and B. Jallow. 2001. Ecosystems and their goods and services. *In* J. J. McCarthy, F. C. Osvaldo, N. A. Leary, D. J. Dokken, and K. S. White, eds. Climate change 2001: Impacts, adaptation, and vulnerability. Contribution of Working Group II to the Third Assessment Report of the Intergovernmental Panel on Climate Change (IPCC). Cambridge University Press, New York, 237–342. http://www.grida.no/publications/other/ipcc_tar/?src=/climate/ipcc_tar/wg2/196.htm.

Glasgow, H. B., and J. M. Burkholder. 2000. Water quality trends and management implications from a five-year study of a eutrophic estuary. Ecol. Appl. 10:1024–1046.

Glooschenko, V., P. Blancher, J. Herskowitz, R. Fulthorpe, and S. Rang. 1986. Association of wetland acidity with reproductive parameters and insect prey of the eastern kingbird (*Tyrannus tyrannus*) near Sudbury, Ontario. Water Air Soil Pollut. 30(3/4):553–567.

Golden, H. E., and E. W. Boyer. 2009. Contemporary estimates of atmospheric nitrogen deposition to the watersheds of New York State, USA. Environ. Monitor. Assess. 155:319–339.

Golet, W. J., and T. A. Haines. 2001. Snapping turtles (*Chelydra serpentina*) as monitors for mercury contamination of aquatic environments. Environ. Monitor. Assess. 71(3):211–220.

Goodale, C. L., and J. D. Aber. 2001. The long-term effects of land-use history on nitrogen cycling in northern hardwood forests. Ecol. Appl. 11(1):253–267.

Goodale, C. L., J. D. Aber, and W. H. McDowell. 2000. The long-term effects of disturbance on organic and inorganic nitrogen export in the White Mountains, New Hampshire. Ecosystems 3:433–450.

Goriup, P. D. 1989. Acidic air pollution and birds in Europe. Oryx 23:82–86.

Gotelli, N. J., and A. M. Ellison. 2002. Nitrogen deposition and extinction risk in the northern pitcher plant, *Sarracenia purpurea*. Ecology 83(10):2758–2765.

Graveland, J. 1998. Effects of acid rain on bird populations. Environ. Rev. 6:41–54.

Gray, J., R. Wu, and Y. Or. 2002. Effects of hypoxia and organic enrichment on the coastal marine environment. Mar. Ecol. Prog. Ser. 238:249–279.

Great Lakes Environmental Research Laboratory. 2014. Harmful algal blooms. http://www.glerl.noaa.gov/res/waterQuality/?targetTab=habs.

Greaver, T., L. Liu, and R. Bobbink. 2011. Wetlands. *In* L. H. Pardo, M. J. Robin-Abbott, and C. T. Driscoll, eds. Assessment of nitrogen deposition effects and empirical critical loads of nitrogen for ecoregions of the United States. General Technical Report NRS-80. U.S. Forest Service, Newtown Square, PA, 193–208.

Green, S. M., R. Machin, and M. S. Cresser. 2008. Effect of long-term changes in soil chemistry induced by road salt applications on N-transformations in roadside soils. Environ. Pollut. 152:20–31.

Grigal, D. F. 2002. Inputs and outputs of mercury from terrestrial watersheds: A review. Environ. Rev. 10:1–39.

Grigal, D. F. 2003. Mercury sequestration in forests and peatlands: A review. J. Environ. Qual. 32(2):393–405.

Grimm, J. W., and J. A. Lynch. 1997. Enhanced Wet Deposition Estimates Using Modeled Precipitation Inputs. Final Report to U.S. Forest Service under Cooperative Agreement #23-721. Environmental Resources Research Institute, The Pennsylvania State University, University Park, PA.

Grimm, J. W., and J. A. Lynch. 2005. Improved daily precipitation nitrate and ammonium concentration models for the Chesapeake Bay Watershed. Environ. Pollut. 135(3):445–455.

Guerold, F., J. P. Boudot, G. Jacquemin, D. Vein, D. Merlet, and J. Rouiller. 2000. Macroinvertebrate community loss as a result of headwater stream acidification in the Vosges Mountains (N-E France). Biodivers. Conserv. 9(6):767–783.

Gunderson, P. 1992. Mass balance approaches for establishing critical loads for N in terrestrial ecosystems. Background Document for UN-ECE Workshop Critical Loads for Nitrogen. Lökeberg, Sweden.

Hachmoller, B., R. A. Matthews, and D. F. Brakke. 1991. Effects of riparian community structure, sediment size, and water-quality on the macroinvertebrate communities in a small, suburban stream. Northwest Sci. 65(3):125–132.

Haeuber, R. 2013. Energy-related environmental science and policy: Challenges and strategies. Paper presented at New York State Energy Research and Development Authority conference Environmental Monitoring, Evaluation, and Protection in New York: Linking Science and Policy. Albany, NY, November 6–7, 2013.

Haines, T. A., and J. P. Baker. 1986. Evidence of fish population responses to acidification in the eastern United States. Water Air Soil Pollut. 31:605–629.

Hall, L. W. 1987. Acidification effects on larval striped bass, *Morone saxatilis*, in Chesapeake Bay tributaries: A review. Water Air Soil Pollut. 35:87–96.

Hallett, R. A., S. W. Bailey, S. B. Horsley, and R. P. Long. 2006. Influence of nutrition and stress on sugar maple at a regional scale. Can. J. For. Res. 36:2235–2246.

Hallett, R., M. Martin, L. Lepine, J. Pontius, J. Siemion, and P. Murdoch. 2010. Assessment of regional forest health and stream and soil chemistry using a multi-scale approach and new methods of remote sensing interpretation in the Catskill Mountains of New York. NYSERDA Report 10-28. New York State Energy Research and Development Authority, Albany, NY.

Halman, J. M., P. G. Schaberg, G. J. Hawley, and C. Eagar. 2008. Calcium addition at the Hubbard Brook Experimental Forest increases sugar storage, antioxidant activity and cold tolerance in native red spruce (*Picea rubens*). Tree Physiol. 28:855–862.

Hames, R. S., K. V. Rosenberg, J. D. Lowe, S. E. Barker, and A. A. Dhondt. 2002. Adverse effects of acid rain on the distribution of the wood thrush *Hylocichla mustelina* in North America. Proc. Nat. Acad. Sci. 99:11235–11240.

Hamilton, M., A. Scheuhammer, and N. Basu. 2011. Mercury, selenium, and neurochemical biomarkers in different brain regions of migrating common loons from Lake Erie, Canada. Ecotoxicology 20(7):1677–1683.

Hammerschmidt, C. R., and W. F. Fitzgerald. 2004. Geochemical controls on the production and distribution of methylmercury in near-shore marine sediments. Environ. Sci. Technol. 38(5):1487–1495.

Hammerschmidt, C. R., and M. B. Sandheinrich. 2005. Maternal diet during oogenesis is the major source of methylmercury in fish embryos. Environ. Sci. Technol. 39:3580–3584.

Handy, R. D., and W. S. Penrice. 1993. The influence of high oral doses of mercuric chloride on organ toxicant concentrations and histopathology in rainbow trout, *Oncorhynchus mykiss*. Comp. Biochem. Physiol. C Pharmacol. Toxicol. Endocrinol. 106:717–724.

Harris, R. C., J. W. M. Rudd, M. Amyot, C. L. Babiarz, K. G. Beaty, P. J. Blanch-field, R. A. Bodaly, B. A. Branfireun, C. C. Gilmour, J. A. Graydon, A. Heyes, H. Hintlemann, J. P. Hurley, C. A. Kelly, D. P. Krabbenhoft, S. E. Lindberg, R. P. Mason, M. J. Paterson, C. L. Podemski, A. Robinson, K. A. Sandilands, G. R. Southworth, V. L. St. Louis, and M. T. Tate. 2007. Whole-ecosystem study shows rapid fish-mercury response to changes in mercury deposition. Proc. Nat. Acad. Sci. 104(42):16586–16591.

Harvey, H., and C. Lee. 1982. Historical fisheries changes related to surface water pH changes in Canada. *In* R. E. Johnson, ed. Acid rain/fisheries. American Fisheries Society, Bethesda, MD, 44–55.

Hasler, A. D. 1969. Cultural eutrophication is reversible. BioScience 19:425–431.

Hauxwell, J., J. Cebrian, C. Furlong, and I. Valiela. 2001. Macroalgal canopies contribute to eelgrass (*Zostera marina*) decline in temperate estuarine ecosystems. Ecology 82:1007–1022.

Havas, M. 1985. Aluminum bioaccumulation and toxicity to *Daphnia magna* in soft water at low pH. Can. J. Fish. Aquat. Sci. 42:1741–1748.

Havas, M. 1986. Effects of acid deposition on aquatic ecosystems. *In* A. Stern, ed. Air pollution. Academic Press, Orlando, FL, 351–389.

Havens, K. E., and R. E. Carlson. 1998. Functional complementarity in plankton communities along a gradient of acid stress. Environ. Pollut. 101(3):427–436.

Hawley, G. J., P. G. Schabery, C. Eager, and C. H. Borer. 2006. Calcium addition at the Hubbard Brook Experimental Forest reduced winter injury to red spruce in a high-injury year. Can. J. For. Res. 36:2544–2549.

Hazlett, P. W., J. M. Curry, and T. P. Weldon. 2011. Assessing decadal change in mineral soil cation chemistry at the Turkey Lakes watershed. Soil Sci. Soc. Am. J. 75:287–305.

Hecky, R. E., R. E. H. Smith, D. R. Barton, S. J. Guildford, W. D. Taylor, M. N. Charlton, and T. Howell. 2004. The nearshore phosphorus shunt: A consequence of ecosystem engineering by dreissenids in the Laurentian Great Lakes. Can. J. Fish. Aquat. Sci. 61:1285–1293.

Hedin, L. O., J. J. Armesto, and A. H. Johnson. 1995. Patterns of nutrient loss from unpolluted, old-growth temperate forests: Evaluation of biogeochemical theory. Ecology 76(2):493–509.

Hefting, M. M., R. Bobbink, and H. De Caluwe. 2003. Nitrous oxide emission and denitrification in chronically nitrate-loaded riparian buffer zones. J. Environ. Qual. 32(4):1194–1203.

Heil, G. W., and M. Bruggink. 1987. Competition for nutrients between *Calluna vulgaris* (L.) Hull and *Molinia caerulea* (L.) Moench. Oecologia 73:105–108.

Heil, G. W., and W. H. Diemont. 1983. Raised nutrient levels change heathland into grassland. Vegetation 53:113–120.

Heinz, G. H., D. J. Hoffman, J. D. Klimstra, K. R. Stebbins, S. L. Konrad, and C. A. Erwin. 2009. Species differences in the sensitivity of avian embryos to methylmercury. Arch. Environ. Contam. Toxicol. 56:129–138.

Heinz, G. H., D. Hoffman, J. D. Klimstra, and K. R. Stebbins. 2010. Predicting mercury concentrations in mallard eggs from mercury in the diet or blood of adult females and from duckling down feathers. Environ. Toxicol. Chem. 29:389–392.

Hemond, H. F. 1990. Wetlands as the source of dissolved organic carbon to surface waters. *In* E. M. Perdue and E. T. Gjessing, eds. Organic acids in aquatic ecosystems. John Wiley & Sons, New York, 301–313.

Hemond, H. F. 1994. Role of organic acids in acidification of fresh waters. *In* C. E. W. Steinberg and R. F. Wright, eds. Acidification of freshwater ecosystems: Implications for the future. John Wiley & Sons Ltd., Chichester, 103–116.

Henriksen, A. 1980. Acidification of freshwaters: A large scale titration. *In* D. Drabløs and A. Tollan, eds. Proceedings of the International Conference on the Ecological Impact of Acid Precipitation. SNSF Project, Sandefjord, Norway, 68–74.

Henriksen, A. 1984. Changes in base cation concentrations due to freshwater acidification. Verh. Int. Verein. Limnol. 22:692–698.

Henriksen, A., and M. Grande. 2002. Lake Langtjern: Fish studies in the Langtjern area, 1966–2000. Report SNO 4537-2002. Norwegian Institute for Water Research, Oslo.

Henriksen, A., and M. Posch. 2001. Steady-state models for calculating critical loads of acidity for surface waters. Water Air Soil Pollut: Focus 1(1–2):375–398.

Herlihy, A. T., D. P. Larsen, S. G. Paulsen, N. S. Urquhart, and B. J. Rosenbaum. 2000. Designing a spatially balanced, randomized site selection process for regional stream surveys: The EMAP mid-Atlantic pilot study. Environ. Monitor. Assess. 63:95–113.

Hill, B. H., C. M. Elonen, T. M. Jicha, A. M. Cotter, A. S. Trebitz, and N. P. Danz. 2006. Sediment microbial enzyme activity as an indicator of nutrient limitation in Great Lakes coastal wetlands. Freshw. Biol. 51:1670–1683.

Hinga, K. R., A. A. Keller, and C. A. Ovlatt. 1991. Atmospheric deposition and nitrogen inputs to coastal waters. Ambio 20(6):256–260.

Hodgkin, E. P., and B. H. Hamilton. 1993. Fertilizers and eutrophication in south-western Australia: Setting the scene. Fertilizer Research 36:95–103.

Hodgkins, G. A., R. W. Dudley, and T. G. Huntington. 2003. Changes in the timing of high river flow in New England over the 20th century. J. Hydrol. 278:244–252.

Högberg, P., H. Fan, M. Quist, D. Binkleys, and C. Oloftamm. 2006. Tree growth and soil acidification in response to 30 years of experimental nitrogen loading on boreal forest. Glob. Change Biol. 12:489–499.

Holt, C., and N. D. Yan. 2003. Recovery of crustacean zooplankton communities from acidification in Killarney Park, Ontario, 1971–2000: pH 6 as a recovery goal. R. Swedish Acad. Sci. 32(3):203–207.

Holt, C. A., N. D. Yan, and K. M. Somers. 2003. pH 6 as the threshold to use in critical load modeling for zooplankton community change with acidification in lakes of south-central Ontario: accounting for morphometry and geography. Can. J. Fish. Aquat. Sci. 60:151–158.

Hongve, D., S. Haaland, G. Riise, I. Blakar, and S. Norton. 2012. Decline of acid rain enhances mercury concentrations in fish. Environ. Sci. Technol. 46(5):2490–2491.

Hornberger, G. M., B. J. Cosby, and R. F. Wright. 1989. Historical reconstructions and future forecasts of regional surface water acidification in southernmost Norway. Water Resour. Res. 25:2009–2018.

Horsley, S. B., R. P. Long, S. W. Bailey, R. A. Hallet, and T. J. Hall. 1999. Factors contributing to sugar maple decline along topographic gradients on the glaciated and unglaciated Allegheny Plateau. *In* S. B. Horsley and R. P. Long, eds. Sugar maple ecology and health: Proceedings of an international symposium. General Technical Report NE-261. U.S. Department of Agriculture, Forest Service, Radnor, PA, 60–62.

Horsley, S. B., R. P. Long, S. W. Bailey, R. A. Hallett, and T. J. Hall. 2000. Factors associated with the decline disease of sugar maple on the Allegheny Plateau. Can. J. For. Res. 30:1365–1378.

Horsley, S. B., S. W. Bailey, T. E. Ristau, R. P. Long, and R. A. Hallett. 2008. Linking environmental gradients, species composition, and vegetation indicators of sugar maple health in the northeastern United States. Can. J. For. Res. 38:1761–1774.

Houle, D., R. Carignan, and R. Ouimet. 2001. Soil organic sulfur dynamics in a coniferous forest. Biogeochemistry 53(1):105–124.

Howarth, R. W., and R. Marino. 2006. Nitrogen as the limiting nutrient for eutrophication in coastal marine ecosystems: Evolving views over three decades. Limnol. Oceanogr. 51:364–376.

Howarth, R. W., G. Billen, D. Swaney, A. Townsend, N. Jaworski, K. Lajtha, J. A. Downing, R. Elmgren, N. Caraco, T. Jordan, F. Berendse, J. Freney, V. Kudeyarov, P. S. Murdoch, and Z. Zhao-Liang. 1996. Regional nitrogen budgets and riverine N & P fluxes for the drainages to the North Atlantic Ocean: Natural and human influences. Biogeochemistry 35:75–139.

Howarth, R., D. Anderson, J. Cloern, C. Elfring, C. Hopkinson, B. Lapointe, T. Malone, N. Marcus, K. McGlathery, A. Sharpley, and D. Walker. 2000a. Nutrient pollution of coastal rivers, bays, and seas. Issues in Ecology no. 7. Ecological Society of America, Washington, DC.

Howarth, R. W., D. P. Swaney, T. J. Butler, and R. Marino. 2000b. Climatic control on eutrophication of the Hudson River Estuary. Ecosystems 3:210–215.

Hoyer, M. V., and J. R. Jones. 1983. Factors affecting the relation between phosphorus and chlorophyll *a* in midwestern reservoirs. Can. J. Fish. Aquat. Sci. 40:192–199.

Hrabik, T. R., and C. J. Watras. 2002. Recent declines in mercury concentration in a freshwater fishery: Isolating the effects of de-acidification and decreased atmospheric mercury deposition in Little Rock Lake. Sci. Total Environ. 297:229–237.

Hudson, J. J., P. J. Dillon, and K. M. Somers. 2003. Long-term patterns in dissolved organic carbon in boreal lakes: The role of incident radiation, precipitation, air temperature, southern oscillation, and acid deposition. Hydrol. Earth Syst. Sci. 7:390–398.

Hunter, M. D., and J. C. Schultz. 1995. Fertilization mitigates chemical induction and herbivore responses within damaged trees. Ecology 76:1226–1232.

Hunter, M. L., J. Jones, J. J., K. E. Gibbs, and J. R. Moring. 1986. Duckling responses to lake acidification: do black ducks and fish compete? Oikos 47:26–32.

Hurd, T. M., A. R. Brach, and D. J. Raynal. 1998. Response of understory vegetation of Adirondack forests to nitrogen additions. Can. J. For. Res. 28:799–807.

Hurd, T. M., D. J. Raynal, and C. R. Schwintzer. 2001. Symbiotic N_2 fixation of *Alnus incana* ssp. *rugosa* in shrub wetlands of the Adirondack Mountains, New York, USA. Oecologia 126:94–103.

Hurd, T. M., K. Gökkaya, B. D. Kiernan, and D. J. Raynal. 2005. Nitrogen sources in Adirondack wetlands dominated by nitrogen-fixing shrubs. Wetlands 25(1):192–199.

Husar, R. B., T. J. Sullivan, and D. F. Charles. 1991. Historical trends in atmospheric sulfur deposition and methods for assessing long-term trends in surface water chemistry. *In* D. F. Charles, ed. Acidic deposition and aquatic ecosystems: Regional case studies. Springer-Verlag, New York, 65–82.

HydroQual Inc. 1995. Analysis of factors affecting historical dissolved oxygen trends in Western Long Island Sound. Job Number: NENG0040. Management Committee of the Long Island Sound Estuary Study, Stamford, Connecticut, and the New England Interstate Water Pollution Control Commission, New York.

ICF International. 2006. Mercury transport and fate through a watershed. STAR Report 4(1). http://www.epa.gov/ncer/publications/starreport/starten.pdf.

[ICSU-UNESCO-UNU] International Council for Science–United Nations Educational, Scientific and Cultural Organization–United Nations University. 2008. Ecosystem change and human well-being: Research and monitoring priorities based on the millennium ecosystem assessment. Paris, International Council for Science.

Ingersoll, C. G., D. R. Mount, D. D. Gulley, T. W. La Point, and H. L. Bergman. 1990. Effects of pH, aluminum, and calcium on survival and growth of eggs and fry of brook trout (*Salvelinus fontinalis*). Can. J. Fish. Aquat. Sci. 47:1580–1592.

[IPCC] Intergovernmental Panel on Climate Change. Climate Change 2007: The physical science basis. Contribution of Working Group I to the Fourth Assessment Report of the Intergovernmental Panel on Climate Change. Solomon, S., D. Qin, M. Manning, Z. Chen, M. Marquis, K. B. Averyt, M. Tignor and H. L. Miller, eds. Cambridge University Press, Cambridge, UK. http://www.ipcc.ch/publications_and_data/publications_ipcc_fourth_assessment_report_wg1_report_the_physical_science_basis.htm.

Ito, M., M. J. Mitchell, and C. T. Driscoll. 2002. Spatial patterns of precipitation quantity and chemistry and air temperature in the Adirondack region of New York. Atmos. Environ. 36:1051–1062.

Ito, M., M. J. Mitchell, C. T. Driscoll, and K. M. Roy. 2005. Factors affecting acid neutralizing capacity in the Adirondack region of New York: A solute mass balance approach. Environ. Sci. Technol. 39(11):4076–4081.

Ittekot, V. 2003. Carbon-silicon interactions. *In* J. M. Melillo, C. B. Field, and B. Moldan, eds. Interactions of the major biogeochemical cycles. Island Press, Washington, DC, 311–322.

Jampeetong, A., and H. Brix. 2009a. Effects of NH_4^+ concentration on growth, morphology and NH_4^+ uptake kinetics of *Salvinia natans*. Ecol. Eng. 35(5):695–702. http://dx.doi.org/10.1016/j.ecoleng.2008.11.006.

Jampeetong, A., and H. Brix. 2009b. Nitrogen nutrition of *Salvinia natans*: Effects of inorganic nitrogen form on growth, morphology, nitrate reductase activity and uptake kinetics of ammonium and nitrate. Aquat. Bot. 90(1):67–73.

Jampeetong, A., and H. Brix. 2009c. Oxygen stress in *Salvinia natans*: Interactive effects of oxygen availability and nitrogen source. Environ. Exp. Bot. 66(2):153–159.

Jaworski, N. A., R. S. Howarth, and L. J. Hetling. 1997. Atmospheric deposition of nitrogen oxides onto the landscape contributes to coastal eutrophication in the northeast United States. Environ. Sci. Technol. 31(7):1995–2004.

Jay, D. A., W. R. Geyer, and D. R. Montgomery. 2000. An ecological perspective on estuarine classification. *In* J. Hobbie, ed. Estuarine synthesis: The next decade. Island Press, New York, 148–176.

Jenkins, A., P. G. Whitehead, B. J. Cosby, R. C. Ferrier, and D. J. Waters. 1990. Modelling long term acidification: A comparison with diatom reconstructions and the implications for reversibility. Phil. Trans. R. Soc. Lond. 327:435–440.

Jentsch, A., J. Kreyling, and C. Beierkuhlein. 2007. A new generation of climate change experiments: Events, not trends. Front. Ecol. Environ. 5:365–374.

Jeremiason, J. D., D. R. Engstrom, E. B. Swain, E. A. Nater, B. M. Johnson, J. E. Almendinger, B. A. Monson, and R. K. Kolka. 2006. Sulfate addition increases methylmercury production in an experimental wetland. Environ. Sci. Technol. 40(12):3800–3806.

Johannsson, O. E., R. Dermott, D. M. Graham, J. A. Dahl, E. S. Millard, D. D. Myles, and J. LeBlanc. 2000. Benthic and Pelagic secondary production in Lake Erie after the invasion of *Dreissena* spp. with implications for fish production. J. Great Lakes Res. 26(1):31–54.

Johnson, A. H., S. B. Andersen, and T. G. Siccama. 1994a. Acid rain and soils of the Adirondacks. I. Changes in pH and available calcium, 1930–1984. Can. J. For. Res. 24:39–45.

Johnson, A. H., A. J. Friedland, E. K. Miller, and T. G. Siccama. 1994b. Acid rain and soils of the Adirondacks. III. Rates of soil acidification in a montane spruce-fir forest at Whiteface Mountain, New York. Can. J. For. Res. 24:663–669.

Johnson, A. H., A. J. Moyer, J. E. Bedison, S. L. Richter, and S. A. Willig. 2008. Seven decades of calcium depletion in organic horizons of Adirondack forest soils. Soil Sci. Soc. Am. J. 72:1824–1830.

Johnson, C., and L. J. Hetling. 1995. A historical review of pollution loadings to the lower Hudson River. Paper presented at the winter meeting of the New York Water Environment Association, January 31, 2015, New York.

Johnson, C. E. 1995. Soil nitrogen status 8 years after whole-tree clear-cutting. Can. J. For. Res. 25:1346–1355.

Johnson, C. E., A. H. Johnson, and T. G. Siccama. 1991. Whole-tree clear-cutting effects on exchangeable cations and soil acidity. Soil Sci. Soc. Am. J. 55:502–508.

Johnson, C. E., C. T. Driscoll, T. G. Siccama, and G. E. Likens. 2000 Element fluxes and landscape position in a northern hardwood forest watershed ecosystem. Ecosystems 3:159–184.

Johnson, D. W., and I. J. Fernandez. 1992. Soil mediated effects of atmospheric deposition on eastern U.S. spruce fir forests. *In* C. Eagar and M. B. Adams, eds. Ecology and decline of red spruce in the eastern United States. Springer-Verlag, New York, 235–270.

Johnson, D. W., and M. J. Mitchell. 1998. Responses of forest ecosystems to changing sulfur inputs. *In* D. G. Maynard, ed. Sulfur in the environment. Marcel Dekker, Inc, New York, 219–262.

Johnson, D. W., H. A. Simonin, J. R. Colquhoun, and F. M. Flack. 1987. *In situ* toxicity tests of fishes in acid waters. Biogeochemistry 3:181–208.

Johnson, D. W., M. S. Cresser, S. I. Milsson, J. Turner, B. Ulrich, D. Binkley, and D. W. Cole. 1991a. Soil changes in forest ecosystems: Evidence for and probable causes. Proc. Royal Soc. Edinburg Section B: Biol. Sci. 97B:81–116.

Johnson, D. W., H. V. Van Miegroet, S. E. Lindberg, D. E. Todd, and R. B. Harrison. 1991b. Nutrient cycling in red spruce forests of the Great Smoky Mountains. Can. J. For. Res. 21:769–787.

Johnson, D. W., W. Cheng, and L. C. Burke. 2000. Biotic and abiotic nitrogen retention in a variety of forested soils. Soil Sci. Soc. Am. J. 64:1503–1514.

Johnson, D. W., R. B. Susfalk, T. G. Caldwell, J. D. Murphy, W. W. Miller, and R. F. Walker. 2004. Fire effects on carbon and nitrogen budgets in forests. Water Air Soil Pollut: Focus 4(2–3):263–275.

Johnson, D. W., D. E. Todd, C. F. Trettin, and P. J. Mulholland. 2008. Decadal changes in potassium, calcium, and magnesium in a deciduous forest soil. Soil Sci. Soc. Am. J. 72:1795–1805.

Johnson, L. B., C. Richards, G. E. Host, and J. W. Arthur. 1997. Landscape influences on water chemistry in Midwestern stream ecosystems. Freshw. Biol. 37:193–208.

Joint, I., J. Lewis, J. Aiken, R. Proctor, G. Moore, W. Higman, and M. Donald. 1997. Interannual variability of PSP outbreaks on the north east UK coast. J. Plankton Res. 19:937–956.

Jones, J. A., F. J. Swanson, B. C. Wemple, and K. U. Snyder. 2000. Effects of roads on hydrology, geomorphology, and disturbance patches in stream networks. Conserv. Biol. 14(1):76–85.

Jongmans, A. G., N. van Breemen, U. Lundström, P. A. W. van Hees, R. D. Finlay, M. Srinivasan, T. Unestam, R. Giesler, P.-A. Melkerud, and M. Olsson. 1997. Rock-eating fungi. Nature 389:682–683.

Joslin, J. D., and M. H. Wolfe. 1994. Foliar deficiencies of mature Southern Appalachian red spruce determined from fertilizer trials. Soil Sci. Soc. Am. J. 58:1572–1579.

Joslin, J. D., P. A. Mays, M. H. Wolfe, J. M. Kelly, R. W. Garber, and P. F. Brewer. 1987. Chemistry of tension lysimeter water and lateral flow in spruce and hardwood stands. J. Environ. Qual. 16:152–160.

Joslin, J. D., J. M. Kelly, and H. van Miegroet. 1992. Soil chemistry and nutrition of North American spruce-fir stands: Evidence for recent change. J. Environ. Qual. 21:12–30.

Juice, S. M., T. J. Fahey, T. G. Siccama, C. T. Driscoll, E. G. Denny, C. Eagar, N. L. Cleavitt, R. Minocha, and A. D. Richardson. 2006. Response of sugar maple to calcium addition to northern hardwood forest. Ecol. Soc. 87(5):1267–1280.

Juillerat, J. I., D. S. Ross, and M. S. Bank. 2012. Mercury in litterfall and upper soil horizons in forested ecosystems in Vermont, USA. Environ. Toxicol. Chem. 31(8):1720–1729.

Justic, D., N. N. Rabalais, and R. E. Turner. 1995a. Stoichiometric nutrient balance and origin of coastal eutrophication. Mar. Pollut. Bull. 30:41–46.

Justic, D., N. N. Rabalais, R. E. Turner, and Q. Dortch. 1995b. Changes in nutrient structure of river-dominated coastal waters: Stoichiometric nutrient balance and its consequences. Estuar. Coast. Shelf Sci. 40:339–356.

Kahl, J. S., S. A. Norton, C. S. Cronan, I. J. Fernandez, L. C. Bacon, and T. A. Haines. 1991. Maine. In D. F. Charles, ed. Acidic deposition and aquatic ecosystems: Regional case studies. Springer-Verlag, New York, 203–235.

Kahl, J. S., T. A. Haines, S. A. Norton, and R. B. Davis. 1993. Recent temporal trends in the acid-base status of surface waters in Maine, USA. Water Air Soil Pollut. 67:281–300.

Kamman, N. C., N. M. Burgess, C. T. Driscoll, H. A. Simonin, W. Goodale, J. Linehan, R. Estabrook, M. Hutcheson, A. Major, A. M. Scheuhammer, and D. A. Scruton. 2005. Mercury in freshwater fish of northeast North America: A geographic perspective based on fish tissue monitoring databases. Ecotoxicology 14:163–180.

Kareiva, P., and M. Marvier. 2011. Conservation science: Balancing the needs of people and nature. Roberts and Company, Greenwood Village, CO.

Karr, J. R., and L. W. Chu. 1999. Restoring life in running rivers: Better biological monitoring. Island Press, Washington, DC.

Kaufmann, P. R., A. T. Herlihy, J. W. Elwood, M. E. Mitch, W. S. Overton, M. J. Sale, J. J. Messer, K. A. Cougan, D. V. Peck, K. H. Reckhow, A. J. Kinney, S. J. Christie, D. D. Brown, C. A. Hagley, and H. I. Jager. 1988. Population descriptions and physico-chemical relationships. Vol. 1 of Chemical characteristics of streams in the Mid-Atlantic and Southeastern United States. EPA/600/3-88/021a. U.S. Environmental Protection Agency, Washington, DC.

Kauppi, P. E., K. Mielikäinen, and K. Kuusela. 1992. Biomass and carbon budget of European forests, 1971 to 1990. Science 256:70–74.

Kenow, K. P., M. W. Meyer, R. Rossmann, A. Gendron-Fitzpatrick, and B. R. Gray. 2011. Effects of injected methylmercury on the hatching of common loon (*Gavia immer*) eggs. Ecotoxicology 20:1684–1693.

Kerekes, J., R. Tordon, A. Nieuwburg, and L. Risk. 1994. Fish-eating bird abundance in oligotrophic lakes in Kejimkujik National Park, Nova Scotia, Canada. Hydrobiologia 279/280:57–61.

Kerin, E. J., C. C. Gilmour, E. Roden, M. T. Suzuki, J. D. Coates, and R. P. Mason. 2006. Mercury methylation by dissimilatory iron-reducing bacteria. Appl. Environ. Microbiol. 72(12):7919–7921.

Kerr, J. G., M. C. Eimers, I. F. Creed, M. B. Adams, F. Beall, D. Burns, J. L. Campbell, S. F. Christopher, T. A. Clair, F. Courchesne, L. Duchesne, I. Fernandez, D. Houle, D. S. Jeffries, G. E. Likens, M. J. Mitchell, J. Shanley, and H. Yao. 2011. The effect of seasonal drying on sulphate dynamics in streams across southeastern Canada and the northeastern USA. Biogeochemistry 111(1–3):393–409.

Khan, R. A. 2003. Health of flatfish from localities in Placentia Bay, Newfoundland, contaminated with petroleum and PCBs. Arch. Environ. Contam. Toxicol. 44:485–492.

Kiernan, B. D., T. M. Hurd, and D. J. Raynal. 2003. Abundance of *Alnus incana* ssp. *rugosa* in Adirondack Mountain shrub wetlands and its influence on inorganic nitrogen. Environ. Pollut. 123:347–354.

Kirby, M., and H. Miller. 2005. Response of a benthic suspension feeder (*Crassostrea virginica Gmelin*) to three centuries of anthropogenic eutrophication in Chesapeake Bay. Estuar. Coast. Shelf Sci. 62:679–689.

Klauda, R. J., R. E. Palmer, and M. J. Lenkevich. 1987. Sensitivity of early life stages of blueback herring to moderate acidity and aluminum in soft freshwater. Estuaries 10(1):44–53.

Klopatek, J. M., M. J. Barry, and D. W. Johnson. 2006. Potential canopy interception of nitrogen in the Pacific Northwest, USA. For. Ecol. Manage. 234:344–354.

Knights, J. S., F. J. Zhao, B. Spiro, and S. P. McGrath. 2000. Long-term effects of land use and fertilizer treatments on sulfur cycling. J. Environ. Qual. 29(6): 1867–1874.

Knoepp, J. D., and J. M. Vose. 2007. Regulation of nitrogen mineralization and nitrification in Southern Appalachian ecosystems: Separating the relative importance of biotic vs. abiotic controls. Pedobiologia 51:89–97.

Kobe, R. K., G. E. Likens, and C. Eagar. 2002. Tree seedling growth and mortality responses to manipulations of calcium and aluminum in a northern hardwood forest. Can. J. For. Res. 32(6):954–966.

Kolka, R. K., C. P. J. Mitchel, J. D. Jeremiason, N. A. Hines, D. F. Grigal, D. R. Engstrom, J. K. Coleman-Wasik, E. A. Nater, Edward B. Swain, B. A. Monson, J. A. Fleck, B. Johnson, J. E. Almendinger, B. A. Branfireun, P. L. Brezonik, and J. B. Cotner. 2011. Mercury cycling in peatland watersheds. *In* R. K. Kolka, S. D. Sebestyen, E. S. Verry, and K. N. Brooks, eds. Peatland biogeochemistry and watershed hydrology at the Marcell Experimental Forest. CRC Press, Boca Raton, FL, 349–370.

Kooijman, A. M., and C. Bakker. 1994. The acidification capacity of wetland bryophytes as influenced by simulated clean and polluted rain. Aquat. Bot. 48(2):133–144.

Kopáček, J., B. Cosby, C. Evans, J. Hruška, F. Moldan, F. Oulehle, H. Šantrůčková, K. Tahovská, and R. Wright. 2013. Nitrogen, organic carbon and sulphur cycling in terrestrial ecosystems: linking nitrogen saturation to carbon limitation of soil microbial processes. Biogeochemistry 115(1–3):33–51.

Kramar, D., W. M. Goodale, L. M. Kennedy, L. W. Carstensen, and T. Kaur. 2005. Relating land cover characteristics and common loon mercury levels using geographic information systems. Ecotoxicology 14:253–262.

Kretser, W., J. Gallagher, and J. Nicolette. 1989. Adirondack Lakes Study, 1984–1987: An evaluation of fish communities and water chemistry. Adirondack Lakes Survey Corp, Ray Brook, NY.

Krug, E. C., and C. R. Frink. 1983. Acid rain on acid soil: A new perspective. Science 221:520–525.

Krug, E. C., P. J. Isaacson, and C. R. Frink. 1985. Appraisal of some current hypotheses describing acidification of watersheds. J. Air Pollut. Control Assoc. 35:109–114.

Krupa, S. V. 2003. Effects of atmospheric ammonia (NH_3) on terrestrial vegetation: A review. Environ. Pollut. 124(179–221).

Larsen, D. P., and N. S. Urquhart. 1993. A framework for assessing the sensitivity of the EMAP design. *In* D. P. Larsen and S. J. Christie, eds. EMAP-Surface Waters 1991 pilot report. U.S. Environmental Protection Agency, Corvallis, OR, 4.1–4.37.

Larsen, D. P., K. W. Thornton, N. S. Urquhart, and S. G. Paulsen. 1994. The role of sample surveys for monitoring the condition of the nation's lakes. Environ. Monitor. Assess. 32:101–134.

Latty, E. F., C. D. Canham, and P. L. Marks. 2004. The effects of land-use history on soil properties and nutrient dynamics in northern hardwood forests of the Adirondack Mountains. Ecosystems 7(2):193–207.

Lawrence, G. B. 2002. Persistent episodic acidification of streams linked to acid rain effects on soil. Atmos. Environ. 36:1589–1598.

Lawrence, G. 2013. The latest on Adirondack soils and streams as acidic deposition continues to decrease. Paper presented at New York State Energy Research and Development Authority conference Environmental Monitoring, Evaluation, and Protection in New York: Linking Science and Policy. Albany, NY, November 6–7.

Lawrence, G. B., and T. G. Huntington. 1999. Soil-calcium depletion linked to acid rain and forest growth in the eastern United States. Water-Resources Investigations Report 98-4267. U.S. Geological Survey, Reston, VA.

Lawrence, G. B., R. D. Fuller, and C. T. Driscoll. 1986. Spatial relationships of aluminum chemistry in the streams of the Hubbard Brook Experimental Forest, New Hampshire. Biogeochemistry 2:115–135.

Lawrence, G. B., M. B. David, and W. C. Shortle. 1995. A new mechanism for calcium loss in forest-floor soils. Nature 378:162–165.

Lawrence, G. B., M. M. David, G. M. Lovett, P. S. Murdoch, D. A. Burns, J. L. Stoddard, B. Baldigo, J. H. Porter, and A. W. Thompson. 1999. Soil calcium status and the response of stream chemistry to changing acidic deposition rates. Ecol. Appl. 9(3):1059–1072.

Lawrence, G. B., J. W. Sutherland, C. W. Boylen, S. A. Nierzwicki-Bauer, B. Momen, B. P. Baldigo, and H. A. Simonin. 2007. Acid rain effects on aluminum mobilization clarified by inclusion of strong organic acids. Environ. Sci. Technol. 41(1):93–98.

Lawrence, G. B., B. P. Baldigo, K. M. Roy, H. A. Simonin, R. W. Bode, S. I. Passy, and S. B. Capone. 2008a. Results from the 2003–2005 Western Adirondack Stream Survey. Final Report 08-22. New York State Energy Research and Development Authority, Albany, NY.

Lawrence, G. B., K. M. Roy, B. P. Baldigo, H. A. Simonin, S. B. Capone, J. W. Sutherland, S. A. Nierzwicki-Bauer, and C. W. Boylen. 2008b. Chronic and episodic acidification of Adirondack streams from acid rain in 2003–2005. J. Environ. Qual. 37:2264–2274.

Lawrence, G. B., W. C. Shortle, M. B. David, K. T. Smith, R. A. F. Warby, and A. G. Lapenis. 2012. Early indications of soil recovery from acidic deposition in U.S. red spruce forests. Soil Sci. Soc. Am. J. 76:1407–1417.

Lawrence, G. B., J. E. Dukett, N. Houck, P. Snyder, and S. Capone. 2013a. Increases in dissolved organic carbon accelerate loss of toxic Al in Adirondack lakes recovering from acidification. Environ. Sci. Technol. 47(13):7095–7100.

Lawrence, G. B., I. J. Fernandez, D. D. Richter, D. S. Ross, P. W. Hazlett, S. Bailey, R. Ouimet, R. Warby, A. E. Johnson, H. Lin, J. Kaste, A. G. Lapenis, and T. J. Sullivan. 2013b. Measuring environmental change in forest ecosystems by repeated soil sampling: A North American perspective. J. Environ. Qual. 42(3):623–639.

Lazarus, B. E., P. G. Schaberg, D. H. DeHayes, and G. J. Hawley. 2004. Severe red spruce winter injury in 2003 creates unusual ecological event in the northeastern United States. Can. J. For. Res. 34:1784–1788.

LeBlanc, D. C. 1992. Spatial and temporal variation in the prevalence of growth decline in red spruce populations of the northeastern United States. Can. J. For. Res. 22:1351–1363.

LeBlanc, G. A. 1995. Trophic-level differences in the bioconcentration of chemicals: Implications in assessing environmental biomagnification. Environ. Sci. Technol. 29:154–160.

Lehman, E. M. 2007. Seasonal occurrence and toxicity of *Microcystis* in impoundments of the Huron River, Michigan. Water Res. 41:795–802.

Lenoir, J., J. C. Gegout, P. A. Marquet, P. de Ruffray, and H. Brisse. 2008. A significant shift in plant species optimum elevations during the 20th century. Science 320:1768–1771.

Lepistö, A., P. G. Whitehead, C. Neal, and B. J. Cosby. 1988. Modelling the effects of acid deposition: Estimation of long-term water quality responses in forested catchments in Finland. Nord. Hydrol. 19:99–120.

Levine, C. R., and R. Yanai. 2012. Assessment of long-term monitoring of nitrogen, sulfur, and mercury deposition and environmental effects in New York State. NYSERDA Report 12-21. New York State Energy Research and Development Authority, Albany, NY.

Levine, S. N., A. D. Shambaugh, S. E. Pomeroy, and M. Braner. 1997. Phosphorus, nitrogen, and silica as controls on phytoplankton biomass and species composition in Lake Champlain (USA-Canada). Intern. Assoc. for Great Lakes Res. 23(2):131–148.

Li, H., and S. G. McNulty. 2007. Uncertainty analysis on simple mass balance model to calculate critical loads for soil acidity. Environ. Pollut. 149:315–326.

Likens, G. E., and D. C. Buso. 2012. Dilution and the elusive baseline. Environ. Sci. Technol. 46(8):4382–4387.

Likens, G. E., F. H. Bormann, R. S. Pierce, and W. A. Reiners. 1978. Recovery of a deforested ecosystem. Science 199:492–496.

Likens, G. E., C. T. Driscoll, and D. C. Buso. 1996. Long-term effects of acid rain: response and recovery of a forest ecosystem. Science 272(5259):244–246.

Likens, G. E., C. T. Driscoll, D. Buso, T. G. Siccama, C. E. Johnson, G. M. Lovett, T. J. Fahey, W. A. Reiners, D. F. Ryan, C. W. Martin, and S. W. Bailey. 1998. The biogeochemistry of calcium at Hubbard Brook. Biogeochemistry 41:89–173.

Likens, G. E., C. T. Driscoll, D. C. Buso, M. J. Mitchell, G. M. Lovett, S. W. Bailey, T. G. Siccama, W. A. Reiners, and C. Alewell. 2002. The biogeochemistry of sulfur at Hubbard Brook. Biogeochemistry 60:235–316.

Lilleskov, E. A., T. J. Fahey, and G. M. Lovett. 2001. Ectomycorrhizal fungal aboveground community change over an atmospheric nitrogen deposition gradient. Ecol. Appl. 11(2):397–410.

Lind, C. J., and J. D. Hem. 1975. Effects of organic solutes on chemical reactions of aluminum. U.S. Geological Survey Water Supply Paper 1827-G.

Linthurst, R. A., D. H. Landers, J. M. Eilers, D. F. Brakke, W. S. Overton, E. P. Meier, and R. E. Crowe. 1986. Population descriptions and physico-chemical relationships. Vol. 1 of Characteristics of lakes in the eastern United States. EPA/600/4-86/007a. U.S. Environmental Protection Agency, Washington, DC.

Lipton, D. W., and R. Hicks. 2003. The cost of stress: Low dissolved oxygen and recreational striped bass (*Morosne saxatilis*) fishing in the Patuxent River. Estuaries 26:310–315.

[LISS] Long Island Sound Study. 2003. Sound health 2003: A report on status and trends in the health of the Long Island Sound. EPA Long Island Sound Office, Stamford, CT. http://www.longislandsoundstudy.net.

Locke, A., and W. G. Sprules. 1994. Effects of lake acidification and recovery on the stability of zooplankton food webs. Ecology 75(2):498–506.

Long, R. P., S. B. Horsley, and P. R. Lilja. 1997. Impact of forest liming on growth and crown vigor of sugar maple and associated hardwoods. Can. J. For. Res. 27:1560–1573.

Long, R. P., S. B. Horsley, and T. J. Hall. 2011. Long-term impact of liming on growth and vigor of northern hardwoods. Can. J. For. Res. 41:1295–1307.

Longcore, J. R., H. Boyd, R. T. Brooks, J. D. Gill, G. M. Haramis, D. K. McNicol, J. R. Newman, K. A. Smith, and F. Stearns. 1993. Acidic depositions: Effects on wildlife and habitats. Technical Review 93-1. The Wildlife Society, Bethesda, MD.

Lovett, G. M. 1992. Atmospheric deposition and canopy interactions of nitrogen. *In* D. W. Johnson and S. E. Lindberg, eds. Atmospheric deposition and forest nutrient cycling. Springer-Verlag, New York, 159–166.

Lovett, G. M. 2013. Critical issues for critical loads. Proc. Nat. Acad. Sci. 110(3):808–809.

Lovett, G. M., and C. L. Goodale. 2011. A new conceptual model of nitrogen saturation based on experimental nitrogen addition to an oak forest. Ecosystems 14(4):615–631.

Lovett, G. M., and H. Rueth. 1999. Soil nitrogen transformation in beech and maple stands along a nitrogen deposition gradient. Ecol. Appl. 9(4):1330–1344.

Lovett, G. M., A. W. Thompson, J. B. Anderson, and J. J. Bowser. 1999. Elevational patterns of sulfur deposition at a site in the Catskill Mountains, New York. Atmos. Environ. 33:617–624.

Lovett, G. M., K. C. Weathers, and W. V. Sobczak. 2000. Nitrogen saturation and retention in forested watersheds of the Catskill Mountains, NY. Ecol. Appl. 10:73–84.

Lovett, G. M., L. M. Christensen, P. M. Groffman, C. G. Jones, J. E. Hart, and M. J. Mitchell. 2002a. Insect defoliation and nitrogen cycling in forests. BioScience 52(4):335–341.

Lovett, G. M., K. C. Weathers, and M. A. Arthur. 2002b. Control of nitrogen loss from forested watersheds by soil carbon:nitrogen ratio and tree species composition. Ecosystems 5:712–718.

Lovett, G. M., K. C. Weathers, M. A. Arthur, and J. C. Schultz. 2004. Nitrogen cycling in a northern hardwood forest: Do species matter? Biogeochemistry 67:289–308.

Lowrance, R. 1992. Groundwater nitrate and denitrification in a coastal plain riparian forest. J. Environ. Qual. 21:401–405.

Lundström, U. S., N. v. Breemen, and D. Bain. 2000. The podzolization process. A review. Geoderma 94:91–107.

Lynam, M. M., and G. J. Keeler. 2005. Artifacts associated with the measurement of particulate mercury in an urban environment: the influence of elevated ozone concentrations. Atmos. Environ. 39:3081–3088.

Lynch, J. A., and E. S. Corbett. 1989. Hydrologic control of sulfate mobility in a forested watershed. Water Resour. Res. 25(7):1695–1703.

MacAvoy, S. W., and A. J. Bulger. 1995. Survival of brook trout (*Salvelinus fontinalis*) embryos and fry in streams of different acid sensitivity in Shenandoah National Park, USA. Water Air Soil Pollut. 85:445–450.

Magill, A. H., J. D. Aber, G. M. Berntson, W. H. McDowell, K. J. Nadelhoffer, J. M. Melillo, and P. Steudler. 2000. Long-term nitrogen additions and nitrogen saturation in two temperate forests. Ecosystems 3:238–253.

Magill, A. H., J. D. Aber, W. S. Currie, K. J. Nadelhoffer, M. E. Martin, W. H. McDowell, J. M. Melillo, and P. Steudler. 2004. Ecosystem response to 15 years of chronic nitrogen additions at the Harvard Forest LTER, Massachusetts, USA. For. Ecol. Manage. 196:7–28.

Mahaffey, K. R. 2005. NHANES 1999–2002 Update on Mercury. *In* Proceedings of the 2005 national forum on contaminants in fish. U.S. Environmental Protection Agency, Washington, DC, II-51. http://water.epa.gov/scitech/swguidance/fishshellfish/techguidance/upload/2008_11_18_fish_forum_2005_proceedings2005.pdf.

Makarewicz, J. C. 2009. Nonpoint source reduction to the nearshore zone via watershed management practices: Nutrient fluxes, fate, transport, and biotic response: Background and objectives. J. Great Lakes Res. 35:3–9.

Makarewicz, J. C., and O. Bertram. 1991. Restoration of the Lake Erie ecosystem. BioScience 41:216–223.

Makarewicz, J. C., G. L. Boyer, W. Guenther, M. Arnold, and T. W. Lewis. 2006. The occurrence of cyanotoxins in the nearshore and coastal embayments of Lake Ontario. Great Lakes Research Review 7:25–29.

Makarewicz, J. C., P. E. D'Aiuto, and I. Bosch. 2007. Elevated nutrient levels from agriculturally dominated watersheds stimulate metaphyton growth. J. Great Lakes Res. 33:437–448.

Mallin, M. A., H. W. Paerl, J. Rudek, and P. W. Bates. 1993. Regulation of estuarine primary production by watershed rainfall and river flow. Mar. Ecol. Prog. Ser. 93:199–203.

Malone, T. C. 1977. Environmental regulation of phytoplankton productivity in the lower Hudson estuary. Estuarine and Coastal Marine Science 5:157–171.

Manera, M., R. Serra, G. Isani, and E. Carpene. 2000. Macrophage aggregates in gilt-head sea bream fed copper, iron and zinc enriched diet. J. Fish. Biol. 57:457–465.

Mann, L. K., D. W. Johnson, D. C. West, D. W. Cole, J. W. Hornbeck, C. W. Martin, H. Riekerk, C. T. Smith, W. T. Swank, L. M. Tritton, and D. H. Van Lear. 1988. Effects on whole-tree and stem-only clearcutting on postharvest hydrologic losses, nutrient capital, and regrowth. For. Sci. 34:412–428.

Martin, C. W., D. S. Noel, and C. A. Federer. 1984. Effects of forest clearcutting in New England on stream chemistry. J. Environ. Qual. 13:204–210.

Martin, P. A., T. V. McDaniel, K. D. Hughes, and B. Hunter. 2011. Mercury and other heavy metals in free-ranging mink of the lower Great Lakes basin, Canada, 1998–2006. Ecotoxicology 20:1701–1712.

Martinson, L., N. Lamersdorf, and P. Warfvinge. 2005. The Solling roof revisited: Slow recovery from acidification observed and modeled despite a decade of "clean-rain" treatment. Environ. Pollut. 135(2):293–302.

Mason, C. F., and C. D. Wren. 2001. Carnivora. *In* R. F. Shore and B. A. Rattner, eds. Ecotoxicology of wild mammals. Wiley, Chichester, England, 315–370.

Mason, J., and H. M. Seip. 1985. The current state of knowledge on acidification of surface waters and guidelines for further research. Ambio 14:45–51.

Mason, R. P., M. L. Abbott, R. A. Bodaly, O. R. Bullock Jr, C. T. Driscoll, D. Evers, S. B. Lindberg, M. Murray, and E. B. Swain. 2005. Monitoring the response to changing mercury deposition. Environ. Sci. Technol. 39:14A–22A.

Mason, R. R., B. E. Wickman, R. C. Beckwith, and H. G. Paul. 1992. Thinning and nitrogen fertilization in a grand fir stand infested with spruce budworm. Part I: Insect response. For. Sci. 38:235–251.

Matuszek, J. E., and G. L. Beggs. 1988. Fish species richness in relation to lake area, pH, and other abiotic factors in Ontario lakes. Can. J. Fish. Aquat. Sci. 45:1931–1941.

McAuley, D. G., and J. R. Longcore. 1988a. Foods of juvenile ring-necked ducks: Relationship to wetland pH. J. Wildl. Manage. 52:177–185.

McAuley, D. G., and J. R. Longcore. 1988b. Survival of juvenile ring-necked ducks on wetlands of different pH. J. Wildl. Manage. 52:169–176.

McClain, M. E., E. W. Boyer, C. L. Dent, S. E. Gergel, N. B. Grimm, P. M. Groffman, S. C. Hart, J. W. Harvey, C. A. Johnston, E. Mayorga, W. H. McDowell, and G. Pinay. 2003. Biogeochemical hot spots and hot moments at the interface of aquatic and terrestrial ecosystems. Ecosystems 6:301–312.

McCormick, J. H., K. M. Jensen, and L. E. Anderson. 1989. Chronic effects of low pH and elevated aluminum on survival, maturation, spawning, and embryo-larval development of the fathead minnow in soft water. Water Air Soil Pollut. 43:293–307.

McDonald, C. P., N. R. Urban, and C. M. Casey. 2010. Modeling historical trends in Lake Superior total nitrogen concentrations. J. Great Lakes Res. 36:715–721.

McDonnell, T. C., B. J. Cosby, and T. J. Sullivan. 2012. Regionalization of soil base cation weathering for evaluating stream water acidification in the Appalachian Mountains, USA. Environ. Pollut. 162:338–344.

McDonnell, T. C., S. Belyazid, T. J. Sullivan, H. Sverdrup, W. D. Bowman, and E. M. Porter. 2014. Modeled subalpine plant community response to climate change and atmospheric nitrogen deposition in Rocky Mountain National Park, USA. Environ. Pollut. 187:55–64.

McHale, M. R., P. S. Murdoch, D. A. Burns, and G. B. Lawrence. 2007. Factors controlling aluminum release after a clearcut in a forested watershed with calcium-poor soils. Biogeochemistry 84:311–331.

McKenney, D. W., J. H. Pedlar, K. Lawrence, K. Campbell, and M. F. Hutchinson. 2007. Potential impacts of climate change on the distribution of North American trees. BioScience 57:939–948.

McKie, B. G., Z. Petrin, and B. Malmqvist. 2006. Mitigation or disturbance? Effects of liming on macroinvertebrate assemblage structure and leaf-litter decomposition in the humic streams of northern Sweden. J. Appl. Ecol. 43(4):780–791.

McLaughlin, S. B., and M. J. Tjoelker. 1992. Growth and physiological changes in red spruce saplings associated with acidic deposition at high elevations in the southern Appalachians, USA. For. Ecol. Manage. 51(1–3):43–51.

McNeil, B. E., J. M. Read, and C. T. Driscoll. 2012. Foliar nitrogen responses to the environmental gradient matrix of the Adirondack Park, New York. Ann. Assoc. Amer. Geogr. 102(1):1–16.

McNicol, D. K. 2002. Relation of lake acidification and recovery to fish, common loon and common merganser occurrence in Algoma Lakes. Water Air Soil Pollut.: Focus 2:151–168.

McNulty, S. G., J. D. Aber, and R. D. Boone. 1991. Spatial changes in forest floor and foliar chemistry of spruce-fir forests across New England. Biogeochemistry 14:13–29.

McNulty, S. G., J. D. Aber, and S. D. Newman. 1996. Nitrogen saturation in a high elevation New England spruce-fir stand. For. Ecol. Manage. 84:109–121.

McNulty, S. G., J. Boggs, J. D. Aber, L. Rustad, and A. Magill. 2005. Red spruce ecosystem level changes following 14 years of chronic N fertilization. For. Ecol. Manage. 219:279–291.

Meinelt, T., R. Kruger, M. Pietrock, R. Osten, and C. Steinberg. 1997. Mercury pollution and macrophage centres in pike (*Esox lucius*) tissues. Environ. Sci. Pollut. Res. 4:32–36.

Melillo, J. M., J. D. Aber, and J. F. Muratore. 1982. Nitrogen and lignin control of hardwood leaf litter decomposition dynamics. Ecology 63:621–626.

Meyer, M. W., P. W. Rasmussen, C. J. Watras, B. M. Fevold, and K. P. Kenow. 2011. Bi-phasic trends in mercury concentrations in blood of Wisconsin common loons during 1992–2010. Ecotoxicology 20(7):1659–1668.

Meyers, T., J. Sickles, R. Dennis, K. Russell, J. Galloway, and T. Church. 2001. Atmospheric nitrogen deposition to coastal estuaries and their watersheds. *In* R. Valigura, R. Alexander, M. Castro, T. Meyers, H. Paerl, P. Stacey, and R. E. Turner, eds. Nitrogen loading in coastal water bodies: An atmospheric perspective. Coastal and Estuarine Studies. American Geophysical Union, Washington, DC, 53–76.

Michalak, A. M., E. J. Anderson, D. Beletsky, S. Boland, N. S. Bosch, T. B. Bridgeman, J. D. Chaffin, K. Cho, R. Confesor, I. Daloglu, J. V. DePinto, M. A. Evans, G. L. Fahnenstiel, L. He, J. C. Ho, L. Jenkins, T. H. Johengen, K. C. Kuo, E. LaPorte, X. Liu, M. R. McWilliams, M. R. Moore, D. J. Posselt, R. P. Richards, D. Scavia, A. L. Steiner, E. Verhamme, D. M. Wright, and M. A. Zagorski. 2013. Record-setting algal bloom in Lake Erie caused by agricultural and meteorological trends consistent with expected future conditions. Proc. Nat. Acad. Sci. 110(16):6448–6452.

Mida, J. L., D. Scavia, G. L. Fahnenstiel, S. A. Pothoven, H. A. Vanderploeg, and D. M. Dolan. 2010. Long-term and recent changes in southern Lake Michigan water quality with implications for present trophic status. J. Great Lakes Res. 36:42–49.

Mierle, G., E. M. Addison, K. S. MacDonald, and D. G. Joachim. 2000. Mercury levels in tissues of otters from Ontario, Canada: Variation with age, sex, and location. Environ. Toxicol. Chem. 19:3044–3051.

Millard, E. S., O. E. Johannsson, M. A. Neilson, and A. H. El-Shaarawi. 2003. Long-term, seasonal and spatial trends in nutrients, chlorophyll *a* and light attenuation in Lake Ontario. *In* M. Munawar, ed. State of Lake Ontario (SOLO): Past, present, and future. Ecovision World Monograph Series. Aquatic Ecosystem Health and Management Society, Burlington, ON, Canada, 97–132.

Millennium Ecosystem Assessment. 2005. Living beyond our means: Natural assets and human well-being (statement from the board). Island Press, Washington, DC.

Miller, E. K., J. D. Blum, and A. J. Friedland. 1993. Determination of soil exchangeable-cation loss and weathering rates using Sr isotopes. Nature 362:438–441.

Millhollen, A. G., M. S. Gustin, and D. Obrist. 2006. Foliar mercury accumulation and exchange for three tree species. Environ. Sci. Technol. 40(19):6001–6006.

Mills, E. L., R. M. Dermott, E. F. Roseman, D. Dustin, E. Mellina, D. B. Conn, and A. P. Spindle. 1993. Colonization, ecology, and population structure of the "quagga" mussel (bivalivia: Dreissenidae) in the lower Great Lakes. Can. J. Fish. Aquat. Sci. 50:2304–2314.

Mills, E. L., C. E. Hoffman, J. P. Gillette, L. G. Rudstam, R. McCullough, D. Bishop, W. Pearsall, S. LaPan, B. Trometer, B. Lantry, R. O'Gorman, and T. Schaner. 2006. 2005 Status of the Lake Ontario ecosystem: A biomonitoring approach. Section 20. NYSDEC Lake Ontario Annual Report 2005. New York State Department of Environmental Conservation, Albany, NY.

Mills, K. H., S. M. Chalanchuk, J. C. Mohr, and I. J. Davies. 1987. Responses of fish populations in Lake 223 to 8 years of experimental acidification. Can. J. Fish. Aquat. Sci. 44 (Suppl. 1):114–125.

Minocha, R., W. C. Shortle, G. B. Lawrence, M. B. David, and S. C. Minocha. 1997. Relationships among foliar chemistry, foliar polyamines, and soil chemistry in red spruce trees growing across the northeastern United States. Plant Soil 191(1):109–122.

Mitchell, M., D. Raynal, and C. Driscoll. 1996a. Biogeochemistry of a forested watershed in the central Adirondack Mountains: Temporal changes and mass balances. Water, Air, and Soil Pollut. 88(3–4):355–369.

Mitchell, M. J., C. T. Driscoll, J. S. Kahl, G. E. Likens, P. S. Murdoch, and L. H. Pardo. 1996b. Climatic control of nitrate loss from forested watersheds in the northeast United States. Environ. Sci. Technol. 30(8):2609–2612.

Mitchell, M. J., K. B. Piatek, S. Christopher, B. Mayer, C. Kendall, and P. J. McHale. 2006. Solute sources in stream water during consecutive fall storms in a northern hardwood forest watershed: A combined hydrological, chemical and isotopic approach. Biogeochemistry 78:217–246.

Mitchell, M. J., G. Lovett, S. Bailey, D. Burns, D. Buso, T. A. Clair, F. Courchesne, L. Duchesne, C. Eimers, I. Fernandez, D. Houle, D. S. Jeffries, G. E. Likens, M. D. Moran, C. Rogers, D. Schwede, J. Shanley, K. C. Weathers, and R. Vet. 2011. Comparisons of watershed sulfur budgets in southeast Canada and northeast US: New approaches and implications. Biogeochemistry 103:181–207.

Mitchell, R. S., and G. S. Tucker. 1997. Revised checklist of New York State plants. Bulletin No. 490. The New York State Museum, Albany, NY.

Mitsch, W. J., and J. G. Gosselink. 2000. Wetlands. John Wiley & Sons, Inc., New York, NY.

Mohseni, O., H. G. Stefan, and J. G. Eaton. 2003. Global warming and potential changes in fish habitat in U.S. streams. Climatic Change 59:389–409.

· Moldan, F., R. F. Wright, R. C. Ferrier, B. I. Andersson, and H. Hultberg. 1998. Simulating the Gårdsjön covered catchment experiment with the MAGIC model. *In* H. Hultberg and R. A. Skeffington, eds. Experimental reversal of acid rain effects. The Gårdsjön Roof Project. Wiley and Sons, Chichester, UK, 351–362.

Momen, B., G. B. Lawrence, S. A. Nierzwicki-Bauer, J. W. Sutherland, L. Eichler, J. P. Harrison, and C. W. Boylen. 2006. Trends in summer chemistry linked to productivity in lakes recovering from acid deposition in the Adirondack region of New York. Ecosystems 9:1306–1317.

Monson, B. A. 2009. Trend reversal of mercury concentrations in piscivorous fish from Minnesota lakes: 1982–2006. Environ. Sci. Technol. 43:1750–1755.

Monson, B. A., D. F. Staples, S. P. Bhavsar, T. M. Holsen, C. S. Schrank, S. K. Moses, D. J. McGoldrick, S. M. Backus, and K. A. Williams. 2011. Spatiotemporal trends of mercury in walleye and largemouth bass from the Laurentian Great lakes region. Ecotoxicology 20:1555–1567.

Monteith, D. T., J. L. Stoddard, C. D. Evans, H. A. de Wit, M. Forsius, T. Hogasen, A. Wilander, B. L. Skjelkvale, D. S. Jeffries, J. Vuorenmaa, B. Keller, J. Kopacek, and J. Vesely. 2007. Dissolved organic carbon trends resulting from changes in atmospheric deposition chemistry. Nature 450(7169):537–540.

Montgomery, D. R., and J. M. Buffington. 1998. Channel processes, classification, and response. *In* R. J. Naiman and R. Bilby, eds. River ecology and management. Springer-Verlag, New York, 13–42.

Moon, J. B., and H. J. Carrick. 2007. Seasonal variation of phytoplankton nutrient limitation in Lake Erie. Aquatic Microbial Ecology 48:61–71.

Moore, T., C. Blodau, J. Turenen, N. Roulet, and P. J. H. Richard. 2004. Patterns of nitrogen and sulfur accumulation and retention in ombrotrophic bogs, eastern Canada. Glob. Change Biol. 11:356–367.

Morgan, C., and N. Owens. 2001. Benefits of water quality policies: the Chesapeake Bay. Ecol. Econ. 39(2):271–284.

Morrice, J. A., J. R. Kelly, A. S. Trebitz, A. M. Cotter, and M. L. Knuth. 2004. Temporal dynamics of nutrients (N and P) and hydrology in a Lake Superior coastal wetland. J. Great Lakes Res. 30(Suppl 1):82–96.

Morris, J. T. 1991. Effects of nitrogen loading on wetland ecosystems with particular reference to atmospheric deposition. Ann. Rev. Ecol. Syst. 22:257–279.

Mörth, C. M., P. Torssander, O. J. Kjonaas, A. O. Stuanes, F. Moldan, and R. Giesler. 2005. Mineralization of organic sulfur delays recovery from anthropogenic acidification. Environ. Sci. Technol. 39(14):5234–5240.

Moss, D. M., M. T. Furse, J. F. Wright, and P. D. Armitage. 1987. The prediction of the macro-invertebrate fauna of unpolluted running-water sites in Great Britain using environmental data. Freshw. Biol. 17:41–52.

Mount, D. R., C. G. Ingersoll, D. D. Gulley, J. D. Fernandez, T. W. LaPoint, and H. L. Bergman. 1988. Effect of long-term exposure to acid, aluminum, and low calcium on adult brook trout (*Salvelinus fontinalis*). I. Survival, growth, fecundity, and progeny survival. Can. J. Fish. Aquat. Sci. 45:1623–1632.

Mulholland, P. J., H. M. Valett, J. R. Webster, S. A. Thomas, L. W. Cooper, S. K. Hamilton, and B. Peterson. 2004. Stream denitrification and total nitrate uptake rates measured using a field ^{15}N tracer addition approach. Limnol. Oceanogr. 49(3):809–820.

Munawar, M., I. F. Munawar, R. Dermott, M. Fitzpatrick, and H. Niblock. 2006. The threat of exotic species to the food web in Lake Ontario. Verh. Int. Verein. Limnol. 29:1194–1198.

Muniz, I. P., and H. Levivestad. 1980. Acidification effects on freshwater fish. *In* D. Drabløs and A. Tollan, eds. Proceedings of the International Conference on the Ecological Impact of Acid Precipitation. SNSF Project, Oslo, Norway, 84–92.

Murdoch, P. S., and J. L. Stoddard. 1992. The role of nitrate in the acidification of streams in the Catskill Mountains of New York. Water Resour. Res. 28(10):2707–2720.

Murdoch, P. S., and J. L. Stoddard. 1993. Chemical characteristics and temporal trends in eight streams of the Catskill Mountains. Water Air Soil Pollut. 67:367–395.

Murdoch, P. S., D. A. Burns, and G. B. Lawrence. 1998. Relation of climate change to the acidification of surface waters by nitrogen deposition. Environ. Sci. Technol. 32:1642–1647.

Murdoch, P. S., J. S. Baron, and T. L. Miller. 2000. Potential effects of climate change on surface-water quality in North America. J. Am. Water Resour. Assoc. 36:347–366.

Murphy, T. P., K. Irvine, J. Guo, J. Davies, H. Murkin, M. Charlton, and S. B. Watson. 2003. New microcystin concerns in the lower Great Lakes. Water Quality Research Journal of Canada 38(1):127–140.

Myers, M. D., M. A. Ayers, J. S. Baron, P. R. Beauchemin, K. T. Gallagher, M. B. Goldhaber, D. R. Hutchinson, J. W. LaBaugh, R. G. Sayre, S. E. Schwarzbach, E. S. Schweig, J. Thormodsgard, C. van Riper III, and W. Wilde. 2007. USGS goals for the coming decade. Science 318:200–201.

Nadelhoffer, K., M. Downs, B. Fry, A. Magill, and J. Aber. 1999. Controls on N retention and exports in a forested watershed. Environ. Monitor. Assess. 55(1):187–210.

Naidoo, R., A. Balmford, R. Costanza, B. Fisher, R. E. Green, B. Lehner, T. R. Malcolm, and T. H. Rickets. 2008. Global mapping of ecosystem services and conservation priorities. Proc. Nat. Acad. Sci. 105(28):9495–9500.

Nalepa, T. F., D. L. Fanslow, and G. A. Lang. 2009. Transformation of the offshore benthic community in Lake Michigan: recent shift from the native amphipod *Diporeia* spp. to the invasive mussel *Dreissena rostriformis bugensis*. Freshw. Biol. 54:466–479.

[NADP] National Atmospheric Deposition Program. 2004. National Atmospheric Deposition Program 2003 annual summary. NADP Data Report 2004-1. Illinois State Water Survey, Champaign, IL.

[NADP] National Acid Deposition Program. 2014. Atmospheric Mercury Network (AMNet). http://nadp.sws.uiuc.edu/amn/.

[NAPAP] National Acid Precipitation Assessment Program. 1991. Integrated assessment report. National Acid Precipitation Assessment Program, Washington, DC.

[NAPAP] National Acid Precipitation Assessment Program. 1998. NAPAP Biennial Report to Congress: An Integrated Assessment. National Acid Precipitation Assessment Program, Silver Spring, MD.

[NOAA] National Oceanic and Atmospheric Administration. 1988. Hudson/Raritan Estuary: Issues, resources, status, and management. Estuary-of-the-Month Seminar Series no. 9. National Oceanographic and Atmospheric Administration, Washington, DC.

[NOAA] National Oceanic and Atmospheric Administration. 1996. South Atlantic region. Vol. 1 of NOAA's Estuarine Eutrophication Survey. National Oceanic and Atmospheric Administration, National Ocean Service, Office of Ocean Resources Conservation and Assessment, Silver Spring, MD. http://www.eutro.org/documents/south%20atlantic%20regional%20report.pdf.

[NOAA] National Oceanic and Atmospheric Administration. 1998. Pacific Coast region. Vol. 5 of NOAA's Estuarine Eutrophication Survey. National Oceanic and Atmospheric Administration, National Ocean Service, Office of Ocean Resources Conservation and Assessment, Silver Spring, MD.

[NRC] National Research Council. 2000. Clean coastal waters: Understanding and reducing the effects of nutrient pollution. National Academy Press, Washington, DC.

National Science and Technology Council. 1998. National Acid Precipitation Assessment Program biennial report to Congress: An integrated assessment; executive summary. U.S. Department of Commerce, National Oceanic and Atmospheric Administration, Silver Spring, MD.

New York State Department of Health. 2014. Regional fish health advisories. http://www.health.ny.gov/environmental/outdoors/fish/health_advisories/regional/.

Newton, R. M., D. A. Burns, V. L. Blette, and C. T. Driscoll. 1996. Effect of whole catchment liming on the episodic acidification of two Adirondack streams. Biogeochemistry 32:299–322.

Nicholls, K. H., G. J. Hopkins, S. J. Standke, and L. Nakamoto. 2001. Trends in total phosphorus in Canadian nearshore waters of Laurentian Great Lakes: 1976–1999. Great Lakes Res. 27:402–422.

Niemi, G. J., D. Wardrop, R. Brooks, S. Anderson, V. Brady, H. Paerl, C. Rakocinski, M. Brouwer, B. Levinson, and M. McDonald. 2004. Rationale for a new generation of ecological indicators for coastal waters. Environ. Health Perspect. 112(9):979–986.

Niemi, G. J., J. R. Kelly, and N. P. Danz. 2007. Environmental Indicators for the coastal region of the North American Great Lakes: Introduction and prospectus. J. Great Lakes Res. 33, Supplement 3(0):1–12.

Nierzwicki-Bauer, S. A., C. W. Boylen, L. W. Eichler, J. P. Harrison, J. W. Sutherland, W. Shaw, R. A. Daniels, D. F. Charles, F. W. Acker, T. J. Sullivan, B. Momen, and P. Bukaveckas. 2010. Acidification in the Adirondacks: Defining the biota in trophic levels of 30 chemically diverse acid-impacted lakes. Environ. Sci. Technol. 44:5721–5727.

Nilsson, J., and P. Grennfelt, eds. 1988. Critical loads for sulphur and nitrogen. Report from a workshop held at Skokloster, Sweden, March 19–24. Nordic Council of Ministers, Copenhagen.

Nixon, S. W. 1986. Nutrient dynamics and the productivity of marine coastal waters. *In* R. Halwagy, D. Clayton, and M. Behbehani, eds. Marine environment and pollution. Alden Press, Oxford, 97–115.

Nixon, S., and B. A. Buckley. 2005. "A strikingly rich zone": Nutrient enrichment and secondary production in coastal marine ecosystems. Estuaries 25:782–796.

Nobre, A. M., J. G. Ferreira, A. Newton, T. Simas, J. D. Icely, and R. Neves. 2005. Managing eutrophication: integration of field data, ecosystem-scale simulations and screening models. Journal of Marine Systems 56(3/4):375–390.

Nodvin, S. C., H. V. Miegroet, S. E. Lindberg, N. S. Nicholas, and D. W. Johnson. 1995. Acidic deposition, ecosystem processes, and nitrogen saturation in a high elevation southern Appalachian watershed. Water Air Soil Pollut. 85:1647–1652.

North, R. L., S. J. Guidford, R. E. H. Smith, S. M. Havens, and M. R. Twiss. 2007. Evidence for phosphorus, nitrogen, and iron colimitation of phytoplankton communities in Lake Erie. Limnol. Oceanogr. 52:315–328.

Norton, J. M., and J. M. Stark. 2011. Regulation and measurement of nitrification in terrestrial systems. Methods in Enzymology 486:343–368.

Norton, S. A., J. J. Akielaszek, T. A. Haines, K. J. Stromborg, and J. R. Longcore. 1982. Bedrock geologic control of sensitivity of aquatic ecosystems in the United States to acidic deposition. National Atmospheric Deposition Program, Fort Collins, CO.

Norton, S. A., R. F. Wright, J. S. Kahl, and J. P. Scofield. 1992. The MAGIC simulation of surface water acidification at, and first year results from, the Bear Brook Watershed Manipulation, Maine, USA. Environ. Pollut. 77:279–286.

NYS Seagrass Taskforce. 2009. Final Report of the New York State Seagrass Task Force: Recommendations to the New York State governor and legislature.

Obrist, D., D. W. Johnson, S. E. Lindberg, Y. Luo, O. Hararuk, R. Bracho, J. J. Battles, D. B. Dail, R. I. Edmonds, R. K. Monson, S. V. Ollinger, S. G. Pallardy, K. S. Pregitzer, and D. E. Todd. 2011. Mercury distribution across 14 U.S. forests. Part I. Spatial patterns of concentrations in biomass, litter, and soils. Environ. Sci. Technol. 45:3974–3981.

Officer, C. B., and J. H. Ryther. 1980. The possible importance of silicon in marine eutrophication. Mar. Ecol. Prog. Ser. 3:83–91.

Olem, H. 1991. Liming acidic surface waters. Lewis Publishers, Chelsea, MI.

Olha, J. 1990. Novel algal blooms: Common underlying causes with particular reference to New York and New Jersey coastal waters. MS thesis, Marine Sciences Research Center, State University of New York, Stony Brook, NY.

Oliver, B. G., E. M. Thurman, and R. L. Malcolm. 1983. The contribution of humic substances to the acidity of colored natural waters. Geochim. Cosmochim. Acta 47:2031–2035.

Ollinger, S. V., R. G. Lathrop, J. M. Ellis, J. D. Aber, G. M. Lovett, and S. E. Millham. 1993. A spatial model of atmospheric deposition for the northeastern US. Ecol. Appl. 3(3):459–472.

Ollinger, S. V., M. L. Smith, M. E. Martin, R. A. Hallett, C. L. Goodale, and J. D. Aber. 2002. Regional variation in foliar chemistry and N cycling among forests of diverse history and composition. Ecology 83(2):339–355.

Ormerod, S. J., and S. J. Tyler. 1987. Dippers (*Cinclus cinclus*) and grey wagtails (*Motacilla cinerea*) as indicators of stream acidity in upland Wales. ICBP Technical Publication No. 6. International Council for Bird Preservation, Cambridge, UK.

Ormerod, S. J., S. J. Tyler, and J. M. S. Lewis. 1985. Is the breeding distribution of dippers influenced by stream acidity? Bird Study 32:32–39.

Ormerod, S. J., N. Allinson, D. Hudson, and S. J. Tyler. 1986. The distribution of breeding dippers (*Cinclus cinclus* [L.]; Aves) in relation to stream acidity in upland Wales. Freshw. Biol. 16:501–507.

O'Shea, M. L., and T. M. Brosnan. 2000. Trends in indicators of eutrophication in western Long Island Sound and the Hudson-Raritan Estuary. Estuaries 23(6):877–901.

Paerl, H. W. 1985. Enhancement of marine primary production by nitrogen-enriched acid rain. Nature 315:747–749.

Paerl, H. 1995. Coastal eutrophication in relation to atmospheric nitrogen deposition: Current perspectives. Ophelia 41:237–259.

Paerl, H. 1997. Coastal eutrophication and harmful algal blooms: importance of atmospheric deposition and groundwater as "new" nitrogen and other nutrient sources. Limnol. Oceanogr. 42:1154–1162.

Paerl, H. W. 2002. Connecting atmospheric nitrogen deposition to coastal eutrophication: research is needed to understand this air-water quality interaction. Environ. Sci. Technol. 36(15):323A–326A.

Paerl, H. W., and D. R. Whitall. 1999. Anthropogenically-derived atmospheric nitrogen deposition, marine eutrophication and harmful algal bloom expansion: Is there a link? Ambio 28:307–311.

Paerl, H. W., W. R. Boynton, R. L. Dennis, C. T. Driscoll, H. S. Greening, J. N. Kremer, N. N. Rabalais, and S. P. Seitzinger. 2001. Atmospheric deposition of nitrogen in coastal waters: biogeochemical and ecological implications. *In* R. W. Valigura, R. B. Alexander, M. S. Castro, T. P. Meyers, H. W. Paerl, P. E. Stacey, and R. E. Turner, eds. Nitrogen loading in coastal water bodies. An atmospheric perspective. American Geophysical Union, Washington, DC, 11–52.

Paerl, H. W., R. L. Dennis, and D. R. Whitall. 2002. Atmospheric deposition of nitrogen: Implications for nutrient over-enrichment of coastal waters. Estuaries 25(4b):677–693.

Paerl, H. W., L. M. Valdes, B. L. Peierls, J. E. Adolf, and L. W. Harding. 2006. Anthropogenic and climatic influences on the eutrophication of large estuarine ecosystems. Limnol. Oceanogr. 51(1, part 2):448–462.

Page, B. D., and M. J. Mitchell. 2008. The influence of basswood (*Tilia americana*) and soil chemistry on soil nitrate concentrations in a northern hardwood forest. Can. J. For. Res. 38:667–676.

Pardo, L. H., C. T. Driscoll, and G. E. Likens. 1995. Patterns of nitrate loss from a chronosequence of clear-cut watersheds. Water Air Soil Pollut. 85:1659–1664.

Pardo, L. H., M. J. Robin-Abbott, and C. T. Driscoll, eds. 2011. Assessment of nitrogen deposition effects and empirical critical loads of nitrogen for ecoregions of the United States. General Technical Report NRS-80. U.S. Forest Service, Newtown Square, PA.

Parker, C. A., and J. E. O'Reilly. 1991. Oxygen depletion in Long Island Sound: A historical perspective. Estuaries 14:248–264.

Parker, D. R., L. W. Zelazny, and T. B. Kinraide. 1989. Chemical speciation and plant toxicity of aqueous aluminum. *In* T. E. Lewis, ed. Environmental chemistry and toxicology of aluminum. American Chemical Society, Chelsea, MI, 117–145.

Parker, K. E. 1988. Common loon reproduction and chick feeding on acidified lakes in the Adirondack Park, New York. Can. J. Zool. 66:804–810.

Passy, S. I. 2006. Diatom community dynamics in streams of chronic and episodic acidification: the roles of environment and time. J. Phycol. 42(2):312–323.

Passy, S. I. 2010. A distinct latitudinal gradient of diatom diversity is linked to resource supply. Ecology 91(1):36–41.

Pearson, J., and R. G. Stewart. 1993. Atmospheric ammonia deposition and its effects on plants. Tansley Review no. 56. New Phytol. 125:283–305.

Perdue, E. M., J. H. Reuter, and R. S. Parrish. 1984. A statistical model of proton binding by humus. Geochim. Cosmochim. Acta 48:1257–1263.

Peterjohn, W. T., and D. L. Correll. 1984. Nutrient dynamics in an agricultural watershed: Observation on the role of a riparian forest. Ecology 65:1466–1475.

Peterson, J., D. L. Schmoldt, D. Peterson, J. M. Eilers, R. Fisher, and R. Bachman. 1992. Guidelines for evaluating air pollution impacts on Class I wilderness areas in the Pacific Northwest. PNW-GTR-299. USDA Forest Service, Portland, OR.

Petrin, Z., H. Laudon, and B. Malmqvist. 2007a. Does freshwater macroinvertebrate diversity along a pH-gradient reflect adaptation to low pH? Freshw. Biol. 52(11):2172–2183.

Petrin, Z., B. McKie, I. Buffam, H. Laudon, and B. Malmqvist. 2007b. Landscape-controlled chemistry variation affects communities and ecosystem function in headwater streams. Can. J. Fish. Aquat. Sci. 64(11):1563–1572.

Pinay, G., L. Roques, and A. Fabre. 1993. Spatial and temporal patterns of denitrification in a riparian forest. J. Appl. Ecol. 30:581–591.

Pinay, G., B. V. J., A. M. Planty-Tabacchi, B. Gumiero, and H. Decamps. 2000. Geomorphic control of denitrification in large river floodplain soils. Biogeochemistry 50:163–182.

Pittman, H. T., W. W. Bowerman, L. H. Grim, T. G. Grubb, and W. C. Bridges. 2011. Using nestling feathers to assess spatial and temporal concentrations of mercury in bald eagles at Voyageurs National Park, Minnesota, USA. Ecotoxicology 20:1626–1635.

Plafkin, J. L., M. T. Barbour, K. D. Porter, S. K. Gross, and R. M. Hughes. 1989. Rapid bioassessment protocols for use in streams and rivers: Benthic macroinvertebrates and fish. EPA 440-89-001. U.S. Environmental Protection Agency, Office of Water Regulations and Standards, Washington, DC.

Poff, N. L., J. D. Allan, M. B. Bain, J. R. Karr, K. L. Prestegaard, B. D. Richter, R. E. Sparks, and J. C. Stromberg. 1997. The natural flow regime: A paradigm for river conservation and restoration. BioScience 47:769–784.

Potyondy, J. P., B. B. Roper, S. E. Hixson, R. L. Leiby, R. L. Lorenz, and C. M. Knopp. 2006. Aquatic ecological unit inventory technical guide: Valley segment and river reach. General Technical Report. USDA Forest Service, Ecosystem Management Coordination Staff, Washington, DC.

Pound, K. L., G. B. Lawrence, and S. I. Passy. 2013. Wetlands serve as natural sources for improvement of stream ecosystem health in regions affected by acid deposition. Glob. Change Biol. 19(9):2720–2728.

Pourmokhtarian, A., C. T. Driscoll, J. L. Campbell, and K. Hayhoe. 2012. Modeling potential hydrochemical responses to climate change and increasing CO_2 at the Hubbard Brook Experimental Forest using a dynamic biogeochemical model (PnET-BGC). Water Resour. Res. 48(7): W07514.

Rabalais, N. N. 2002. Nitrogen in aquatic ecosystems. Ambio 31:102–112.

Rago, P. J., and J. G. Wiener. 1986. Does pH affect fish species richness when lake area is considered? Trans. Am. Fish. Soc. 115:438–447.

Rapp, L., and K. Bishop. 2009. Surface water acidification and critical loads: Exploring the F-factor. Hydrol. Earth Syst. Sci. 13:2191–2201.

Rattner, B. A., G. M. Haramis, D. S. Chu, C. M. Bunck, and C. G. Scanes. 1987. Growth and physiological condition of black ducks reared on acidified wetlands. Can. J. Zool. 65:2953–2958.

Raulund-Rasmussen, K., O. K. Borggaard, J. C. B. Hansen, and M. Olsson. 1998. Effect of natural organic soil solutes on weathering rates of soil minerals. Eur. J. Soil Sci. 49:397–406.

Raynal, D. J., M. J. Mitchell, C. T. Driscoll, and K. M. Roy. 2004. Effects of atmospheric deposition of sulfur, nitrogen, and mercury on Adirondack Ecosystems. Report 04-03. New York State Energy Research and Development Authority, Albany, NY.

Redbo-Torstensson, P. 1994. The demographic consequences of nitrogen fertilization of a population of sundew, *Drosera rotundifolia*. Acta Bot. Neerl. 43(2):175–188.

Reinfelder, J. R., N. S. Fisher, S. N. Luoma, J. W. Nichols, and W.-X. Wang. 1998. Trace element trophic transfer in aquatic organisms: A critique of the kinetic model approach. Sci. Total Environ. 219:117–135.

Resh, V. H., R. H. Norris, and M. T. Barbour. 1995. Design and implementation of rapid assessment approaches for water resource monitoring using benthic macroinvertebrates. Aust. J. Ecol. 20:108–121.

Reuss, J. O. 1983. Implications of the calcium-aluminum exchange system for the effect of acid precipitation on soils. J. Environ. Qual. 12(4):591–595.

Reuss, J. O., and D. W. Johnson. 1986. Acid deposition and the acidification of soil and water. Springer-Verlag, New York.

Riegman, R. 1992. *Phaeocystis* blooms and eutrophication of the continental coastal zones of the North Sea. Mar. Biol. 112:479–484.

Riegman, R. 1998. Species composition of harmful algal blooms in relation to macronutrient dynamics. *In* D. M. Anderson, A. D. Cembella, and G. M. Hallegraeff, eds. Physiological ecology of harmful algal blooms. North Atlantic Treaty Organization Series Vol. G 41. Springer, Berlin, 475–488.

Riggan, P. J., R. N. Lockwood, P. M. Jacks, C. G. Colver, F. Weirich, L. F. DeBano, and J. A. Brass. 1994. Effects of fire severity on nitrate mobilization in watersheds subject to chronic atmospheric deposition. Environ. Sci. Technol. 28:369–375.

Rimmer, C. C., K. P. McFarland, D. C. Evers, E. Miller, K., Y. Aubry, D. Busby, and R. J. Taylor. 2005. Mercury concentrations in Bicknell's thrush and other insectivorous passerines in montane forests of northeastern North America. Ecotoxicology 14:223–240.

Rimmer, C. C., E. K. Miller, K. P. McFarland, R. J. Taylor, and S. D. Faccio. 2009. Mercury bioaccumulation and trophic transfer in the terrestrial food web of a montane forest. Ecotoxicology 19:697–709.

Risch, M. R., J. F. DeWild, D. P. Krabbenhoft, R. K. Kolka, and L. Zhang. 2012. Litterfall mercury dry deposition in the eastern USA. Environ. Pollut. 161:264–290.

Riva-Murray, K., R. W. Bode, P. J. Phillips, and G. L. Wall. 2002. Impact source determination with biomonitoring data in New York State: Concordance with environmental data. Northeast. Nat. 9(2):127–162.

Riva-Murray, K., L. C. Chasar, P. M. Bradley, D. A. Burns, M. E. Brigham, M. J. Smith, and T. A. Abrahamsen. 2011. Spatial patterns of mercury in macroinvertebrates and fishes from streams of two contrasting forested landscapes in the eastern United States. Ecotoxicology 20(7):1530–1542.

Riva-Murray, K., P. M. Bradley, L. C. Chasar, D. T. Button, M. E. Brigham, B. C. Scudder Eikenberry, C. A. Journey, and M. A. Lutz. 2013a. Influence of dietary carbon on mercury bioaccumulation in streams of the Adirondack Mountains of New York and the Coastal Plain of South Carolina, USA. Ecotoxicology 22(1):60–71.

Riva-Murray, K., P. M. Bradley, B. C. Scudder Eikenberry, C. D. Knightes, C. A. Journey, M. E. Brigham, and D. T. Button. 2013b. Optimizing stream water mercury sampling for calculation of fish bioaccumulation factors. Environmental Science and Technology 47(11):5904–5912.

Rochefort, L., D. H. Vitt, and S. E. Bayley. 1990. Growth, production and decomposition dynamics of *Sphagnum* under natural and experimentally acidified conditions. Ecology 71(5):1986–2000.

Rochelle, B. P., M. R. Church, and M. B. David. 1987. Sulfur retention at intensively studied sites in the U.S. and Canada. Water Air Soil Pollut. 33(1–2):73–83.

Rolfhus, K. R., B. D. Hall, B. A. Monson, A. M. Paterson, and J. D. Jeremiason. 2011. Assessment of mercury bioaccumulation within the pelagic food web of lakes in the western Great Lakes region. Ecotoxicology 20:1520–1529.

Rood, B. E., J. F. Gottgens, J. J. Delfino, C. D. Earle, and T. L. Crisman. 1995. Mercury accumulation trends in Florida Everglades and Savannas Marsh flooded soils. Water, Air, and Soil Pollut. 80(1–4):981–990.

Rose, C., and R. P. Axler. 1998. Uses of alkaline phosphatase activity in evaluating phytoplankton community phosphorous deficiency. Hydrobiologia 361:145–156.

Rosenbrock, H. H. 1960. An automatic method for finding the greatest or least value of a function. Computer Journal 3:175–184.

Rosenzweig, C., W. Solecki, A. DeGaetano, M. O'Grady, S. Hassol, and P. Grabhorn, eds. 2011. Responding to climate change in New York State: The ClimAID integrated assessment for effective climate change adaptation. Synthesis report. New York State Research and Development Authority, Albany, NY.

Ross, D. S., G. B. Lawrence, and G. Fredriksen. 2004. Mineralization and nitrification patterns at eight northeastern USA forested research sites. For. Ecol. Manage. 188:317–335.

Rosseland, B. O., and M. Staurnes. 1994. Physiological mechanisms for toxic effects and resistance to acidic. water: an ecophysiological and ecotoxicological approach. *In* C. E. W. Steinberg and R. F. Wright, eds. Acidification of freshwater ecosystems: Implications for the future. John Wiley & Sons Ltd., New York, NY, 227–246.

Roué-LeGall, A., M. Lucotte, J. Carreau, R. Canuel, and E. Garcia. 2005. Development of an ecosystem sensitivity model regarding mercury levels in fish using a preference modeling methodology: Application to the Canadian boreal system. Environ. Sci. Technol. 39(24):9412–9423.

Roy, K. M., R. P. Curran, J. W. Barge, D. M. Spada, D. J. Bogucki, E. B. Allen, and W. A. Kretser. 1996. Watershed protection for Adirondack wetlands: A demonstration-level GIS characterization of subcatchments of the Oswegatchie/Black River watershed. Final report for State Wetlands Protection, Program U.S. Environmental Protection Agency. New York State Adirondack Park Agency, Ray Brook, NY.

Roy, K. M., J. Dukett, N. Houck, and G. B. Lawrence. 2013. A long-term monitoring program for evaluating changes in water quality in selected Adirondack waters. Program Summary Report 2012. NYSERDA Report 13-26. New York State Energy Research and Development Authority, Albany, NY.

Rutkiewicz, J., D.-H. Nam, T. Cooley, K. Neumann, I. B. Padilla, W. Route, S. Strom, and N. Basu. 2011. Mercury exposure and neurochemical impacts in bald eagles across several Great Lakes states. Ecotoxicology 20:1669–1676.

Ruzycki, E. M., R. P. Axler, J. R. Henneck, N. R. Will, and G. E. Host. 2011. Estimating mercury concentrations and loads from four western Lake Superior watersheds using continuous in-stream turbidity monitoring. Aquatic Ecosystem Health & Management 14(4):422–432.

Rysgaard, S., N. Risgaard-Petersen, N. P. Sloth, K. Jensen, and L. P. Nielsen. 1994. Oxygen regulation of nitrification and denitrification in sediments. Limnol. Oceanogr. 39:1643–1652.

Ryther, J. H., and W. M. Dunstan. 1971. Nitrogen, phosphorus, and eutrophication in the Coastal Marine Environment. Science 171:1008–1013.

Sanchez-Andrea, I., K. Knittel, R. Amann, R. Amils, and J. L. Sanz. 2012. Quantification of Tinto River sediment microbial communities: Importance of sulfate-reducing bacteria and their role in attenuating acid mine drainage. Appl. Environ. Microbiol. 78(13):4638–4645.

Sandheinrich, M. B., and J. G. Wiener. 2011. Methylmercury in freshwater fish: Recent advances in assessing toxicity of environmentally relevant exposures. *In* W. N. Beyer and J. P. Meador, eds. Environmental contaminants in biota: Interpreting tissue concentrations. CRC Press, Boca Raton, 169–190.

Saunders, P. A., W. H. Shaw, and P. A. Bukaveckas. 2000. Differences in nutrient limitation and grazer suppression of phytoplankton in seepage and drainage lakes of the Adirondack region, NY, U.S.A. Freshw. Biol. 43:391–407.

Scavia, D., and S. B. Bricker. 2006. Coastal eutrophication assessment in the United States. Biogeochemistry 79:187–208.

Scheffe, R., J. Lynch, A. Reff, J. Kelly, B. Hubbell, T. Greaver, and J. T. Smith. 2014. The Aquatic Acidification Index: A new regulatory metric linking atmospheric and biogeochemical models to assess potential aquatic ecosystem recovery. Water Air Soil Pollut. 225(2):1–15.

Schelske, C. L. 1991. Historical nutrient enrichment of Lake Ontario: paleolimnological evidence. Can. J. Fish. Aquat. Sci. 48:1529–1538.

Schelske, C. L., E. F. Stoermer, D. J. Conley, J. A. Robbins, and R. M. Glove. 1983. Early eutrophication of the lower Great Lakes: New evidence from biogenic silica in sediments. Science 222:320–322.

Schelske, C. L., E. F. Stoermer, G. L. Fahnenstiel, and M. Haibach. 1986. Phosphorus enrichment, silica utilization, and biogeochemical silica depletion in the Great Lakes. Can. J. Fish. Aquat. Sci. 43:407–415.

Schelske, C. L., E. F. Stoermer, and W. F. Kenney. 2006. Historic Low-level phosphorus enrichment in the Great Lakes inferred from biogenic silica accumulation in sediments. Limnol. Oceanogr. 51(1, Part 2):728–748.

Scheuhammer, A. M. 1991. Effects of acidification on the availability of toxic metals and calcium to wild birds and mammals. Environ. Pollut. 71:329–375.

Scheuhammer, A. M., M. W. Meyer, M. B. Sandheinrich, and M. W. Murray. 2007. Effects of environmental methylmercury on the health of wild birds, mammals, and fish. Ambio 36(1):12–18.

Scheuhammer, A. M., N. Basu, N. M. Burgess, J. E. Elliott, G. D. Campbell, M. Wayland, L. Champoux, and J. Rodrigue. 2008. Relationships among mercury, selenium, and neurochemical parameters in common loons (*Gavia immer*) and bald eagles (*Haliaeetus leucocephalus*). Ecotoxicology 17:93–101.

Schiff, S. L., J. Spoelstra, R. G. Semkin, and D. S. Jeffries. 2005. Drought induced pulses of SO_4^{2-} from a Canadian shield wetland: Use of $\delta^{34}S$ and $\delta^{18}O$ in SO_4^{2-} to determine sources of sulfur. Appl. Geochem. 20:691–700.

Schimel, J. P., and J. Bennett. 2004. Nitrogen mineralization: challenges of a changing paradigm. Ecology 85(3):597–602.

Schindler, D. W. 1988. Effects of acid rain on freshwater ecosystems. Science 239:232–239.

Schindler, D. W., K. H. Mills, D. F. Malley, M. S. Findlay, J. A. Schearer, I. J. Davies, M. A. Turner, G. A. Lindsey, and D. R. Cruikshank. 1985. Long-term ecosystem stress: Effects of years of experimental acidification. Can. J. Fish. Aquat. Sci. 37:342–354.

Schlegel, H., R. G. Amundson, and A. Hüttermann. 1992. Element distribution in red spruce (*Picea rubens*) fine roots: Evidence for aluminum toxicity at Whiteface Mountain. Can. J. For. Res. 22:1132–1138.

Schnitzer, M., and S. I. M. Skinner. 1963. Organo-metallic interactions in soils: 2. Reactions between different forms of iron and aluminum and the organic matter of a podzol Bh horizon. Soil Sci. 96:181–186.

Schoch, N. 2006. The Adirondack Cooperative Loon Program: Loon conservation in the Adirondack Park. Adir. J. Environ. Stud. Fall/Winter 2006:18–22.

Schoch, N., M. Glennon, D. Evers, M. Duron, A. Jackson, C. Driscoll, X. Yu, and J. Simonin. 2011. Long-term monitoring and assessment of mercury based on integrated sampling efforts using the common loon, prey fish, water, and sediment. NYSERDA Report 12-06. New York State Research and Development Authority, Albany, NY.

Schofield, C. L. 1993. Habitat suitability for brook trout (*Salvelinus fontinalis*) reproduction in Adirondack lakes. Water Resour. Res. 29:875–879.

Schofield, C. L., and C. T. Driscoll. 1987. Fish species distribution in relation to water quality gradients in the North Branch of the Moose River Basin. Biogeochemistry 3:63–85.

Schofield, C. L., and C. Keleher. 1996. Comparison of brook trout reproductive success and recruitment in an acidic Adirondack lake following whole lake liming and watershed liming. Biogeochemistry 32:323–337.

Schofield, C. L., and J. R. Trojnar. 1980. Aluminum toxicity to brook trout (*Salvelinus fontinalis*) in acidified waters. *In* T. Y. Toribara, M. W. Miller, and P. E. Morrows, eds. Polluted rain. Plenum Press, New York, 341–362.

Schultz, A. M., M. H. Begemann, D. A. Schmidt, and K. C. Weathers. 1993. Longitudinal trends in pH and aluminum chemistry of the Coxing Kill, Ulster County, New York. Water Air & Soil Pollut. 69:113–125.

Schwindt, A. R., J. W. Fournie, D. H. Landers, C. B. Schreck, and M. L. Kent. 2008. Mercury concentrations in salmonids from western U.S. national parks and relationships with age and macrophage aggregates. Environ. Sci. Technol. 42:1365–1370.

Scudlark, J. R., and T. M. Church. 1993. Atmospheric input of inorganic nitrogen to Delaware Bay. Estuaries 16(4):747–759.

Seitzinger, S. P. 1988. Denitrification in freshwater and coastal marine ecosystems: Ecological and geochemical significance. Limnol. Oceanogr. 33:702–724.

Seitzinger, S. P., R. V. Styles, E. W. Boyer, R. Alexander, G. Billen, R. W. Howarth, B. Mayer, and N. Van Breemen. 2002. Nitrogen retention in rivers: Model development and application to watersheds in the northeastern U.S.A. Biogeochemistry 57/58:199–237.

Seitzinger, S., J. A. Harrison, J. K. Bohlke, A. F. Bouwman, R. Lowrance, B. Peterson, C. Tobias, and G. Van Drecht. 2006. Denitrification across landscapes and waterscapes: A synthesis. Ecol. Appl. 16(6):2064–2090.

Sellers, P., C. A. Kelly, and J. W. M. Rudd. 2001. Fluxes of methylmercury to the water column of a drainage lake: The relative importance of internal and external sources. Limnol. Oceanogr. 46:623–631.

Selvendiran, P., C. T. Driscoll, J. T. Bushey, and M. R. Montesdeoca. 2008a. Wetland influence on mercury fate and transport in a temperate forested watershed. Environmental Pollut. 154(1):46–55.

Selvendiran, P., C. T. Driscoll, M. R. Montesdeoca, and J. T. Bushey. 2008b. Inputs, storage, and transport of total and methyl mercury in two temperate forest wetlands. J. Geophys. Res. 113.

Selvendiran, P., C. T. Driscoll, M. R. Montesdeoca, H. D. Choi, and T. M. Holsen. 2009. Mercury dynamics and transport in two Adirondack lakes. Limnol. Oceanogr. 54(2):413–427.

Shortle, W. C., and K. T. Smith. 1988. Aluminum-induced calcium deficiency syndrome in declining red spruce. Science 240:1017–1018.

Shortle, W. C., K. T. Smith, R. Minocha, G. B. Lawrence, and M. B. David. 1997. Acidic deposition, cation mobilization, and biochemical indicators of stress in healthy red spruce. J. Environ. Qual. 26(3):871–876.

Simmons, J. A., J. B. Yavitt, and T. J. Fahey. 1996. Watershed liming effects on the forest floor N cycle. Biogeochemistry 32:221–244.

Simonin, H. A., W. A. Kretser, D. W. Bath, M. Olson, and J. Gallagher. 1993. *In situ* bioassays of brook trout (*Salvelinus fontinalis*) and blacknose dace (*Rhinichthys atratulus*) in Adirondack streams affected by episodic acidification. Can. J. For. Res. 50:902–912.

Simonin, H. A., J. J. Loukmas, L. C. Skinner, and K. M. Roy. 2008. Lake variability: Key factors controlling mercury concentrations in New York State fish. Environ. Pollut. 154:107–115.

Skjelkvåle, B. L., J. L. Stoddard, and T. Andersen. 2001. Trends in surface water acidification in Europe and North America (1989–1998). Water Air Soil Pollut. 130:787–792.

Smith, A. J., R. W. Bode, and G. S. Kleppel. 2007. A nutrient biotic index (NBI) for use with benthic macroinvertebrate communities. Ecol. Indicat. 7(2):371–386.

Smith, R. A., G. E. Schwarz, and R. B. Alexander. 1997. Regional interpretation of water-quality monitoring data. Water Resour. Res. 33(12):2781–2798.

Smith, V. H., G. D. Tilman, and J. C. Nekola. 1999. Eutrophication: Impacts of excess nutrient inputs on freshwater, marine, and terrestrial ecosystems. Environ. Pollut. 100:179–196.

Smithwick, E. A. H., D. M. Eissenstat, G. M. Lovett, R. D. Bowden, L. E. Rustad, and C. T. Driscoll. 2013. Root stress and nitrogen deposition: consequences and research priorities. New Phytol. 197(3):712–719.

Snucins, E., and J. M. Gunn. 2000. Interannual variation in the thermal structure of clear and colored lakes. Limnol. Oceanogr. 45:1647–1654.

Sobczak, W. V., S. Findlay, and S. Dye. 2003. Relationship between DOC variability and nitrate removal in and upland stream: An experimental approach. Biogeochemistry 62(3):309–327.

Sorensen, J. A., L. W. Kallemeyn, and M. Sydor. 2005. Relationship between mercury accumulation in young-of-the-year yellow perch and water-level fluctuations. Environ. Sci. Technol. 39:9237–9243.

Spranger, T., J.-P. Hettelingh, J. Slootweg, and M. Posch. 2008. Modelling and mapping long-term risks due to reactive nitrogen effects: An overview of LRTAP convention activities. Environ. Pollut. 154:482–487.

St. Louis, V. L., L. Breebaart, and J. C. Barlow. 1990. Foraging behaviour of tree swallows over acidified and nonacidic lakes. Can. J. Zool. 68(11):2385–2392.

St. Louis, V. L., J. W. M. Rudd, C. A. Kelly, K. G. Beaty, N. S. Bloom, and R. J. Flett. 1994. Importance of wetlands as sources of methylmercury to boreal forest ecosystems. Can. J. Fish. Aquat. Sci. 51(5):1065–1076.

St. Louis, V. L., J. W. M. Rudd, C. A. Kelly, K. G. Beaty, R. J. Flett, and N. T. Roulet. 1996. Production and loss of methylmercury and loss of total mercury from boreal forest catchments containing different types of wetlands. Environ. Sci. Technol. 30(9):2719–2729.

Stacey, P. E., J. S. Greening, J. N. Kremer, D. Peterson, and D. A. Tomasko. 2001. Contribution of atmospheric nitrogen deposition to U.S. estuaries: summary and conclusions. *In* R. W. Valigura, R. B. Alexander, M. S. Castro, T. P. Meyers, H. W. Paerl, P. E. Stacey, and R. E. Turner, eds. Nitrogen loading in coastal water bodies: An atmospheric perspective. American Geophysical Union, Washington, DC, 187–226.

Stanford, G., and H. E. Epstein. 1974. Nitrogen mineralization water relations in soils. Soil Science Sci. Soc. Am. Proc. 38(1):103–107.

Stark, J. M. 1996. Modeling the temperature response of nitrification. Biogeochemistry 35:433–445.

Stark, J. M., and M. K. Firestone. 1996. Kinetic characteristics of ammonium-oxidizer communities in a California oak woodland-annual grassland. Soil Biol. Biogeochem. 28:1307–1317.

Stauffer, R. E. 1990. Granite weathering and the sensitivity of alpine lakes to acid deposition. Limnol. Oceanogr. 35(5):1112–1134.

Stauffer, R. E., and B. D. Wittchen. 1991. Effects of silicate weathering on water chemistry in forested, upland, felsic terrain of the USA. Geochim. Cosmochim. Acta 55:3253–3271.

Steinnes, E., O. Ø. Hvatum, B. Bølviken, and P. Varskog. 2005. Atmospheric supply of trace elements studied by peat samples from ombrotrophic bogs. J. Environ. Qual. 34:192–197.

Stemberger, R. S., D. P. Larsen, and T. M. Kincaid. 2001. Sensitivity of zooplankton for regional lake monitoring. Can. J. Fish. Aquat. Sci. 58:2222–2232.

Sterner, R. W., and J. J. Elser. 2002. Ecological stoichiometry: The biology of elements from molecules to the biosphere. Princeton University Press, Princeton, NJ.

Sterner, R. W., E. Anagnostou, S. Brovold, G. S. Bullerjahn, J. C. Finlay, S. Kumar, R. M. L. McKay, and R. M. Sherrell. 2007. Increasing stoichiometric imbalance in North America's largest lake: Nitrification in Lake Superior. Geophys. Res. Lett. 34:L10406.

Stevens, C. J., N. B. Dise, O. J. Mountford, and D. J. Gowing. 2004. Impact of nitrogen deposition on the species richness of grasslands. Science 303:1876–1878.

Stevens, R. J., R. J. Laughlin, and J. P. Malone. 1997. Measuring the contributions of nitrification and denitrification to the flux of nitrous oxide from soil. Soil Biol. Biogeochem. 29:139–151.

Stoddard, J. L. 1991. Trends in Catskill stream water quality: evidence from historical data. Water Resour. Res. 27:2855–2864.

Stoddard, J. L. 1994. Long-term changes in watershed retention of nitrogen: its causes and aquatic consequences. *In* L. A. Baker, ed. Environmental chemistry of lakes and reservoirs. American Chemical Society, Washington, DC, 223–284.

Stoddard, J. L., and J. H. Kellogg. 1993. Trends and patterns in lake acidification in the state of Vermont: Evidence from the Long-Term Monitoring project. Water Air Soil Pollut. 67:301–317.

Stoddard, J. L., and P. S. Murdoch. 1991. Catskill Mountains: An overview of the impact of acidifying pollutants on aquatic resources. *In* D. F. Charles, ed. Acidic deposition and aquatic ecosystems: Regional case studies. Springer-Verlag, New York, 237–271.

Stoddard, J. L., N. S. Urquhart, A. D. Newell, and D. Kugler. 1996. The Temporally Integrated Monitoring of Ecosystems (TIME) project design: 2. Detection of regional acidification trends. Water Resour. Res. 32(8):2529–2538.

Stoddard, J. L., C. T. Driscoll, J. S. Kahl, and J. H. Kellogg. 1998. A regional analysis of lake acidification trends for the northeastern U.S., 1982–1994. Environ. Monitor. Assess. 51:399–413.

Stoddard, J., J. S. Kahl, F. A. Deviney, D. R. DeWalle, C. T. Driscoll, A. T. Herlihy, J. H. Kellogg, P. S. Murdoch, J. R. Webb, and K. E. Webster. 2003. Response of surface water chemistry to the Clean Air Act Amendments of 1990. EPA 620/R-03/001. U.S. Environmental Protection Agency, Office of Research and Development, National Health and Environmental Effects Research Laboratory, Research Triangle Park, NC.

Stoddard, J. L., A. T. Herlihy, D. V. Peck, R. M. Hughes, T. R. Whittier, and E. Tarquinio. 2008. A process for creating multi-metric indices for large scale aquatic surveys. J. North Am. Benthol. Soc. 27:878–891.

Stoermer, E. F., and J. P. Smol, eds. 1999. The diatoms: Applications for the environmental and earth sciences. Cambridge University Press, New York.

Stolte, W., T. McCollin, A. Noordeloos, and R. Riegman. 1994. Effects of nitrogen source on the size distribution within marine phytoplankton populations. J. Exp. Mar. Biol. Ecol. 184:83–97.

Strayer, D. L., and H. Malcom. 2006. Long-term demography of a zebra mussel (*Dreissena polymorpha*). Freshw. Biol. 51:117–130.

Sullivan, T. J. 1990. Historical changes in surface water acid-base chemistry in response to acidic deposition. State of the Science SOS/T 11. National Acid Precipitation Assessment Program, Washington, DC.

Sullivan, T. J. 2000. Aquatic effects of acidic deposition. Lewis Publ./CRC Press, Boca Raton, FL.

Sullivan, T. J. 2012. Combining ecosystem service and critical load concepts for resource management and public policy. Water 4:905–913.

Sullivan, T. J. In revision. Air quality related values (AQRVs) in national parks: Effects from ozone; visibility reducing particles; and atmospheric deposition of acids, nutrients and toxics. Report prepared for the National Park Service.

Sullivan, T. J., and B. J. Cosby. 1995. Testing, improvement, and confirmation of a watershed model of acid-base chemistry. Water Air Soil Pollut. 85:2607–2612.

Sullivan, T. J., and J. Jenkins. 2014. The science and policy of critical loads of pollutant deposition to protect sensitive ecosystems in NY. Ann. NY Acad. Sci. 1313:57–68.

Sullivan, T. J., and T. C. McDonnell. 2012. Application of critical loads and ecosystem services principles to assessment of the effects of atmospheric sulfur and nitrogen deposition on acid-sensitive aquatic and terrestrial resources. Pilot Case Study: Central Appalachian Mountains. Report prepared for the U.S. Environmental Protection Agency, in association with Systems Research and Applications Corporation E&S Environmental Chemistry, Inc, Corvallis, OR.

Sullivan, T. J., D. F. Charles, J. P. Smol, B. F. Cumming, A. R. Selle, D. R. Thomas, J. A. Bernert, and S. S. Dixit. 1990. Quantification of changes in lakewater chemistry in response to acidic deposition. Nature 345:54–58.

Sullivan, T. J., B. J. Cosby, C. T. Driscoll, D. F. Charles, and H. F. Hemond. 1996a. Influence of organic acids on model projections of lake acidification. Water Air Soil Pollut. 91:271–282.

Sullivan, T. J., B. McMartin, and D. F. Charles. 1996b. Re-examination of the role of landscape change in the acidification of lakes in the Adirondack Mountains, New York. Sci. Total Environ. 183:231–248.

Sullivan, T. J., J. M. Eilers, B. J. Cosby, and K. B. Vaché. 1997. Increasing role of nitrogen in the acidification of surface waters in the Adirondack Mountains, New York. Water Air Soil Pollut. 95(1–4):313–336.

Sullivan, T. J., B. J. Cosby, J. R. Webb, K. U. Snyder, A. T. Herlihy, A. J. Bulger, E. H. Gilbert, and D. Moore. 2002. Assessment of the effects of acidic deposition on aquatic resources in the southern Appalachian Mountains. Report prepared for the Southern Appalachian Mountains Initiative (SAMI). E&S Environmental Chemistry, Inc., Corvallis, OR.

Sullivan, T. J., B. J. Cosby, J. A. Laurence, R. L. Dennis, K. Savig, J. R. Webb, A. J. Bulger, M. Scruggs, C. Gordon, J. Ray, H. Lee, W. E. Hogsett, H. Wayne, D. Miller, and J. S. Kern. 2003. Assessment of air quality and related values in Shenandoah National Park. Technical report NPS/NERCHAL/NRTR-03/090. U.S. Department of the Interior, National Park Service, Northeast Region, Philadelphia, PA. http://www.nature.nps.gov/air/Pubs/pdf/SHEN_Assess_Sullivan2003.pdf.

Sullivan, T. J., B. J. Cosby, A. T. Herlihy, J. R. Webb, A. J. Bulger, K. U. Snyder, P. Brewer, E. H. Gilbert, and D. L. Moore. 2004. Regional model projections of future effects of sulfur and nitrogen deposition on streams in the southern Appalachian Mountains. Water Resour. Res. 40: W02101.

Sullivan, T. J., C. T. Driscoll, B. J. Cosby, I. J. Fernandez, A. T. Herlihy, J. Zhai, R. Stemberger, K. U. Snyder, J. W. Sutherland, S. A. Nierzwicki-Bauer, C. W. Boylen, T. C. McDonnell, and N. A. Nowicki. 2006a. Assessment of the extent to which intensively-studied lakes are representative of the Adirondack Mountain region. Final Report 06-17. New York State Energy Research and Development Authority, Albany, NY.

Sullivan, T. J., I. J. Fernandez, A. T. Herlihy, C. T. Driscoll, T. C. McDonnell, N. A. Nowicki, K. U. Snyder, and J. W. Sutherland. 2006b. Acid-base characteristics of soils in the Adirondack Mountains, New York. Soil Sci. Soc. Am. J. 70:141–152.

Sullivan, T. J., B. J. Cosby, A. T. Herlihy, C. T. Driscoll, I. J. Fernandez, T. C. McDonnell, C. W. Boylen, S. A. Nierzwicki-Bauer, and K. U. Snyder. 2007a. Assessment of the extent to which intensively-studied lakes are representative of the Adirondack region and response to future changes in acidic deposition. Water Air Soil Pollut. 185:279–291.

Sullivan, T. J., J. R. Webb, K. U. Snyder, A. T. Herlihy, and B. J. Cosby. 2007b. Spatial distribution of acid-sensitive and acid-impacted streams in relation to watershed features in the southern Appalachian Mountains. Water Air Soil Pollut. 182:57–71.

Sullivan, T. J., B. J. Cosby, J. R. Webb, R. L. Dennis, A. J. Bulger, and F. A. Deviney Jr. 2008. Streamwater acid-base chemistry and critical loads of atmospheric sulfur deposition in Shenandoah National Park, Virginia. Environ. Monitor. Assess. 137:85–99.

Sullivan, T. J., B. J. Cosby, C. T. Driscoll, T. C. McDonnell, and A. T. Herlihy. 2011. Target loads of atmospheric sulfur deposition protect terrestrial resources in the Adirondack Mountains, New York against biological impacts caused by soil acidification. J. Environ. Stud. Sci. 1(4):301–314.

Sullivan, T. J., B. J. Cosby, C. T. Driscoll, T. C. McDonnell, A. T. Herlihy, and D. A. Burns. 2012. Target loads of atmospheric sulfur and nitrogen deposition for protection of acid sensitive aquatic resources in the Adirondack Mountains, New York. Water Resour. Res. 48(1).

Sullivan, T. J., G. B. Lawrence, S. W. Bailey, T. C. McDonnell, and G. T. McPherson. 2013. Effects of acidic deposition and soil acidification on sugar maple trees in the Adirondack Mountains, New York. NYSERDA Report No. 13-04. New York State Energy Research and Development Authority, Albany, NY.

Sundareshwar, P. V., J. T. Morris, E. Koepfler, and B. Fornwalt. 2003. Phosphorus limitation of coastal ecosystem processes. Science 299:563–565.

Sverdrup, H., and P. Warfvinge. 1993. Report in ecological engineering. Vol. 2 of The effect of soil acidification on the growth of trees, grass and herbs as expressed by the (Ca+ Mg+ K)/Al ratio. Lund University, Lund, Sweden.

Sverdrup, H., T. C. McDonnell, T. J. Sullivan, B. Nihlgård, S. Belyazid, B. Rihm, E. Porter, W. D. Bowman, and L. Geiser. 2012. Testing the feasibility of using the ForSAFE-VEG model to map the critical load of nitrogen to protect plant biodiversity in the Rocky Mountains region, USA. Water Air Soil Pollut. 23:371–387.

Tank, J. L., and W. K. Dodds. 2003. Nutrient limitation of epilithic and epixylic biofilms in ten North American streams. Freshw. Biol. 48:1031–1049.

Tessier, J. T., and D. J. Raynal. 2003. Use of nitrogen to phosphorus ratios in plant tissue as an indicator of nutrient limitation and nitrogen saturation. J. Appl. Ecol. 40(3):523–534.

Tett, P., L. Gilpin, H. Svendsen, C. P. Erlandsson, U. Larsson, S. Kratzer, E. Fouilland, C. Janzen, J. Lee, C. Grenz, A. Newton, J. G. Ferreira, T. Fernandes, and S. Scory. 2003. Eutrophication and some European waters of restricted exchange. Cont. Shelf Res. 23:1635–1671.

Thomas, R. Q., C. D. Canham, K. C. Weathers, and C. L. Goodale. 2010. Increased tree carbon storage in response to nitrogen deposition in the US. Nature Geosci. 3:13–17.

Thorne, R. S. J., W. P. Williams, and C. Gordon. 2000. The macroinvertebrates of a polluted stream in Ghana. J. Freshw. Ecol. 15:209–217.

Tipping, E., E. J. Smith, A. J. Lawlor, S. Hughes, and P. A. Stevens. 2003. Predicting the release of metals from ombrotrophic peat due to drought-induced acidification. Environ. Pollut. 123:239–253.

Tober, J., M. Griffin, and I. Valiela. 2005. Growth and abundance of *Fundulus heteroclitus* and *Menidia menidia* in estuaries of Waquoit Bay, Massachusetts, exposed to different rates of nitrogen loading. Aquatic Ecology 34:299–306.

Togersen, T., E. DeAngelo, and J. O'Donnell. 1997. Calculations of horizontal mixing rates using 222Rn and the controls on hypoxia in western Long Island Sound, 1991. Estuaries 20:328–345.

Tomassen, H. B. M., A. J. P. Smolders, L. P. M. Lamers, and J. G. M. Roelofs. 2003. Stimulated growth of *Betula pubescens* and *Molinia caerulea* on ombrotrophic bogs: Role of high levels of atmospheric nitrogen deposition. J. Ecol. 91:357–370.

Tomassen, H. B., A. J. Smolders, J. Limpens, L. P. Lamers, and J. G. Roelofs. 2004. Expansion of invasive species on ombrotrophic bogs: Desiccation or high N deposition? J. Appl. Ecol. 41(1):139–150.

Townsend, A. R., R. W. Howarth, F. A. Bazzaz, M. S. Booth, C. C. Cleveland, S. K. Collinge, A. P. Dobson, P. R. Epstein, D. R. Keeney, M. A. Mallin, C. A. Rogers, P. Wayne, and A. H. Wolfe. 2003. Human health effects of a changing global nitrogen cycle. Front. Ecol. Environ. 1(5):240–246.

Trebitz, A. S., J. C. Brazner, A. M. Cotter, M. L. Knuth, J. A. Morrive, G. S. Peterson, M. E. Sierszen, J. A. Thompson, and J. R. Kelly. 2007. Water quality in

Great Lakes coastal wetlands: Basin-wide patterns and responses to an anthropogenic disturbance gradient. J. Great Lakes Res. 33(3):67–85.

Treseder, K. K. 2004. The meta-analysis of mycorrhizal response to nitrogen, phosphorus, and atmospheric CO_2 in field studies. New Phytol. 164:347–355.

Tritton, L. M., C. W. Martin, J. W. Hornbeck, and R. S. Pierce. 1987. Biomass and nutrient removals from commercial thinning and whole-tree clearcutting of central hardwoods. Environ. Manage. 11:659–666.

Trombulak, S. C., and C. A. Frissell. 2000. Review of ecological effects of roads on terrestrial and aquatic communities. Conserv. Biol. 14(1):18–30.

Turner, R. E., N. Qureshi, N. N. Rabalais, Q. Dortch, D. Justić, R. F. Shaw, and J. Cope. 1998. Fluctuating silicate:nitrate ratios and coastal plankton food webs. Proc. Nat. Acad. Sci. 95(22):13048–13051.

Turner, R. E., D. Stanley, D. Brock, J. Pennock, and N. N. Rabalais. 2001. A comparison of independent N-loading estimates for U.S. estuaries. *In* R. Valigura, R. Alexander, M. Castro, T. Meyers, H. Paerl, P. Stacey, and R. E. Turner, eds. Nitrogen loading in coastal water bodies: An atmospheric perspective. American Geophysical Union, Washington, DC, 107–118.

Turner, R. K., S. Georgiou, I. Gren, F. Wulff, S. Barrett, Söderqvist, T., I. J. Bateman, C. Folke, S. Langaas, T. Żylicz, K. Karl-Goran Mäler, and A. Markowska. 1999. Managing nutrient fluxes and pollution in the Baltic: An interdisciplinary simulation study. Ecol. Econ. 30:333–352.

Turner, R. S., R. B. Cook, H. van Miegroet, D. W. Johnson, J. W. Elwood, O. P. Bricker, S. E. Lindberg, and G. M. Hornberger. 1990. Watershed and lake processes affecting chronic surface water acid-base chemistry. State of the Science SOS/T 10. National Acid Precipitation Assessment Program, Washington DC.

Turner, R. S., R. B. Cook, H. Van Miegroet, D. W. Johnson, J. W. Elwood, O. P. Bricker, S. E. Lindberg, and G. M. Hornberger. 1991. Watershed and lake processes affecting surface water acid-base chemistry. *In* P. M. Irving, ed. Aquatic processes and effects. Vol. 2 of Acidic deposition: State of science and technology report no. 10. National Acid Precipitation Assessment Program, Washington, DC, 10-1–10-167.

Turnquist, M. A., C. T. Driscoll, K. L. Schulz, and M. A. Schlaepfer. 2011. Mercury concentrations in snapping turtles (*Chelydra serpentina*) correlate with environmental and landscape characteristics. Ecotoxicology 20:1599–1608.

Twilley, R. R., W. M. Kemp, K. W. Staver, J. C. Stevenson, and W. R. Boynton. 1985. Nutrient enrichment of estuarine submersed vascular plant communities. 1. Algal growth and effects on production of plants and associated communities. Mar. Ecol. Prog. Ser. 23:179–191.

Tyler, S. J., and S. J. Ormerod. 1992. A review of the likely causal pathways relating the reduced density of breeding dippers *Cinclus cinclus* to the acidification of upland streams. Environ. Pollut. 78(1/3):49–55.

Ueno, Y., S. Nagata, T. Tsutsumi, A. Hasegawa, M. F. Watanabe, H.-D. Park, G.-C. Chen, G. Chen, and S.-Z. Yu. 1996. Detection of microcystins, a blue-green algal hepatotoxin, in drinking water sampled in Haimen and Fusui, endemic areas of primary liver cancer in China, by highly sensitive immunoassay. Carcinogenesis 17(6):1317–1321.

Ulrich, B., R. Mayer, and T. K. Khanna. 1980. Chemical changes due to acid precipitation in a loess-derived soil in central Europe. Soil Sci. 130:193–199.

Urban, N., M. Auer, S. Green, X. Lu, D. Apul, K. Powell, and L. Bub. 2005. Carbon cycling in Lake Superior. J. Geophys. Res. 110:C06590.

Urquhart, N. S., S. G. Paulsen, and D. P. Larsen. 1998. Monitoring for regional and policy-relevant trends over time. Ecol. Appl. 8:246–257.

USDA Natural Resources Conservation Service. 2014. *Drosera rotundiflora* L.: Roundleaf sundew. http://plants.usda.gov/core/profile?symbol=DRRO.

[U.S. EPA] U.S. Environmental Protection Agency. 1993. Air quality criteria for oxides of nitrogen. EPA/600/8-91/049aF-cF.3v. Office of Health and Environmental Assessment, Environmental Criteria and Assessment Office, Research Triangle Park, NC.

[U.S. EPA] U.S. Environmental Protection Agency. 1994. The Long Island Sound Study: The Comprehensive conservation and management plan. U.S. Environmental Protection Agency, Washington, DC.

[U.S. EPA] U.S. Environmental Protection Agency. 1995. Acid Deposition Standard Feasibility Study report to Congress. EPA 430-R-95-001a. U.S. EPA, Office of Air and Radiation, Acid Rain Division.

[U.S. EPA] U.S. Environmental Protection Agency. 1998. Long Island Sound Study: Phase III—Actions for hypoxia management. Report No. EPA 902-R-98-002. U.S. Environmental Protection Agency, Stamford, CT.

[U.S. EPA] U.S. Environmental Protection Agency. 1999a. The benefits and costs of the Clean Air Act, 1990–2010. EPA-410-R-99-001. U.S. Environmental Protection Agency, Washington, DC.

[U.S. EPA] U.S. Environmental Protection Agency. 1999b. Deposition of air pollutants to the great waters. U.S. Government Printing Office, Washington, DC.

[U.S. EPA] U.S. Environmental Protection Agency. 2001. Water quality criterion for the protection of human health: Methylmercury. EPA-823-R-01-001. Office of Science and Technology, Office of Water, U.S. EPA, Washington, DC.

[U.S. EPA] U.S. Environmental Protection Agency. 2004. Air quality criteria for particulate matter. Volumes I and II. EPA/600/P-99/002aF. National Center for Environmental Assessment-RTP, Office of Research and Development, Research Triangle Park, NC.

[U.S. EPA] U.S. Environmental Protection Agency. 2008. Integrated science assessment for oxides of nitrogen and sulfur: Ecological criteria. EPA/600/R-

08/082F. National Center for Environmental Assessment, Office of Research and Development, Research Triangle Park, NC.

[U.S. EPA] U.S. Environmental Protection Agency. 2009. Risk and exposure assessment for review of the secondary national ambient air quality standards for oxides of nitrogen and oxides of sulfur: Final. EPA-452/R-09-008a. Office of Air Quality Planning and Standards, Health and Environmental Impacts Division, Research Triangle Park, NC.

[U.S. EPA] U.S. Environmental Protection Agency. 2010. Guidance for implementing the January 2001 methylmercury water quality criterion. EPA 823-R-10-001. U.S. Environmental Protection Agency, Office of Water, Washington, DC.

[U.S. EPA] U.S. Environmental Protection Agency. 2012a. Great Lakes monitoring. http://www.epa.gov/glindicators/water/trophicb.html.

[U.S. EPA] U.S. Environmental Protection Agency. 2012b. State of the Lakes Ecosystem Conference (SOLEC). http://www.epa.gov/solec/.

[U.S. EPA] U.S. Environmental Protection Agency. 2015. Mercury emissions: The global context. http://www2.epa.gov/international-cooperation/mercury-emissions-global-context, accessed February 27, 2015.

Vadeboncoeur, M. A. 2010. Meta-analysis of fertilization experiments indicates multiple limiting nutrients in northeastern deciduous forests. Can. J. For. Res. 40(9):1766–1780. 10.1139/x10-127.

Valiela, I., and J. E. Costa. 1988. Eutrophication of Buttermilk Bay, a Cape Cod coastal embayment: Concentrations of nutrients and watershed nutrient budgets. Environ. Manage. 12:539–553.

Valiela, I., J. E. Costa, and K. Foreman. 1990. Transport of groundwater-borne nutrients from watersheds and their effects on coastal waters. Biogeochemistry 10:177–197.

Valigura, R. A., R. B. Alexander, D. A. Brock, M. S. Castro, T. P. Meyers, H. W. Paerl, P. E. Stacey, and D. Stanley. 2000. An assessment of nitrogen inputs to coastal areas with an atmospheric perspective. American Geophysical Union, Washington, DC.

Valle-Levinson, A., R. E. Wilson, and R. L. Swanson. 1995. Physical mechanisms leading to hypoxia and anoxia in western Long Island Sound. Environmental International 21:657–666.

van Breemen, N., J. Mulder, and C. T. Driscoll. 1983. Acidification and alkalinization of soils. Plant Soil 75:283–308.

van Breemen, N., R. Giesler, M. Olsson, R. Finlay, U. Lundström, and A. G. Jongmans. 2000. Mycorrhizal weathering: A true case of mineral plant nutrition? Biogeochemistry 49(1):53–67.

van Breemen, N., E. W. Boyer, C. L. Goodale, N. A. Jaworski, K. Paustian, S. P. Seitzinger, K. Lajtha, B. Mayer, D. Van Dam, R. W. Howarth, and K. J. Nadelhoffer. 2002. Where did all the nitrogen go? Fate of nitrogen inputs to large watersheds in the northeastern U.S.A. Biogeochemistry 57/58:267–293.

van Egmond, K., T. Bresser, and L. Bouwman. 2002. The European nitrogen case. Ambio 31:72–78.

Van Metre, P. C. 2012. Increased atmospheric deposition of mercury in reference lakes near major urban areas. Environmental pollution 162:209–215.

Van Miegroet, H., D. W. Cole, and N. W. Foster. 1992a. Nitrogen distribution and cycling. *In* D. W. Johnson and S. E. Lindberg, eds. Atmospheric deposition and forest nutrient cycling. Springer-Verlag, New York, 178–196.

Van Miegroet, H., D. W. Johnson, and D. W. Cole. 1992b. Analysis of N cycles in polluted vs unpolluted environment. *In* D. W. Johnson and S. E. Lindberg, eds. Atmospheric deposition and forest nutrient cycling. Springer-Verlag, New York, 199–202.

van Miegroet, H. V., D. W. Johnson, T. P. Burt, A. L. Heathwaite, and S. T. Trudgill. 1993. Nitrate dynamics in forest soils. *In* Nitrate: Processes, patterns and management. John Wiley & Sons, Ltd., New York, 75–97.

van Sickle, J., and M. R. Church. 1995. Nitrogen bounding study. methods for estimating the relative effects of sulfur and nitrogen deposition on surface water chemistry. EPA/600/R-95/172. U.S. Environmental Protection Agency, National Health and Environment Effects Research Laboratory, Corvallis, OR.

van Sickle, J., J. P. Baker, H. A. Simonin, B. P. Baldigo, W. A. Kretser, and W. E. Sharpe. 1996. Episodic acidification of small streams in the northeastern United States: Fish mortality in field bioassays. Ecol. Appl. 6(2):408–421.

Vanderploeg, H. A., T. F. Nalepa, D. J. Jude, E. L. Mills, K. T. Holeck, J. R. Liebig, I. A. Grigorovich, and H. Ojaveer. 2002. Dispersal and emerging ecological impacts of Ponto-caspian species in the Laurentian Great Lakes. Can. J. Fish. Aquat. Sci. 59:1209–1228.

Vertucci, F. A., and J. M. Eilers. 1993. Issues in monitoring wilderness lake chemistry: A case study in the Sawtooth Mountains, Idaho. Environ. Monitor. Assess. 28:277–294.

Vitousek, P. M., and R. W. Howarth. 1991. Nitrogen limitation on land and in the sea: How can it occur? Biogeochemistry 13:87–115.

Vitousek, P. M., H. A. Mooney, J. Lubchenco, and J. M. Melillo. 1997. Human domination of Earth's ecosystems. Science 277:494–499.

Vitt, D. H., K. Wieder, L. A. Halsey, and M. Turetsky. 2003. Response of *Sphagnum fuscum* to nitrogen deposition: a case study of ombrogenous peatlands in Alberta, Canada. Bryologist 106(2):235–245.

Wall, D. H., and J. C. Moore. 1999. Interaction underground: Soil biodiversity, mutualism, and ecosystem processes. BioScience 49:109–117.

Waller, K., C. Driscoll, J. Lynch, D. Newcomb, and K. Roy. 2012. Long-term recovery of lakes in the Adirondack region of New York to decreases in acidic deposition. Atmos. Environ. 46(0):56–64.

Warby, R. A. F., C. E. Johnson, and C. T. Driscoll. 2009. Continuing acidification of organic soils across the northeastern USA: 1984–2001. Soil Sci. Soc. Am. J. 73(1):274–284.

Wargo, P. M., J. Tilley, G. Lawrence, M. David, K. Vogt, D. Vogt, and Q. Holifield. 2003. Vitality and chemistry of roots of red spruce in forest floors of stands with a gradient of soil Al/Ca ratios in the northeastern United States. Can. J. For. Res. 33(4):635–652.

Warner, K. A., E. E. Roden, and J.-C. Bonzongo. 2003. Microbial mercury transformation in anoxic freshwater sediments under iron-reducing and other electron-accepting conditions. Environ. Sci. Technol. 37(10):2159.

Watmough, S. A., M. C. Eimers, J. Aherne, and P. J. Dillon. 2004. Climate effects on stream nitrate at 16 forested catchments in south central Ontario. Environ. Sci. Technol. 38:2383–2388.

Watras, C. J., K. A. Morrison, R. J. M. Hudson, T. M. Frost, and T. K. Kratz. 2000. Decreasing mercury in northern Wisconsin: Temporal patterns in bulk precipitation and a precipitation-dominated lake. Environ. Sci. Technol. 34(19):4051–4057.

Watts, S. H., and S. P. Seitzinger. 2001. Denitrification rates in organic and mineral soils from riparian sites: a comparison of N_2 flux and acetylene inhibition methods. Soil Biol. Biogeochem. 22:331–335.

Wayland, M., and D. K. McNicol. 1990. Status report on the effects of acid precipitation on common loon reproduction in Ontario: The Ontario Lakes loon survey. Canadian Wildlife Service, Ottawa.

Weathers, K. C., G. M. Lovett, G. E. Likens, and R. Lathrop. 2000. The effect of landscape features on deposition to Hunter Mountain, Catskill Mountains, New York. Ecol. Appl. 10:528–540.

Webber, H. M., and T. A. Haines. 2003. Mercury effects on predator avoidance behavior of a forage fish, golden shiner (*Notemigonus crysoleucas*). Environ. Toxicol. Chem. 22(7):1556–1561.

Webster, K. E., A. D. Newell, L. A. Baker, and P. L. Brezonik. 1990. Climatically induced rapid acidification of a softwater seepage lake. Nature 347:374–376.

Webster, K. L., I. F. Creed, N. S. Nicholas, and H. V. Miegroet. 2004. Exploring interactions between pollutant emissions and climatic variability in growth of red spruce in the Great Smoky Mountains National Park. Water Air Soil Pollut. 159:225–248.

Welsh, B. 1991. Anoxia and hypoxia in Long Island Sound, Chesapeake Bay, and Mobile Bay: A comparative assessment. *In* K. R. Hinga, D. W. Stanley, C. J. Klein, D. T. Lucid, and M. J. Katz, eds. The National Estuarine Eutrophication Project: Workshop proceedings. National Oceanic and Atmospheric Administration and the University of Rhode Island Graduate School of Oceanography, Rockville, MD, 35–40.

Welsh, B. L., and F. C. Eller. 1991. Mechanisms controlling summertime oxygen depletion in western Long Island Sound. Estuaries 14:265–278.

Welsh, B. L., R. I. Welsh, and M. L. DeGiacomo-Cohen. 1994. Quantifying hypoxia and anoxia in Long Island Sound. *In* K. R. Dyer and R. J. Orth, eds. Changes in fluxes in estuaries: Implications from science to management.

ECSA22/ERF Symposium, Institute of Marine Studies, University of Plymouth. Olsen and Olsen Publishers, Fredensborg, Denmark, 131–137.

Weseloh, D. V. C., D. J. Moore, C. E. Hebert, S. R. deSolla, B. M. Braune, and D. J. McGoldrick. 2011. Current concentrations and spatial and temporal trends in mercury in Great Lakes herring gull eggs, 1974–2009. Ecotoxicology 20(7):1644–1658.

Western Regional Air Partnership. 2013. West-wide Jump Start Air Quality Modeling Study. Managed by the Western Governors' Association. http://www.wrapair2.org/WestJumpAQMS.aspx.

Whitall, D., M. Castro, and C. Driscoll. 2004. Evaluation of management strategies for reducing nitrogen loadings to four US estuaries. Sci. Total Environ. 333:25–36.

Whitall, D., S. Bricker, J. Ferreira, A. M. Nobre, T. Simas, and M. Silva. 2007. Assessment of eutrophication in estuaries: Pressure-state-response and nitrogen source apportionment. Environ. Manage. 40:678–690.

White, A. F., and A. E. Blum. 1995. Effects of climate on chemical weathering in watersheds. Geochim. Cosmochim. Acta 59:1729–1747.

White, A. F., and S. L. Brantley. 1995. Chemical weathering rates of silicate minerals. Mineralogical Society of America, Washington, DC.

Whitehead, P. G., B. Reynolds, G. M. Hornberger, C. Neal, B. J. Cosby, and P. Paricos. 1988. Modeling long term stream acidification trends in upland Wales at Plynlimon. Hydrol. Process. 2:357–368.

Whittier, T. R., S. B. Paulsen, D. P. Larsen, S. A. Peterson, A. T. Herlihy, and P. R. Kaufmann. 2002. Indicators of ecological stress and their extent in the population of northeastern lakes: a regional-scale assessment. BioScience 52:235–247.

Wiedinmyer, C., and H. Friedli. 2007. Mercury emission estimates from fires: An initial inventory for the United States. Environ. Sci. Technol. 41:8092–8098.

Wiener, J. G., D. P. Krabbenhoft, G. H. Heinz, and A. M. Scheuhammer. 2003. Ecotoxicology of mercury. *In* D. J. Hoffman, B. A. Rattner, G. A. Burton, and J. Cairns, eds. Handbook of Ecotoxicology (2nd ed.). CRC Press, Boca Raton, FL, 409–463.

Wiener, J. G., B. C. Knights, M. B. Sandheinrich, J. D. Jeremiason, M. E. Brigham, D. R. Engstrom, L. G. Woodruff, W. F. Cannon, and S. J. Balogh. 2006. Mercury in soils, lakes, and fish in Voyageurs National Park (Minnesota): Importance of atmospheric deposition and ecosystem factors. Environ. Sci. Technol. 40(20):6261–6268.

Wiener, J. G., R. A. Bodaly, S. S. Brown, M. Lucotte, M. C. Newman, D. B. Porcella, R. J. Reash, and E. B. Swain. 2007. Monitoring and evaluating trends in methylmercury accumulation in aquatic biota. *In* R. C. Harris, D. P. Krabbenhoft, R. P. Mason, M. W. Murray, R. J. Reash, and T. Saltman, eds. Ecosystem responses to mercury contamination: Indicators of change. SETAC, CRC Press, Taylor and Francis Group, Boca Raton, FL, 87–122.

Wiener, J. G., M. B. Sandheinrich, S. P. Bhavsar, J. R. Bohr, D. C. Evers, B. A. Monson, and C. S. Schrank. 2012. Toxicological significance of mercury in yellow perch in the Laurentian Great Lakes region. Environ. Pollut. 161:350–357.

Wigington, P. J., Jr, T. D. Davies, M. Tranter, and K. N. Eshleman. 1990. Episodic acidification of surface waters due to acidic deposition. PB-92-100486/XAB. U.S. National Acid Precipitation Assessment Program, State of Science/Technology Report 12.

Wigington, P. J., Jr, J. P. Baker, D. R. DeWalle, W. A. Kretser, P. S. Murdoch, H. A. Simonin, J. Van Sickle, M. K. McDowell, D. V. Peck, and W. R. Barchet. 1996a. Episodic acidification of small streams in the northeastern United States: Episodic response project. Ecol. Appl. 6(2):374–388.

Wigington, P. J., Jr, D. R. DeWalle, P. S. Murdoch, W. A. Kretser, H. A. Simonin, J. Van Sickle, and J. P. Baker. 1996b. Episodic acidification of small streams in the northeastern United States: Ionic controls of episodes. Ecol. Appl. 6(2):389–407.

Wigley, T. M. I. 1999. The science of climate change: Global and U.S. perspectives. Pew Center on Global Climate Change, Arlington, VA.

Williams, E. L., L. M. Walter, T. C. W. Ku, K. K. Baptist, J. M. Budai, and G. W. Kling. 2007. Silicate weathering in temperate forest soils: Insights from a field experiment. Biogeochemistry 82(111–126):111–126.

Wobeser, G. A., N. O. Nielsen, and B. Schiefer. 1976. Mercury and mink. II. Experimental methyl mercury intoxication. Canadian Journal of Comparative Medicine 40:34–45.

Wolfe, A. P., J. S. Baron, and R. J. Cornett. 2001. Anthropogenic nitrogen deposition induces rapid ecological changes in alpine lakes of the Colorado Front Range (USA). J. Paleolimnol. 25:1–7.

Wolfe, M. F., S. Schwarzbach, and R. A. Sulaiman. 1998. Effects of mercury on wildlife: A comprehensive review. Environ. Toxicol. Chem. 17(2):146–160.

Wolfe, M. F., T. Atkeson, W. Bowerman, K. Burger, D. C. Evers, M. W. Murray, and E. Zillioux. 2007. Wildlife indicators. *In* R. C. Harris, D. P. Krabbenhoft, R. P. Mason, M. W. Murray, R. J. Reash, and T. Saltman, eds. Ecosystem responses to mercury contamination: Indicators of change. SETAC, CRC Press, Taylor and Francis Group, Boca Raton, FL, 123–189.

Wood, C. M., D. G. McDonald, C. G. Ingersoll, D. R. Mount, O. E. Johannsson, S. Landsberger, and H. L. Bergman. 1990. Effects of water acidity, calcium, and aluminum on whole body ions of brook trout (*Salvelinus fontinalis*) continuously exposed from fertilization to swim-up: A study by instrumental neutron activation analysis. Can. J. Fish. Aquat. Sci. 47:1593–1603.

Wood, P. B., J. H. White, A. Steffer, J. M. Wood, C. F. Facemire, and F. Percival. 1996. Mercury concentrations in tissues of Florida bald eagles. J. Wildl. Manage. 60:178–185.

Woodwell, G. M. 1970. Effects of pollution on the structure and physiology of ecosystems: Changes in natural ecosystems caused by many different types of disturbances are similar and predictable. Science 168:429–433.

Worall, F., and T. P. Burt. 2007. Flux of dissolved organic carbon from UK rivers. Global Biogeochem. Cycles 21. GB10139.

Wren, C. D., D. B. Hunter, J. F. Leatherland, and P. M. Stokes. 1987. The effects of polychlorinated biphenyls and methylmercury, singly and in combination, on mink. I: Uptake and toxic responses. Arch. Environ. Contam. Toxicol. 16:441–447.

Wright, R. F., and B. J. Cosby. 1987. Use of a process-oriented model to predict acidification at manipulated catchments in Norway. Atmos. Environ. 21:727–730.

Wright, R. F., B. J. Cosby, G. M. Hornberger, and J. N. Galloway. 1986. Comparison of paleolimnological with MAGIC model reconstructions of water acidification. Water Air Soil Pollut. 30:367–380.

Wright, R. F., B. J. Cosby, R. C. Ferrier, A. Jenkins, A. J. Bulger, and R. Harriman. 1994. Changes in the acidification of lochs in Galloway, southwestern Scotland, 1979–1988: The MAGIC model used to evaluate the role of afforestation, calculate critical loads, and predict fish status. J. Hydrol. 161:257–285.

Wright, R. F., C. Beier, and B. J. Cosby. 1998. Effects of nitrogen deposition and climate change on nitrogen runoff at Norwegian boreal forest catchments: The MERLIN model applied to Risdalsheia (RAIN and CLIMEX projects). Hydrol. Earth Syst. Sci. 2(4):399–414.

Wu, W., and C. T. Driscoll. 2010. Impact of climate change on three-dimensional dynamic critical load functions. Environ. Sci. Technol. 44:720–726.

Yanai, R. D., M. A. Arthur, T. G. Siccama, and C. A. Federer. 2000. Challenges of measuring forest floor organic matter dynamics: Repeated measures from a chronosequence. For. Ecol. Manage. 138:273–283.

Yanai, R. D., R. P. Phillips, M. A. Arthur, T. G. Siccama, and E. N. Hane. 2005. Spatial and temporal variation in calcium and aluminum in northern hardwood forest floors. Water Air Soil Pollut. 160:109–118.

Yeakley, J. A., D. C. Coleman, B. L. Haines, B. D. Kloeppel, J. L. Meyer, W. T. Swank, B. W. Argo, J. M. Deal, and S. F. Taylor. 2003. Hillslope nutrient dynamics following upland riparian vegetation disturbance. Ecosystems 6:154–167.

Yu, X., C. T. Driscoll, M. Montesdeoca, D. C. Evers, M. Duron, K. A. Williams, N. Schoch, and N. C. Kamman. 2011. Spatial patterns of mercury in biota of Adirondack, New York lakes. Ecotoxicology 20(7):1543–1554.

Zananski, T. J., T. M. Holsen, P. K. Hopke, and B. S. Crimmins. 2011. Mercury temporal trends in top predator fish of the Laurentian Great Lakes. Ecotoxicology 20(7):1568–1576.

Zhai, J., C. T. Driscoll, T. J. Sullivan, and B. J. Cosby. 2008. Regional application of the PnET-BGC model to assess historical acidification of Adirondack Lakes. Water Resour. Res. 44:W01421.

Zhu, G., M. S. Jetten, P. Kuschk, K. F. Ettwig, and C. Yin. 2010. Potential roles of anaerobic ammonium and methane oxidation in the nitrogen cycle of wetland ecosystems. Appl. Microbiol. Biotechnol. 86(4):1043–1055.

Zhu, G., S. Wang, W. Wang, Y. Wang, L. Zhou, B. Jiang, H. J. M. Op den Camp, N. Risgaard-Petersen, L. Schwark, Y. Peng, M. M. Hefting, M. S. M. Jetten, and C. Yin. 2013. Hotspots of anaerobic ammonium oxidation at land–freshwater interfaces. Nature Geosci. 6(2):103–107.

Zillioux, E. J., D. B. B. Porcella, and B. J. M. 1993. Mercury cycling and effects in freshwater wetland ecosystems. Environ. Toxicol. Chem. 12:2245–2264.

About the Author

Timothy Sullivan holds a BA in history from Stonehill College (1972); an MA in biology from Western State College, Colorado (1977); and a PhD in biological sciences from Oregon State University (1983) through an interdisciplinary program that included areas of focus in ecology, zoology, and environmental chemistry. He did his postdoctoral research at the Center for Industrial Research in Oslo, Norway, on the acidification of surface and ground water, episodic hydrologic processes, and Al biogeochemistry. His expertise includes the effects of air pollution on aquatic and terrestrial resources, watershed analysis, critical loads, ecosystem services, nutrient cycling, aquatic acid-base chemistry, episodic processes controlling surface water chemistry, and environmental assessment. He has been president of E&S Environmental Chemistry, Inc., since 1988. He began studies of the effects of air pollution on sensitive aquatic and terrestrial resources in New York in the 1980s and has worked more or less continuously on that research for nearly 30 years. He has served as project manager and/or lead author for a wide variety of projects that have synthesized the science of complex air and water pollution effects for diverse audiences. He was project manager of the effort to draft a scientific summary and integrated science assessment of the effects of nitrogen and sulfur oxides on terrestrial, transitional, and aquatic ecosystems for the U.S. Environmental Protection Agency in support of its review of National Ambient Air Quality Standards. He is the author of the National Acid Precipitation Assessment Program State of Science and Technology Report on past changes in surface-water acid-base chemistry throughout the United States from acidic deposition. He served as project manager for preparation of air quality reviews for national parks throughout California and co-authored similar reviews for the Pacific Northwest and the Rocky Mountain and Great Plains regions. He has summarized air pollution effects at all 272 inventory and monitoring national parks in the United States and has managed dozens of air and water pollution modeling and assessment studies throughout the United States for the National Park Service, the U.S. Forest Service, the U.S. Department of Energy, and the U.S. Environmental Protection Agency. He has authored a book on the aquatic effects of acidic deposition (Lewis Publ. 2000) and more recently co-authored a book on sampling, analysis, and quality assurance protocols for studying air pollution effects on freshwater ecosystems (CRC Press, 2015). He has also published more than 125 peer-reviewed journal articles, book chapters, and technical reports describing the results of his research.